RESUPINATE APHYLLOPHORALES OF INDIA

About the Authors

Dr. Hakimi Meibodi is the Vice Chancellor, Faculty of Natural Resources and Desertification at the Yazd University, Yazd, Iran. He is also a Faculty in Natural Resources and Desert Studies, University of Yazd, Iran. Dr. Hakimi did his Bachelors and Masters Degrees in Plant Science and did his doctoral work in Mycology from University of Pune. He is engaged in 5 research projects and has more than 25 research publications to his credit that pan the journals at the National and International level.

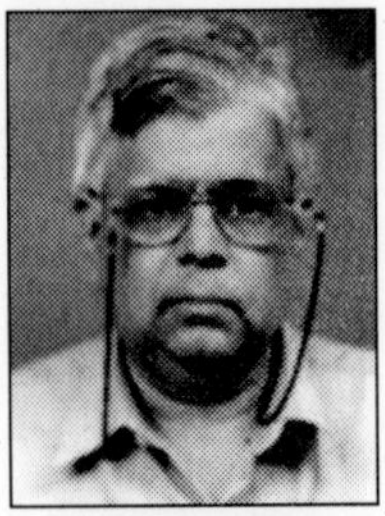

Late Prof. J. G. Vaidya was a distinguished professor in Department of Botany at the University of Pune. His deep engagement and perseverance in Mycology, as an area of study, achieved him wide recognition, both at National and International levels. Heparticipated in about 80 National conferences, 14 International Conferences, 56 University Refresher courses. Through his lifetime, he was invited to serve as a resource person in different institutions and had opportunities to work with esteemed mycologists as Dr. Stapler. It would be significant to highlight his specific contribution as the only expert in Aphyllophorales identification from the State of Maharashtra

Dr. Kiran Ranadive is an Assistant Professor of Botany at the Waghire College. He is also a Contributory teacher at Department of Botany, University of Pune, Pune. Kiran has a Masters and Bachelors in Botany as well as Education and is also a recipient of the prestigious **CSIR-UGC NET** Fellowship. Kiran has got Ph.D. in Aphyllophorales. He has developed the website **www.fungifromindia.com** on fungi as a part of his doctoral research. He has also contributed to the **Vernacular Encyclopedia for Children** or the **Marathi Kumar Vishvakosh**

Dr. Jamaluddin is an Ex Director and Head of Forest Pathology Division, Tropical Forestry Research Institute at Jabalpur. He has a Masters and a D.Phil. Degree and a diploma in French. He has also served as a Member of Several Academic Boards and Societies, providing the necessary support towards planning and accomplishment of academic and research initiatives. He has been felicitated with several awards for his contribution to academic field.

Dr. P.K. Jite is an Associate Professor at the Department of Botany, University of Pune, Pune. She has managed about 7 major research projects on Mycorrhiza. She is working with 5 doctoral and M.Phil. Students, who are on the verge of completion and is guiding 2 post-doctoral students in Botany. She has been actively engaged in academics and has participated in more than 35 conferences and has about 35 National and International publications to her credit. She continues to be actively engaged in the academic field.

RESUPINATE APHYLLOPHORALES OF INDIA

Dr. M.H. Hakimi
Prof. J.G. Vaidya
Dr. K. R. Ranadive
Dr. Jamaluddin
Dr. P. K. Jite

SCIENTIFIC PUBLISHERS (INDIA)

Published by:

Scientific Publishers (India)
5 A, New Pali Road, P.O. Box 91
Jodhpur 342 001 (India)

Branch Office
Scientific Publishers (India)
4806/24, Ansari Road, Daryaganj
New Delhi - 110 002 (India)

E-mail: info@scientificpub.com
Website: www.scientificpub.com

Print : 2021

ISBN: 978-81-7233-840-4

Typeset by Rajesh Ojha
Printed in India

PREFACE

The Western Ghats by and large, is the richest area in the context of composition of wood rotting mycobiots and there exists the endemic angiospermic taxa in this region (Kulkarni, 1979). It was suggested that since a long time, the subtropical vegetations in the Western Ghats of India are related (Palaeontologically or Geologically) with that of tropical Africa, South east Asia and with the part of tropical America which was revealed in the studies of lichen flora (Awasthi, 1965) and in rust flora (Ajrekar, 1912; Sathe, 1969).

The present work has great Mycological significance and will update the knowledge of Resupinate Aphyllophorales of Maharashtra, which presently lacks in the scenario of Indian Mycology. This will be of help for further studies and investigations. Taking into consideration the taxonomic study of Wood Rotting Fungi, this kind of approach is studied for the first time for understanding thoroughly the compilation of the Resupinate Aphyllophorales from a particular region. This work will facilitate further understanding of important pathogens from the vegetation of Maharashtra. The major work deals with the species studied from the family Corticiaceae with non-poroid hymenium; Polyporaceae and Hymenochaetaceae with poroid hymenium. The external morphology of the basidiocarps of non-poroid and poroid Aphyllophorales differ from each other. The morphology of non-poroid Aphyllophorales was studied by Prof. Talbot and he Published a review "Micro morphology of the lower Hymenomycete" (1954), while the morphology of poroid Aphyllophorales were Published by Gilbertson and Ryvarden (1986) in "North American Polypores, Vol. 1" and by Ryvarden (1991) in" Genera of Polypores. Nomenclature and taxonomy".

The present investigation was therefore undertaken to reveal the floristic affinities of the resupinate Aphyllophorales from Western Ghats of India with that of Tropical Africa, South-East Asia and Tropical America. The investigation was undertaken along with a thorough understanding of resupinate Aphyllophorales with respect to its taxonomic identity, floristic affinities, distribution patterns, detailed macro and micro morphological characters, enzymatic patterns, along with their culture characters. In the present study a total of 110 fungal species from the 46 respective hosts were identified out of 355 collected specimens. In the present study a total of 110 species of resupinate Aphyllophorales (27 families and 54 genera) from the 46 respective hosts were identified out of 355 colected specimens throughout the Western Ghats.

Our aim is to provide an entry point for the students in to the World of Aphyllophorales. The work is supplemented with an index of genera, index of hosts/ substrata along with fungi recorded thereon and the list of references. The present publication would prove useful to Mycologist, Plant pathologists, Plant Biotechnologists and also to research students and teachers alike in the disciplines of Mycology and Plant Pathology. Through this book readers are going to get a comprehensive guide of Resupinate Aphyllophorales. Our special thanks to the Scientific Publishers (India), Jodhpur for taking efforts to publish the same. We are especially grateful to Dr. Jamaluddin for his constant encouragement and critical comments to improve the quality of the book. Any suggestions for improving the book will be kindly received and acknowledged.

Authors

ABBREVIATIONS

AM	:	Amyloid
ASL	:	Above sea level.
BR	:	Brown Rot
cm.	:	Centimeter.
cms.	:	Centimeters.
DEX	:	Dextrinoid
diam.	:	Diameter.
Dist	:	District.
DP	:	District Pune
DR	:	District Ratnagiri
DS	:	District Satara
DST	:	Dead Standing Tree
e.g.	:	For example.
EF	:	Effused
EFR	:	Effuso Reflexed
FBT	:	Fallen Branches of Tree
Fig.	:	Figure.
g.	:	Grams.
H_2O_2	:	Hydrogen peroxide.
hrs.	:	Hours.
KOH	:	Potassium hydroxide.
LST	:	Live Standing Tree
mm.	:	Millimeter.
NAM	:	Non Amyloid

No.	:	Number.
nov.	:	Novel/new.
NPO	:	Non - Poroid.
NTI	:	New To India
NTM	:	New to Maharashtra
p-cresol	:	Para - cresol.
PDA	:	Potato Dextrose Agar.
PO	:	Poroid.
PU	:	Pune
PUCC. B.	:	Pune University Culture Collection, Botany.
RS	:	Rainy Season
RSU	:	Rsupinate
RT	:	Ratnagiri
SA	:	Satara
Sp.	:	Species.
VH	:	Vaidya - Hakimi.
WR	:	White Rot
WS	:	Winter Season
α-napthal	:	Alpha - napthal.
μm.	:	mili - micron.

Plate 1

Aleurodiscus aberrans

A. cremeus

Amylocorticium cebennese

A. olivaceoalbum

Amylocystis sericeomolli

Amylosporus bracei

Amylostereum laevigatum

Anomoporia dumontii

Auriporia aurulenta

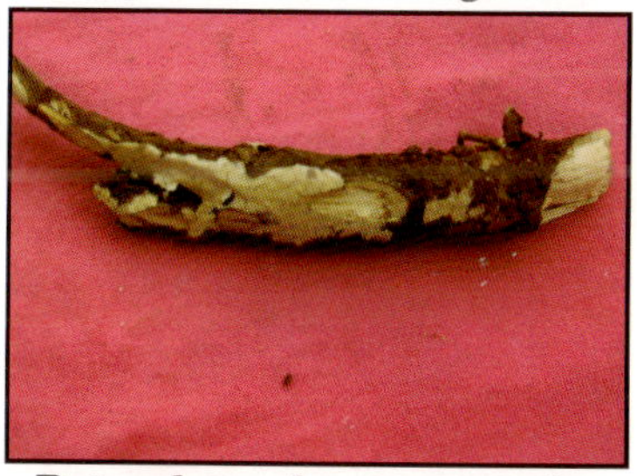

Botryohypochnus anomalus

Candelabrochaete verruculosa

Cejpomyces terrigenus

Cerrena unicolor

Cystostereum murraii

Diplomitoporus hondurensis

Plate 2

Earliella scabrosa

Fibriciellum silvae-ryae

Flavodon flavous

Gloeocystidiellum citrinum

G. convolvens

G. fistulatum

G. flammeum

G. furfuraceum

G. insidiosum

G. irpiscescens

G. kenyense

G. lactescens

G. lacticolor

G. leucoxanthum

G. luridum

Plate 3

G. luteocystidiatum

G. percuriosum

G. sulcatum

G. turpe

Grammothelopsis puiggarii

Hymenochaeta attenuata

H. cinnamomea

H. rubiginosa

H. tabacina

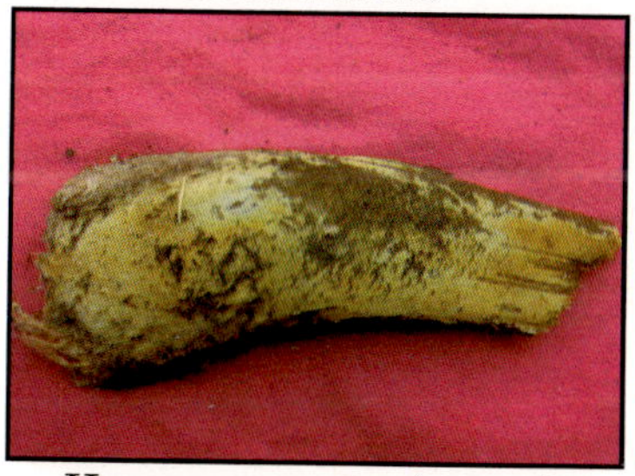

Hymenogramme javensis

Hyphoderma radula

Hyphodontia alienate

Hypochnicium cymosum

H. eichleri

H. globosum

Plate 4

Junghuhnia crustacean

Kavinia himantia

Laeticorticium simplicibasidium

Laxitextum lutescens

Lepidomyces subcalceus

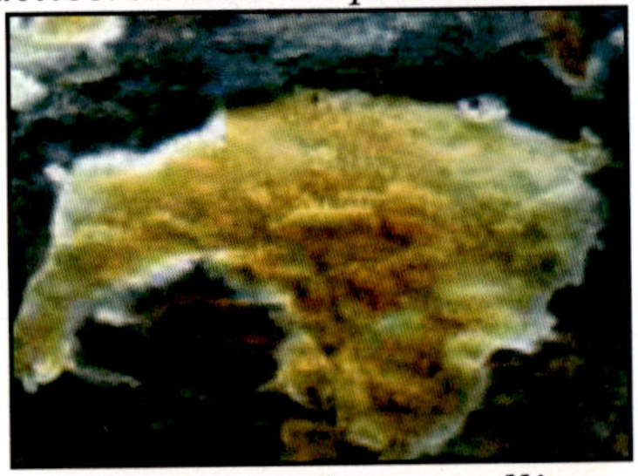

Leucogyrophana mollis

Lopharia fulva

Metulodontia flavidoalba

Oxyporus cervino-gilvus

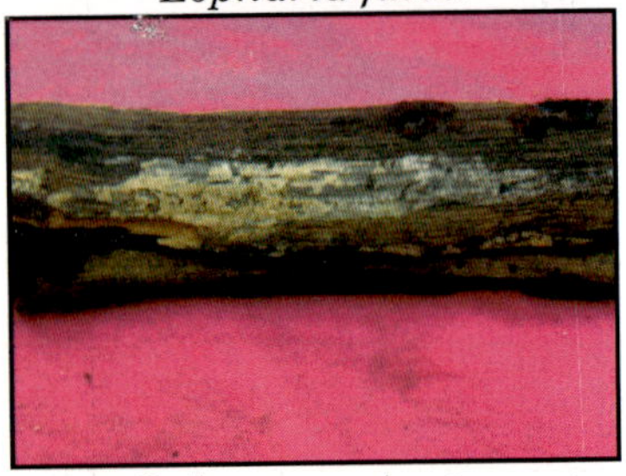

O. latamarginatus

Peniophora aurantiaca

P. farinose

P. gladiola

P. laurentii

P. nuda

Plate 5

P. pithya

P. violaceolivida

Perenniporia gomezii

P. medulla-panis

Phanerochaeta cacaina

P. calotricha

P. joseferroirae

P. laevis

P. sanguinea

P. sordida

P. tuberculata

Phellinus carterii

P. cesatii

P. chryseus

P. inermis

Plate 6

P. lamaensis

P. luctuosus

P. punctatus

P. purpureogilva

P. sonora

P. sonora

Phlebia rufa

P. subcretacea

Phlebiopsis galochroa

P. gigantea

P. peniophoroides

P. roumeguerei

Pseduoxnasma verrucisporum

Pycnoporellus alboluteus

Ramaricium polyporoideum

Plate 7

Rigidoporus ulmarius

Schizopora carneolutea

S. paradoxa

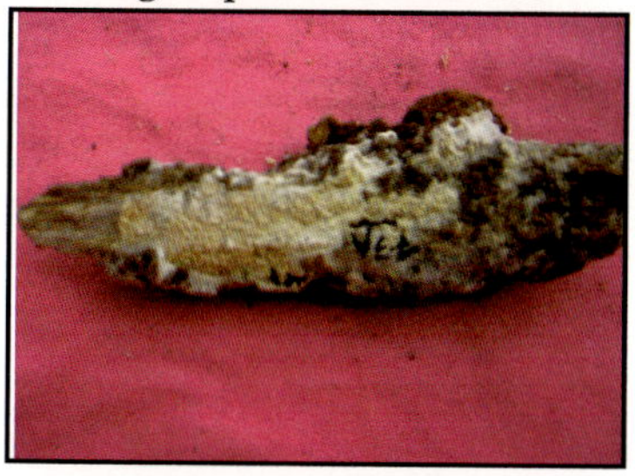
S. roseotigen

S. trichiliae

Scopoloides hydnoides

Scopuloides rimosa

Scytinostromella cerina

Steccherinum setulosum

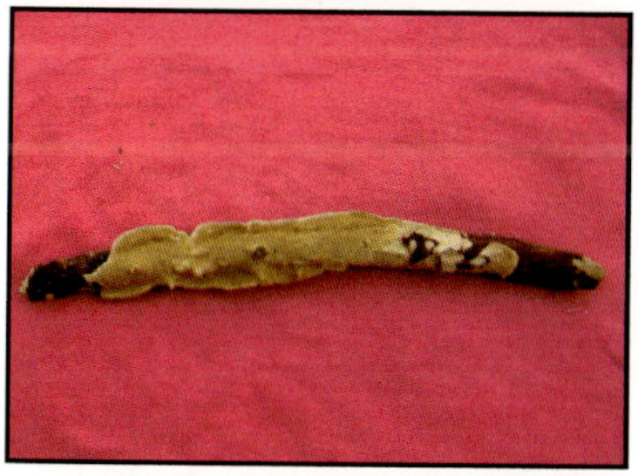
Trichaptum versatile

Tubulicrinis ellipsoideus

Tyromyces undosus

Vararia spheari cospora

Vuilleminia acerina

Wolfiporia cocos

Plate 8

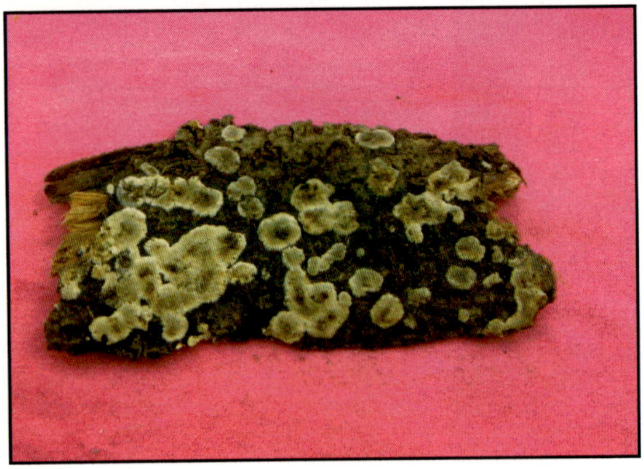

Wrightoporia africana

W. avellanea

W. cremea

Xylobolus apricans

X. frustulatus

Plate 9

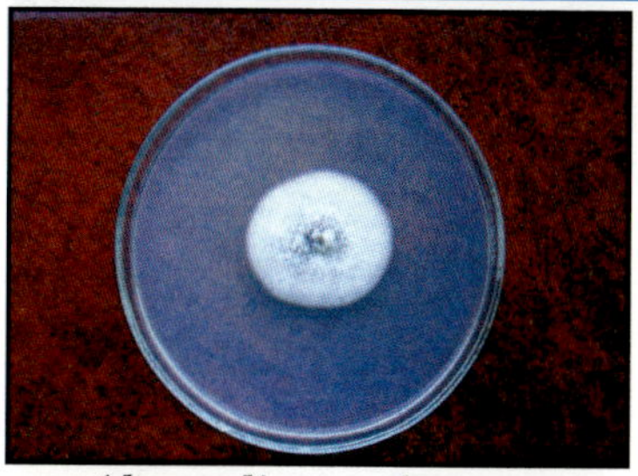
Aleurodiscus aberrans

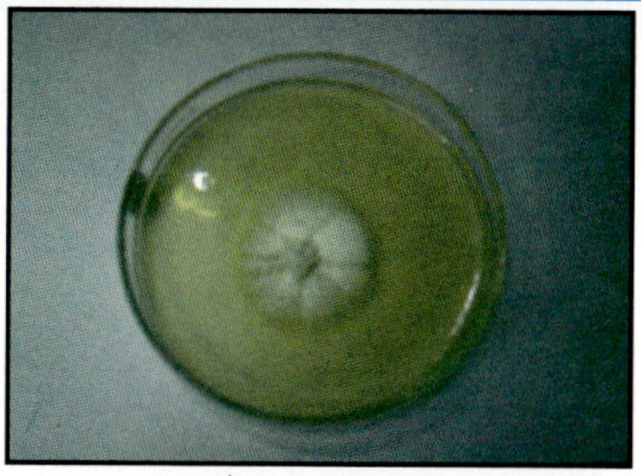
A. cremeus

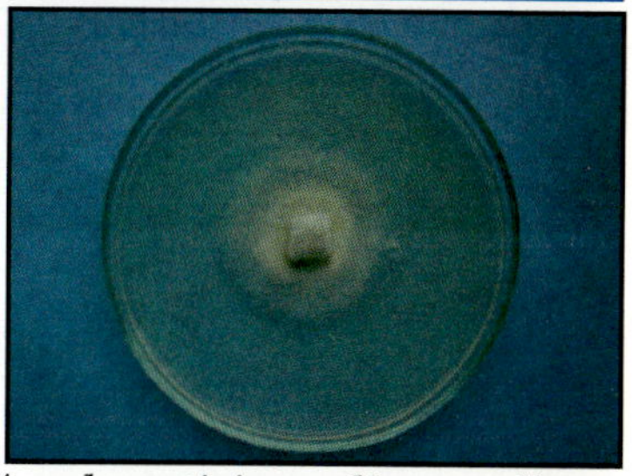
Amylocorticium olivaceoalbum

Amylosporus bracei

Anomoporia dumontii

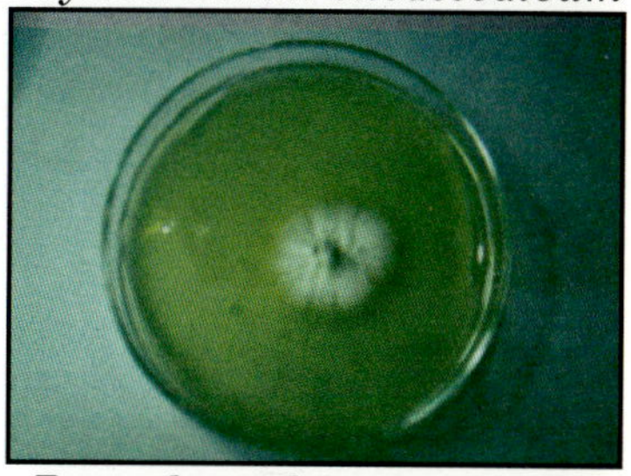
Botryohypochnus anomalus

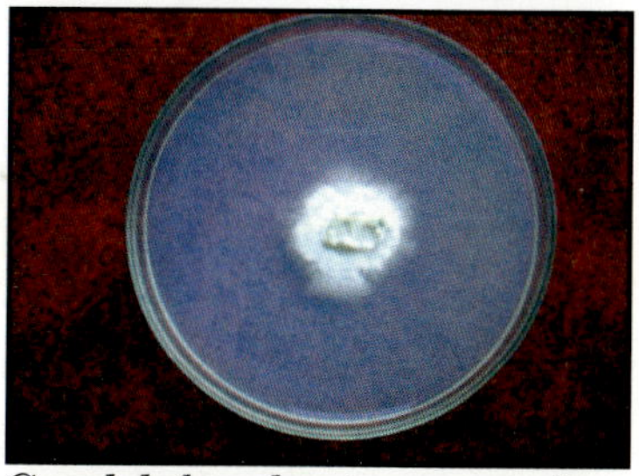
Candelabrochaete verruculosa

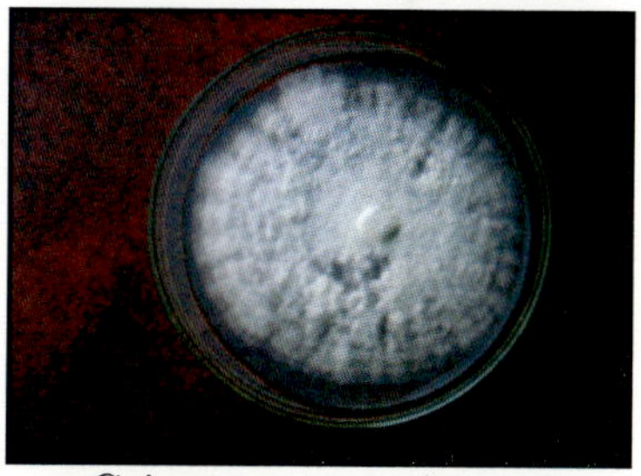
Cejpomyces terrigenus

Cerrena unicolor

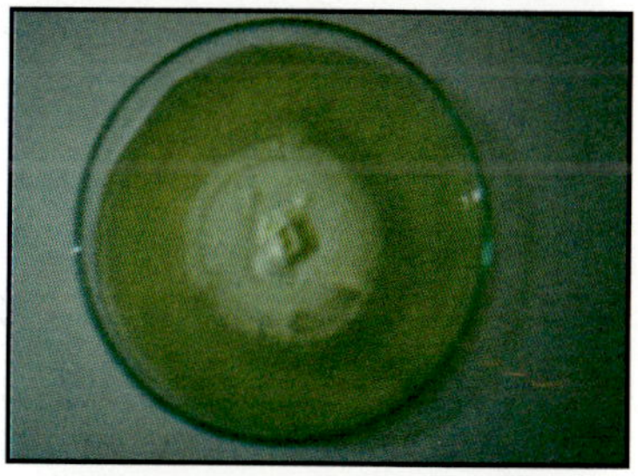
Cystostereum murrayi

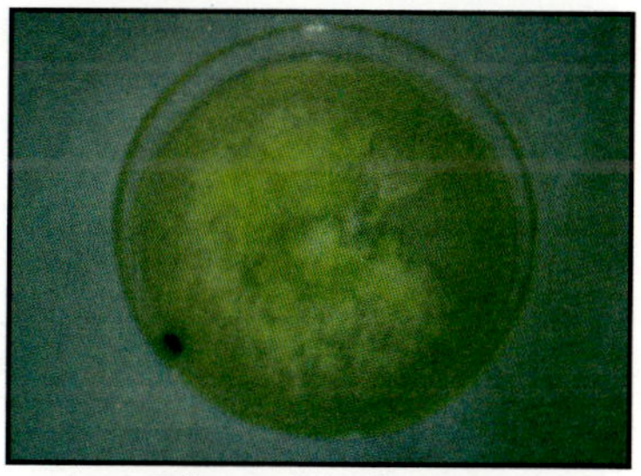
Flavodon flavus

Gloeocystidiellum citrinum

G. lactescens

G. lacticolor

G. luridum

Plate 10

G. luteocystidiatum

Hymenochaete cinnamomea

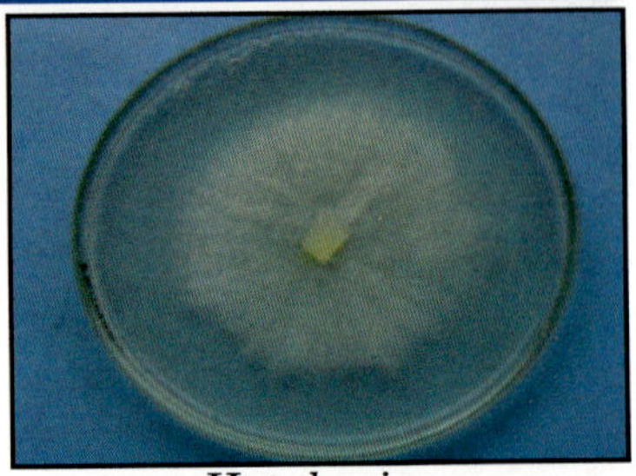

H. tabacina

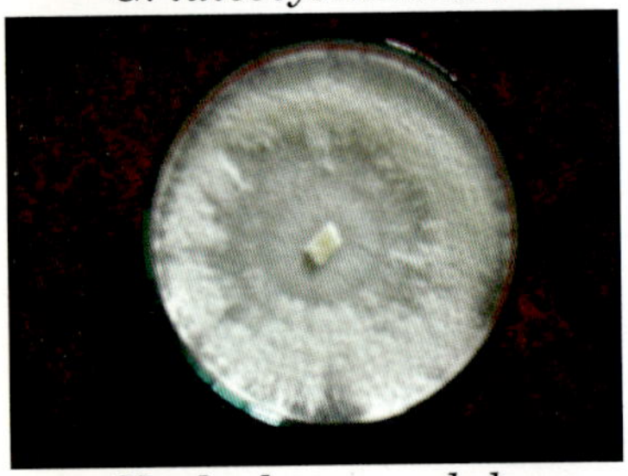

Hyphoderma radula

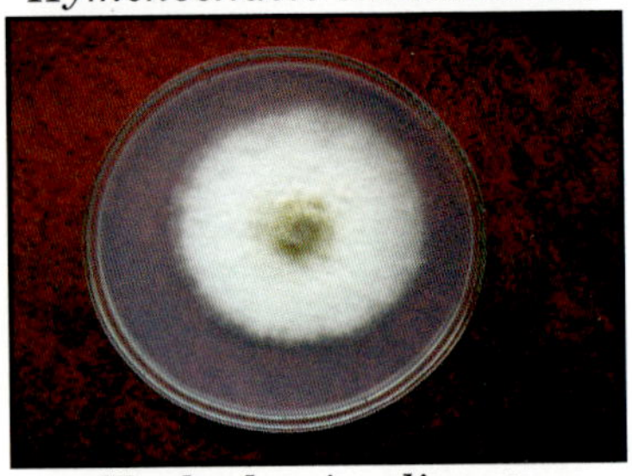

Hyphodontia alienata

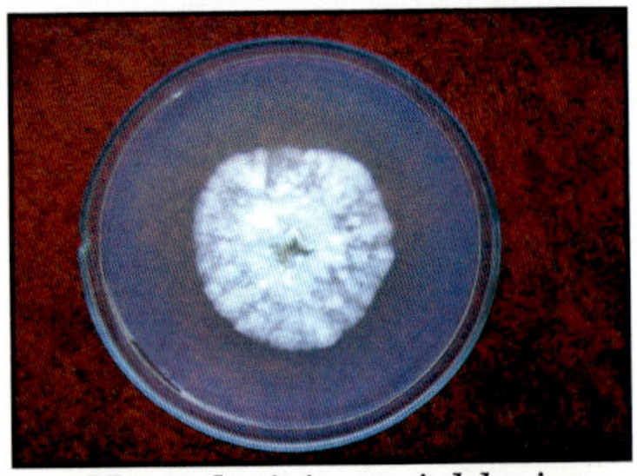

Hypochnicium eichlerie

H. globosum

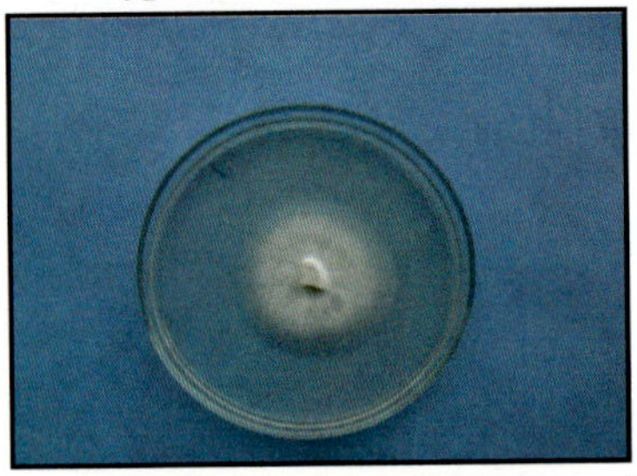

Kavinia himantia

Laeticorticium simplicibasidium

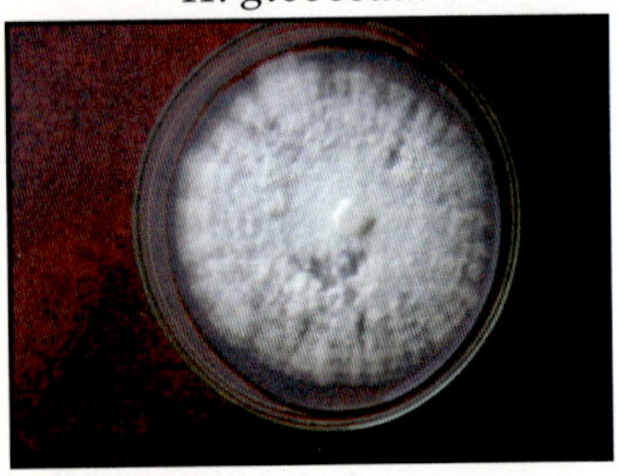

Laxitextum lutescens

Leucogyrophana mollis

Lopharia fulva

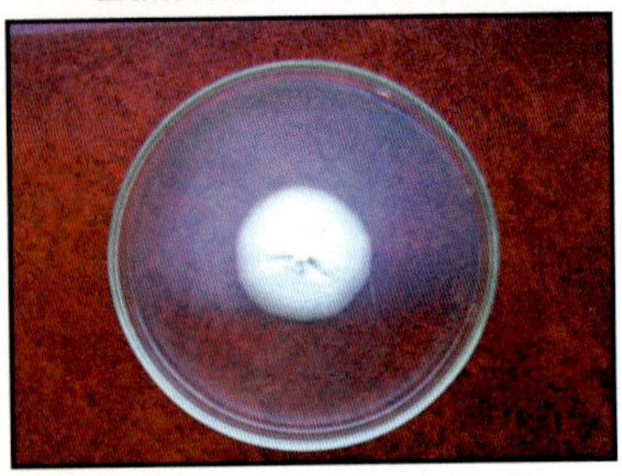

Metulodontia flavidoalba

Oxyporus latemarginatus

Peniophora aurantiaca

Plate 11

Peniophora gladiola

P. nuda

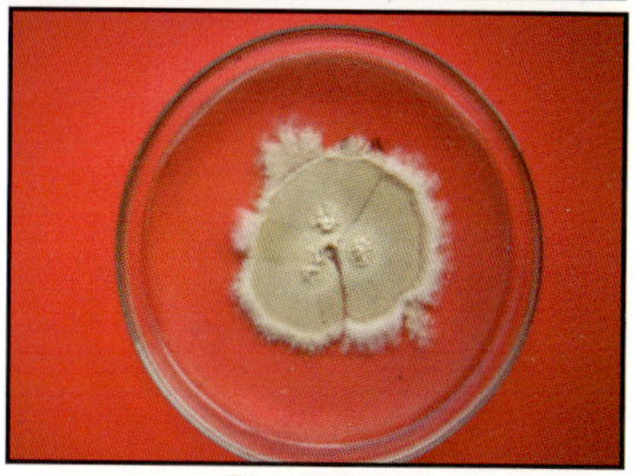

P. pithya

P. violaceolivida

Perenniporia medulla-panis

Phanerochaete cacaina

P. calotricha

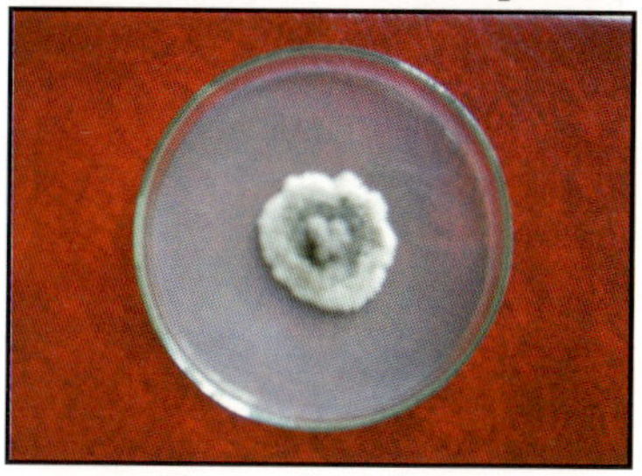

P. joseferroirae

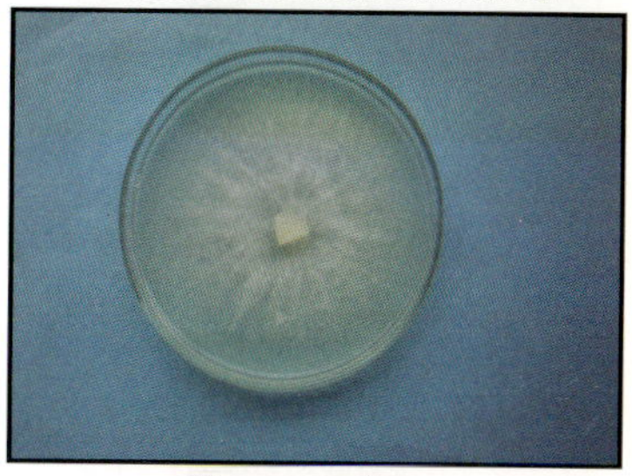

P. sanguinea

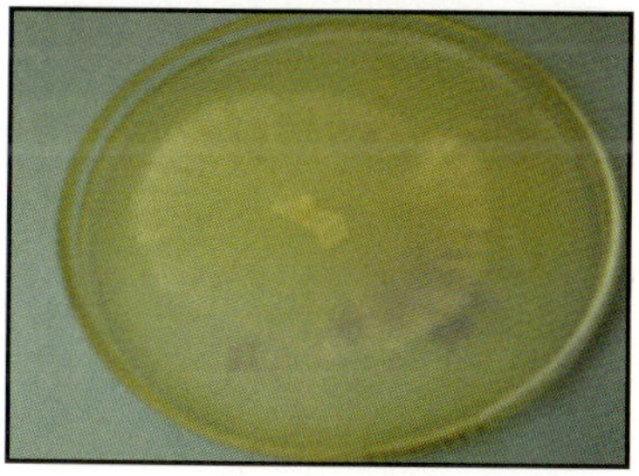

P. sordida

Phellinus punctatus

P. purpureogilva

P. cesatii

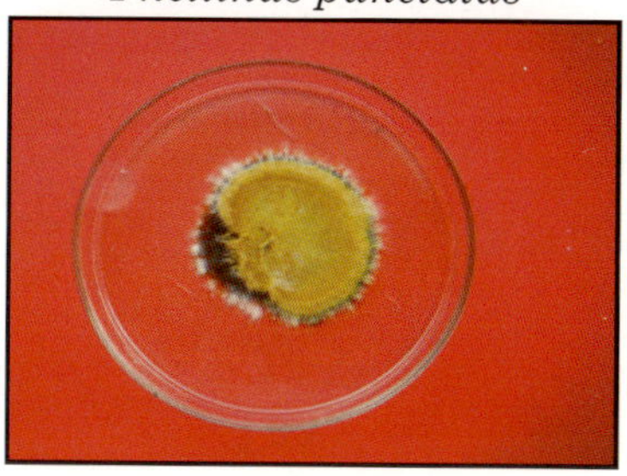

P. lamensis

P. sonorae

Plate 12

Phlebia rufa

Phlebiopsis gigantea

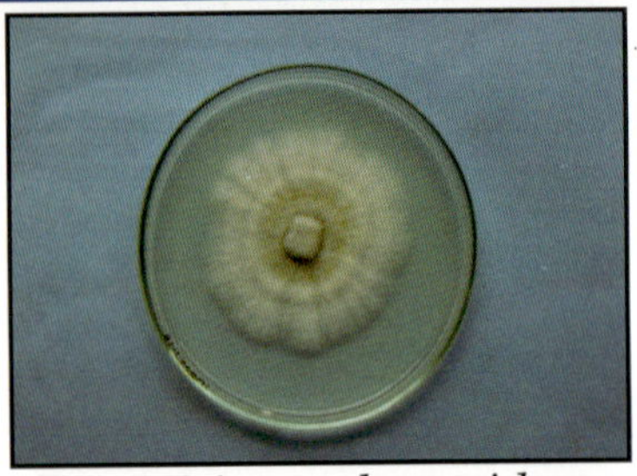

Ramaricium polyporoideum

Schizopora carneolutea

Schizopora roseotigens

Scopuloides hydnoides

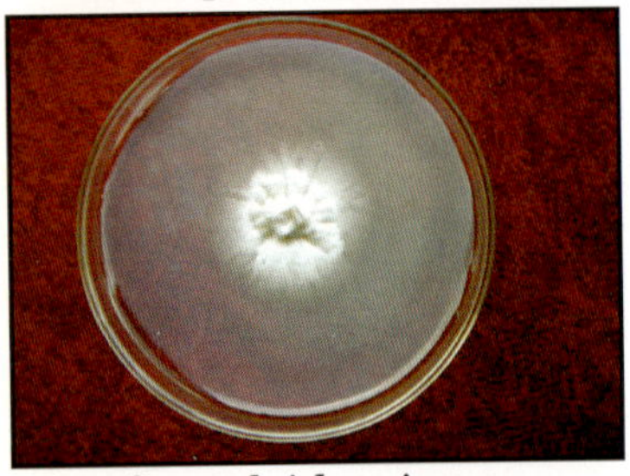

Scopuloides rimosa

Trichaptum versatile

Tubulicrinis ellipsoideus

Vararia sphearicospora

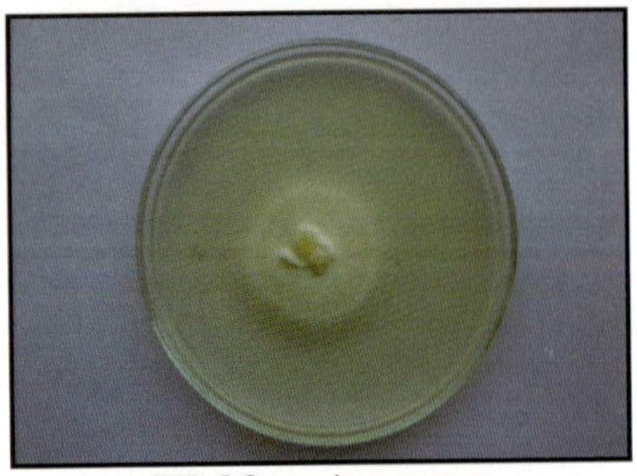

Wolfiporia cocos

Wrightoporia avellanea

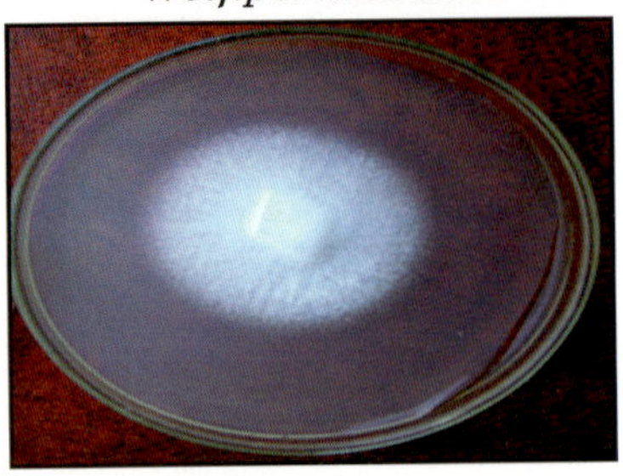

Wrightoporia cremea

CONTENTS

INTRODUCTION

India is rich in fungal flora, especially have a diverse group of resupinate Aphyllophorales fungi. The aim of the present publication is to characterize the resupinate Aphyllophorales from the the Western Ghats of India. The lignicolous fungi cause extensive damage to the forest ecosystem. Hence these mycobiota are of economic importance and majority belongs to wood rotting Basidiomycetes. Around 7-12% of the earth's surface is covered by vegetation biomes. For a healthy environment, our vegetation resources should be 23% or so. The reasons behind the gradual depletion are many e.g., the devegetation due to illicit cutting, commercial plantation, construction of large water bodies, promotion of unsustainable tourism, conversion of vegetation area for the construction of rail, road and communication lines. Indian forest Policy (1988) pays attention for maintenance of 33% forest cover in India.

Among the total biomass production on earth, about 100 x 10^9 tons annually is that of wood perennials (Basham, 1975). Of this approximately 20 x 10^9 tons are lignin. We experience a great loss of commercial timber due to partial attention paid towards the woody tissues, as compared to that of herbaceous plants (Cooke and Rayner, 1984). Now-a-days, wood is has become an important energy source because renewable energy sources like charcoal, are becoming difficult to procure, as well as these are highly expensive commodity. As such wood is finding an, important place towards the domestic and aesthetic purposes, besides being used extensively as fire wood. Hence, there is a lot of illicit cutting of trees leading to devegetation, which leads to removal of literally all wood residue for fulfilling commercial needs, making wood almost an extinct commodity. To avoid this, there is an urgent need to declare our vegetations, at large, as preserved and protected.

Wood decay is mainly due to damage caused to a live tree and in turn this decay becomes a major cause of destruction of trees. Decay usually begins as wounds to branches, trunks or roots and is caused by wood decay fungi and woods borers, where a casual succession of microorganism is responsible for the process of further decay. These wood decaying fungi under favorable conditions produce basidiocarps that remain dry and dormant over a long period of low precipitation. Disease in managed and protected vegetations often occurs when air-borne basidiospores colonize freshly cut stump (infection courts) surfaces during thinning operations. The fungus then grows through the stump and infects other trees at points of stems grafts or contact. This disease goes on for many years inside the tree and is unseen. Strength is lost as wood is decomposed by micro-organisms. Final results are seen as big hollow cores in old trees which come crashing down in storms.

In India Aphyllophorales have been worked by many researchers e.g. Bagchee *et al.* (1950, 1953 and 1954), Bakshi (1958, 1971), Bakshi *et al.* (1970), Thind *et al.* (1956 and 1978), Rehill and Bakshi (1965, 1966). Thind (1973, 1975), Sathe and Rahalker (1976), Rattan (1977), Thind and Dhanda (1978); Anjali Roy and others (1979, 1987), Harsh (1982); Natarajan *et al.* (1980, 1985, 1998); Ganesh and Leelavathy (1986), Vaidya and his colleagues (1987, 1990, 1991, 1993 and 1994), Rabba (1994), Sharma (1995) and Nanda (1996) pertaining mainly to Himalayas and some other parts of Central Peninsula.

This study presents observations on resupinate Aphyllophorales in the investigation areas; a significant, well-preserved large patch of moist tropical vegetation, also distributed elsewhere in India (Western Ghats, Eastern Ghats, Orrisa, Himalayan foothills, North-East India and Nicobar). The investigation areas forms almost the northernmost resort of evergreen vegetation flora and fauna in the Western Ghats expect for a few scattered secondary pockets of evergreen vegetation.

Samples of resupinate Aphyllophorales were collected in the sexual stages from above mentioned localities on different hosts like representing forest species medicinal, horticultural and agricultural plants and their taxonomic study were made.

Taxonomy of Resupinate Aphyllophorales

Aphyllophorales which have resupinate, effused or effuse-reflexed type of fruit bodies are collectively considered as resupinate Aphyllophorales (Patouillard, 1900). They are roughly characterized by non-septate basidia, persistent gymnocarpous and non-putresent fruit bodies which are usually not lamellate (Bourdot and Galzin, 1928; Donk, 1933; Rattan, 1977; Gilbertson, 1979). The basidium bearing structures may be variable like even, tuberculate, meruloid, raduloid, warted, toothed or poroid. The shape of basidium is another variable character which may be spherical, obovoid, clavate or cylindrical. The basidia may show constrictions in the middle or may formed laterally [Pleurobasidia] or terminally on the fertile hyphae. The number of spore-bearing sterigmata is usually four, but may vary from one to eight (Donk, 1930).

The delimitations of the resupinate Aphyllophorales from the order Aphyllophorales from the Agaricales and the Termellales has not yet been definitely established (Bondarzew and Singer, 1941; Oberwinkler, 1972).

The literature on the resupinate Aphyllophorales is vast and can be conveniently divided into three major parts according to its period:

PART-1: This part mainly consists of the work of Linneaus (1953), Persoon (1925), Fries (1849) and Cooke (1886). Their work mainly deals with the broad external characters of these fungi, on the basis of which they are broadly classified into different groups.

PART -2: This part is of major activities, Corner (1932-1953) and Cunningham (1947-1965) brought out the significance of the hyphal organization in the identification of the fungi, belonging to the resupinate Aphyllophorales.

PART- 3 : This part consists of current works of Eriksson (1958); Donk (1964); Anjali Roy (1972-1987); Pegler (1973); Buchana (1992-2000); Hjortstam (1973-1983); Eriksson and Ryvarden (1973, 1975, 1976); Ryvarden (1973-2000); Gilbertson (1977-1980); Rajchenberg (1983-2003); Wu (1990-2006); Wu *et al.* (1990-2004) etc., which deals in details with the external and internal characters, chemical reactions, hyphal structures of the basidiocarps under natural and culture conditions. This leads to an understanding of the phyllogenetic affinities among the members of resupinate Aphyllophorales.

Historical account

The term resupinate Aphyllophorales was proposed by Patouillard, for Basidiomycetes having macroscopic basidiocarps in which the hymenophore is flattened (Thelephoraceae), tooth-like (Hydnaceae) or has the hymenium lining tubes (Polyporaceae) or some times on lamellae, the poroid or lamellate hymenophores being tough and not fleshy as in the Agaricales. Traditionally resupinate Aphyllophorales has had a core of three families based on hymenophore shape, as described above, but recent detailed microscopic studies of basidocarp structure has shown these groupings to be unnatural and thus the taxonomy of resupinate Aphyllophorales is at present in a state of flux. Donk (1964), who recognized 22 families is now followed (Kirk *et al.,* 2001).

Much of the literature of resupinate Aphyllophorales is based on the traditional family of groupings and as under the current re-arrangements, one family may exhibit several different types of hymenophore (e.g. Gomphaceae has effuse, clavarioid, hydnoid and flux cantharelloid hymenophores). Reference to the literature is complicated because information about the genera of any one family may occur in apparently unrelated monographs.

The resupinate Aphyllophorales were not recognized as distinct group by the early workers like Linneaus (1753). Linneaus in "Species plantarum, " (1753), which is considered the stating point of all botanical and now also fungal nomenclature (Stafleu, 1983) used *Boletus* for all fungi with tubes or pores. He described twelve species belonging to the family Polyporaceae. Persoon (1801) was the first to segregate the lamellate and poroid fungi. The fungi now classified in the

Aphyllophorales were placed by Persoon (1801) in the order Hymenochaetales. This order contained all the Basidiomycetes except the Gasteromycetes and some Ascomycetes with which the *Tremellales* were placed in Helvelloidei. The subdivision of the Hymenothecii was based on the hymenial configuration. The division Agaricoidei contained species with a lamellate (*Amanita* and *Agaricus*) or veined hymenophores (*Merulius*). The species with a tabulate or poroid hymenophore were classified under the division Boletoideli and included the *Boletus* and *Polyporus*. The division Hydnoidei had a toothed hymenophore and contained *Sistotrema* and *Hydnum*. Species with an even to papillate or warted hymenium were classified in the division Gymnodermatae, which included the genus *Thelephora*. Finally the species with fleshy elongate basidiocarps with a pileus and stipe were placed in the Clavae-formei with the genus *Clavaria*. Persoon (1801) in his monumental work of "Synopsis Methodica Fungorum", which marks the beginning of the taxonomy of the Hymenomycetes, mentioned over 70 species of the Polyporaceous fungi.

Fries (1821) in his "Systema Mycologicum" accepted two genera for the polypore. His concepts were based on the type of hymenophore (basidiocarps). Fries created the sub-order 'Pileati', which included the genera such as *Agaricus, Schizophyllum, Daedales, Merulius,* and *Favolus* as the sub-genera of the genus *Polyporus,* Later, *Favolus* was raised by him (Fries, 1828) to the generic status. Fries recognized eight genera in total. But soon the workers realized the necessity of segregating Fries artificial and heterogeneous group into more natural ones. Berkeley (1839) was probably the best amongst the old masters who did his observations without the microscopic aids, which were available to his successors. Karsten (1881-1889) and Bresadola (1897) have used microscopic characters in their description of the fungi. Up to the end of the 19th Century, all these classifications were based on macro morphological features of the sporophore. Patouillard (1900) was the first mycologist who used microscopical characters for the delimitation of higher taxa. In his "Essai Taxonomique", Patouillard made groupings in polypores on the basis of characters like detailed hyphal morphology, structure of the pileus and characters of basidia, spores and cystidia. He divided the Basidiomycetes into Homobasidiomycetes and Heterobasidiomycetes. The Heterobasidiomycetes were further subdivided according to the septation of the basidia. Species with transversely septate basidia were classified in the Auriculariaceae, species with longitudinally septate basidia in the Tremellaceae, and species with aseptate basidia in the Tulsnellaceae and Caloceraceae. The Homobasidiomycetes had non-septate basidia and were divided into four families - the parasitic Exobasidiaceae, the gymnocarpous Aphyllophoraceae, the hemiangiocarpous Agaricaceae and the angiocarpous Gasteromycetaceae.

Patoulillard divided the Aphyllophoraceae into two tribes:

(a) **The Clavariales:** With an erect basidiocarps which could be simple, branched or dendroid but never pileate or with amphigenous hymenium.

(b) **The Porohydnales:** With a resupinate or pileate, sessile or stipitate basidiocarp and hymenium underneath the cap. The Porohydnales are subdivided into four sub tribes, based on the form of hymenophore which is cupulate in the Cyphelles, even to warted in the Odonties, poroid in the Pores, and toothed in the Hydnes.

British mycologists Berkeley (1839) described about five hundred and sixty polypores. It was rather difficult to survey the group and no comprehensive flora had been written for any country. Prof. W.A. Murrill (1903 - 1915) felt the need for a manual of the American species and was a pioneer of the long series "Polyporaceae of North America" (1903 - 1908).

Patouillard's system was also adopted by Bourdot and Galzin (1928), in their classical manual. They realised that several species which they considered to be closely related, had to be placed in different groups. Donk (1930, 1933) also fundamentally used the Patouillardian system but his generic concepts were mainly based on microscopical characters. Donk (1960) in his work considered the taxonomic status of all published genera until then. It has resulted in establishment of many monotypic genera. His work is based on Patouillard, Murrill and Berkeley, who studied resupinate Aphyllophorales from different parts of the world.

Corner (1932a,b) distinguished three possible types of hyphae which may be present in the basidiocarps, namely, generative, skeletal and binding hyphae. Corner introduced the concept of 'hyphal system' and thus opened a new era in the field of modern taxonomy. The major significance of Corner's findings were first realized by Cunningham who applied Corner's system in his paper "Notes on classification of Polyporaceae" (1946). Later, in a series of publication he emphasized the value of thorough analysis of hyphal systems in to better understanding of a species. Talbot (1951) in earlier work studied many specimens of Aphyllophorales from tropical countries which initiated him to publish the classical work of "Resupinate Aphyllophorales of South Africa" (1914).

Burt (1918) study on resupinate Aphyllphorales in Florida, Jamaica, Burt (1920) study on resupinate Aphyllphorales in Mexico, Burt (1924) study on resupinate Aphyllphorales in District of Columbia, Florida, Alabama and Mexico. Talbot (1954) published a review paper entitled "Micro-morphology of Lower Hymenomycetes". This paper concerned with macro and micromorphological characters of Aphyllophorales, is still considered as an important literature in understanding the sexual, accessory and hyphal configuration of basidiocarp of Aphyllophorales.

Donk's series on resupinate Hymenomycetes (1954, 1956a, 1956b, 1957, 1958), Lentz (1955) study on resupinate Aphyllophorales in South Africa. Lowe monographed the American species of *Fomes* (1957), *Poria* (1966), *Tyromyces* (1975).

Eriksson's studies on resupinate Aphyllophorales of the Muddus National Park in Sweden (1958), greatly altered the generic delimitation within the Corticiaceae. Boidin (1958b, 1959a, 1959b) published three essays on the genus *Stereum* and redescribed a number of genera, Slysh (1960) study on resupinate Aphyllophorales in Cuba, Alabama and Florida, Cunningham (1963) study on resupinate Aphyllophorales in South Africa, Sri Lanka, Australia, Java, South America, Africa and Malaya.

Lowe (1963b) study on *Anthrodia and Loweporus* in Cuba, Jamaica, Mexico and Brazil, Lemke (1964 a,b) study on *Dendrothele* in British Columbia and Trinid and Reid (1965) monographed the stipitate steroid fungi of the world, Welden (1965) study on *Vararia* in west Indies, Cunningham (1965) study on resupinate Aphyllophorales in Peru, South and Central America West Indies Philippines, East Indies and Australia. Rehil and Bakshi (1965, 1966) study on resupinate Aphyllophorales from North western Hymalayas, Reeves and Welden (1967) study on Hymenochaete in Alabama.

Aleurodiscus by Lemke and Parmasto (1968), Thind and Rattan (1968, 1970, 1971 a,b,c, 1972, 1973a,b) study on resupinate Aphyllophorales from North western Hymalayas, Bakshi (1971) study on *Gloeoporus* in Madagascar, Brazil and Cuba, Bakshi (1971) study on resupinate Aphyllophorales in North western Hymalayas, Boquiren (1971) study on resupinate Aphyllophorales in west Africa.

Mass-Geesteranus revised many of the hydnoid fungi of the eastern old world (1971), Julich (1972) study on resupinate Aphyllophorales in Venezuela, Liberta (1973) study on resupinate Aphyllophorales in Florida, Hawaii, Argentina and west Indies. "Systematic Survey of the Corticiaceae" together with the voluminous work of Eriksson and Ryvarden on the 'Corticiaceae of North Europe' (1973, 1975, 1976) were important contributions to our knowledge of this group of Basidiomycetes.

Welden (1975) study on resupinate Aphyllophorales in Cuba, Jamaica, Mexico and south Africa, Rattan (1977) study on *Scytinostroma* in Pakistan, Rattan (1977) study on *Xylobolus* in North western Hymalayas, Kotlaba and Pouzar (1978) studied resupinate form *Phellinus* in Florida, Alabama and Jamaica, Ryvarden (1978) study on resupinate Aphyllophorales in eastern central Africa, Ryvarden and Johansen (1980) study on resupinate Aphyllophorales in east Africa, Eriksson, Rajchenberg (1982-2005) study on resupinate Aphyllophorales in Argentina, Nakasonek (1990) study on wood-inhabiting Corticiaceae and selected Hymenochaetes in North America, Wu (1990-2006) study on resupinate Aphyllophorales in Taiwan, Wu *et al.* (1990, 1995, 2003, 2004) study on resupinate Aphyllophorales in Taiwan, Decock *et al.* (1997-2000) study on *Phellinus* and *Perenniporia*, Buchanan (1992, 2000) study on resupinate Polypore in New Zealand, Nakasone (2003) studies on fourteen speciemen hydnaceous Hymenocetes in Ecuador, Java, Madagascar, Tunisia, Venezuela and Vitenam, Slippers *et al.* (2003) study on genus *Amylostereum* in South Africa, Dai *et al.* (2004, 2007) study on poroid and non-poroid resupinate Aphyllophorales in western China.

Many regional floristic studies were carried out during the last 50 years and these provided a basis for a more sound. Donk (1964) reviewed all the progress and proposed a new conspectus for the families of "Aphyllophorales". Parmasto (1968) discussed inter-relationships in Corticiaceae and related families.

Taxonomy of Aphyllophorales and resupinate Aphyllophorales is still in a state of flux, as a result, Gilbertson (1980) made an effort to develop a system based on phylogeny. Generic monographs on Aphyllophorales published by Gilbertson (1977), Ginns (1982), Julich (1984) and also type species studies have revealed a lot of ambiguity in confirming the status of old traditional species.

Culture studies

Long and Harsch (1918) and Fritz (1923) were among the first to describe a number of wood inhabiting resupinate Aphyllophorales in pure cultures, particularly polypores. The descriptions are mainly based on such macroscopic characters as colour and texture of the mycelial mat (Long and Harsch, 1918) and the importance of microscopic character of cultures (Fritz, 1923). Campbell (1938) described cultures of 32 species of *Fomes*, paying more attention to microscopic details. Similar details have also been worked out by Davidson *et al.* (1942), Cartwright and Findlay (1958) also gave an important contribution to the knowledge of Wood Rotting Basidiomycetes in pure culture.

Nobles (1948) described 126 species of wood inhabiting Basidiomycetes in pure cultures. She studied morphological and physiological characters like the colour change of the agar, type of rot, odour, host relationship, etc. She was the first to place special emphasis on the characters of advancing (marginal) zone of a culture. Some important studies have been published by Nobles (1948, 1958a, 1958b), Boidin (1958a, 1958b) and Sarkar (1959), and also by many more workers. Nobles (1965) published the most important descriptive key of 149 species, which include the results of above-mentioned studies.

Recently culture studies have been carried out by Lombard and Gilbertson on *Poria* (1965, 1966), David on *Trametes* (1967), Boidin and co-workers on *Aleurodiscus* and related genera (1968) and on *Vararia* (1975). The tropical species received attention by Bakshi *et al.* (1970) and Sen (1973) who described a number of polypore species from India. Buchanan (1992) described cultural resupinate Aphyllophorales from New Zealand. Rabba (1994) described several of *Phellinus* species and Nanda (1996) described 48 species of poroid and non-poroid resupinate Aphyllophorales from India. Wu (1996) described several cultural resupinate Polypores from Taiwan. Chang *et al* (1998) study on culture *Antrodia, Antrodiella* and *Phellinus* from Taiwan. Rajchanberg (2003) described 6 culture of resupinate Polypores from New Zealand, Argentina and Australia. Chang *et al.* (2004) study on cultural *Antrodia* in Taiwan.

The type of rot is considered to be an important taxonomic character, (Davidson *et al.,* 1938; Nobles, 1948; Boidin, 1951; Domanski, 1965). Thus, possibility of detecting the enzymes responsible for the decomposition of lignin in pure culture is also important. The various chemical tests to detect the extra-cellular enzymes especially laccase and peroxidase are established by different workers.

The culture characters were first laid down in a species code by Davidson *et al.* (1942). The first three symbols in this species code were letters, giving information about colour, oxidase reaction and growth rate respectively. Later on Nobles (1948) used eleven columns referring various culture characters. This species code was modified by Nobles (1965) and constructed a species code of 53 numbers. Ultimately the culture studies are well organized and detailed descriptive analytical key of 550 species of wood-inhabiting Aphyllophorales and resupinate Aphyllophorales were given by Stalpers (1978) which is followed here to describe the culture of Western Ghats of Maharashtra.

In recent studies besides morphological and cultural studies, emphasis is laid on microscopical characters and chemical tests as suggested by Donk (1933), Bondartsev and Singer (1941), Ryvarden (1977 a,b), Ryvarden and Johansen (1980), Gilbertson and Ryvarden (1986). This gave a correct and scientific basis for the identification of the genera and species of the resupinate Aphyllophorales and therefore to its taxonomy.

Studies on resupinate Aphyllophorales were initiated along with the launch of studies on Indian Fungi. The first Indian record on the resupinate Aphyllophorales can be traced back to 1901 when Hennings reported Flavodon flavous from Maharashtra and Bengal. Later Lennings (1901) published Fungi India Orientalis.

During the first quarter of the present century, Massee (1901, 1906, 1908, 1910) published several accounts of Indian Fungi based on the collections sent to the Kew Herbarium by several workers, notably by Sir E.J. Butler (1905a,b,c.d, 1918). Several Indian resupinate Aphyllophorales were also reported by Llyod (1898 - 1925), Sydow *et al.* (1906-1916) etc. Theissen (1913) reported many poroid resupinate Aphyllophorales collected from the Bombay presidency by Blatter.

Dr. S.R. Bose (1919-1928) was the first Indian mycologist who gave a comprehensive account of the Indian polypores collected by him from Bengal and its surroundings. Sundaramani and Madurajan (1925) reported some members of polyporaceae from Madras, and by the end of the first quarter (up to 1925) there were several reports on the resupinate Aphyllophorales published in different journals.

Butler and Bisby (1931) for the first time made a compilation of the Indian Fungi in their classic work "The Fungi of India". The important work stimulated the study of Indian Fungi including resupinate Aphyllophorales. Our knowledge about the Indian resupinate Aphyllophorales increased greatly between 1925 to date by the contribution of Bagchee *et al.* (1950, 1953, 1954, 1969), Bakshi (1958, 1971), Bakshi *et al.* (1963, 1970), Puri (1956), Thind *et al.* (1956-1978), Ramakrishan (1959), Rehill *et al.* (1965), Welden (1965), Reeves *et al.* (1967), Thind (1973, 1975), Sathe and Rahalker (1976), Rattan (1977), Thind and Dhanda (1978); Anjali Roy

(1979-1987), Harash, 1982; Natarajan *et al.* (1980, 1985); Vaidya and his colleagues (1987, 1990, 1991, 19993 and 1994), Rabba (1994), Sharma (1995), Nanda (1996), Natarajan *et al.* (1998).

Most of the above mentioned work was carried out on poroid resupinate Aphyllophorales. Bakshi's (1971) monograph for Polyporaceae serve the purpose of identification of poroid Aphyllophorales while there was no monograph available for resupinate members of Aphyllophorales, especially for Corticiaceae (non-poroid Aphyllophorales) and related families except the one published by Natarajan *et al.* (1998). Infact the study of non-poroid resupinate Aphyllophorales had largely remained neglected and there were only few incidental reports of their occurrence. First serious attempt was made by Bagchee and Bakshi (1954) who described 14 species. Six more species were described by Thind and Adalakha (1956), Reid *et al.* (1959), Rehill and Bakshi (1965, 1966) in their generic monographs recognised one species of *Peniophora,* 7 species of *Corticium* and 18 species of *Stereum* as validly reported from the area, while 4 more species were added later on by Thind and Rattan (1971). In addition, there are other scattered reports of a few species in such genera as *Pellicularia, Hymenochaete,* and *Coniophora.* Thind and Rattan (1968 - 1973), described 59 species (under Thelephoraceae) including 36 new records and 6 new species while Thind and Khara (1968), Khara (1978), recorded 24 more species (under Hydnaceae) including one new species, from the North Western Himalayas. Anjali Roy (1972) has discussed the modification of the three kinds of Hyphae and their distribution within the basidiocarps. She considers the regenerating hyphae as the basic type which is derived from the secondary mycelium giving rise to the skeletal and binding hyphae and other structures found in the basidiocarps.

However, in comparison with the work on non-poroid resupinate Aphyllophorales in the other states of India, there is very little or scanty information available on this group in the state of Maharashtra, particularly from Western Ghats. The few done includes the work by Natarajan and Kolandavelu on Resupinate Aphyllophorales from south India (1985, 1998), Charu, D. Naik on wood rotting fungi from Karnala and Kankeshwar (1990), Dr. Arvind Rabba on the genus *Phellinus* Quel. From Maharashtra (1994), Dr. Manjeet K. Nanda on wood Rotting Fungi from Bhimashankar (1996).

Thus, it is a first attempt to study the taxonomy of resupinate Aphyllophorales, collected especially from selected regions of Western Ghats of Maharashtra which deals with the detailed morphological and cultural characters of non-poroid (Corticiaceae) and poroid (Polyporaceae) resupinate Aphyllophorales.

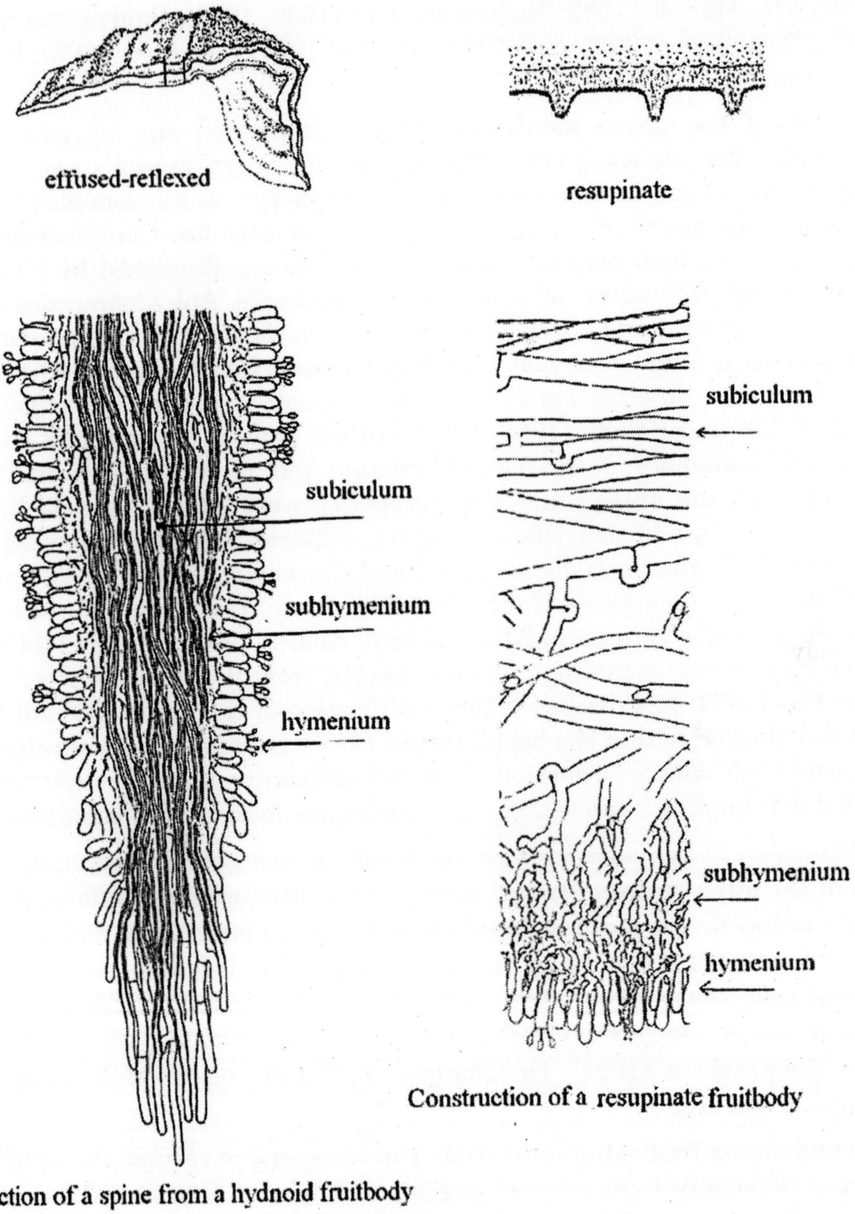

Fig. 1. *(Redrawn from The Corticiaceae of North Europe by Hjortstam, Larsson & Ryvarden 1987)*

Morphology of Resupinate Aphyllophorales (Fig. 1 & 2)

Till the end of 19th century all the study of fungi was based on external morphological features of basidiocarps (Persoon, 1801; Fries, 1821). Patouillard

(1900) was a pioneer worker to bring about a change in this trend of research by introducing microscopical characters in taxonomic study. The present work deals with the species studied from the family Corticiaceae with non-poroid hymenium; Polyporaceae and Hymenochaetaceae with poroid hymenium. The external morphology of the basidiocarps of non-poroid and poroid Aphyllophorales differ from each other. The morphology of non-poroid Aphyllophorales was studied by Prof. Talbot and he Published a review "Micro morphology of the lower Hymenomycete" (1954), while the morphology of poroid Aphyllophorales were Published by Gilbertson and Ryvarden (1986) in "North American Polypores, Vol. I" and by Ryvarden (1991) in" Genera of Polypores. Nomenclature and taxonomy".

(A) Non-Poroid Resupinate Aphyllophorales : Corticiaceae

Corticiaceae is a family within the Homobasidio-mycetes. It is an assembly of species with similar habit. Corticiaceous fungi are characterized by simple fruitbodies. They are more or less effused and have an even, meruliod, poroid or warted to denticulate surface. Many species have very delicate fruitbodies, barely visible for the untrained eye. The colour is usually some shade of white, gray or yellow, sometimes more brightly coloured in red, green or blue. However inconspicuous, the colour is never the less, an essential distinguishing features. The fruitbody is usually soft to tough, seldom hard.

Most "Corticiums" are wood-destroying organisms gaining their energy-yield from the decomposing of wood-substrate such as cellulose or lignin. A smaller group, are constitutes of the soil-microflora only using the wood as support for their fruitbodies. Some species are even suspected of being mycorrhizal.

Micro-morphology of Non-Poroid Resupinate Aphyllophorales (Fig. 2)

Talbot (1954) studied the different criteria for external morphology such as texture and margin of basidiocarps.

I. TEXTURE

The character of texture has been used successfully in the taxonomy of resupinate Hymenomycetes by Bourdot and Galzin (1928) in the genera *Corticium* and *Peniophora*. Texture is a physical character but it may be modified by chemical changes. This occurs in matrices which become gelatinous, mucilaginous or which are coriaceous, suberose or ligneous, the clonsistency of the pileus and cuticle may be correlated with the type of hyphae and their direction in a particular plane (Ames, 1913; Lohwag, 1940). In resupinate Hymenomycetes the texture depends mostly on the compactness of the hymenium in relation to that of the intermediate and basal layers, but it is also influenced by the thickness, and particularly the moistness of the fungus.

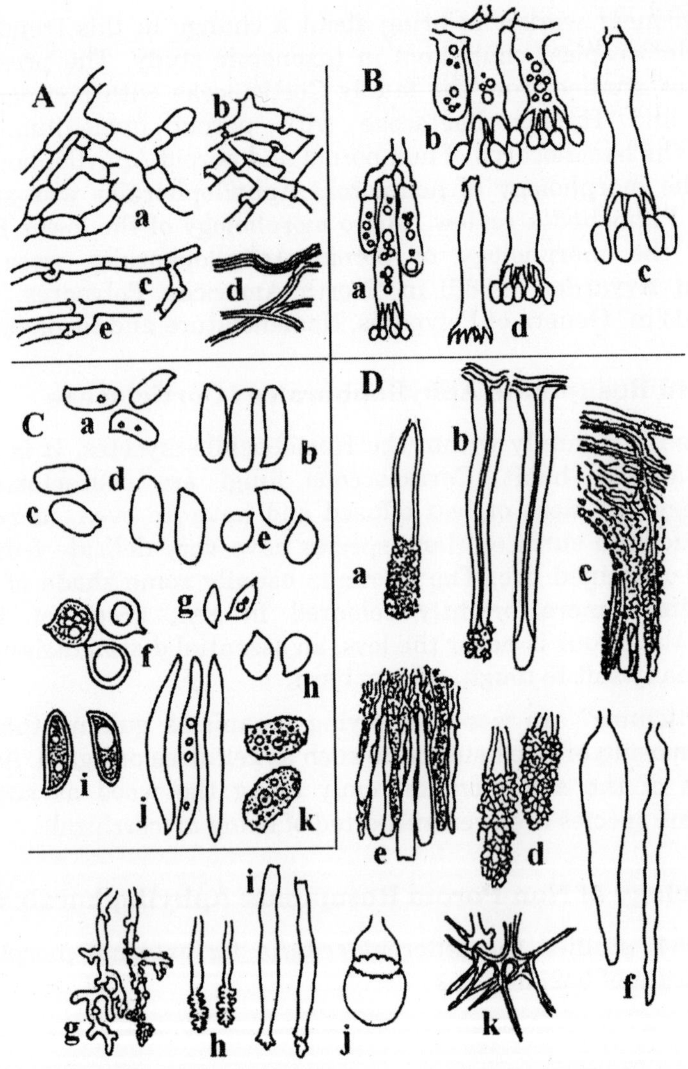

Fig. 2. Micromorphology of Non-Poroids - (A) Types of hyphae and septation: (a) Simple septa, (b) Short-celled hyphae, (c) Scattered clamps, (d) Skeletal hyphae, (e) Clamped septa; **(B) Types of Basidia** - (a) Terminal, (b) Pleural, (c) Constricted waisted, (d) 6-spored basidia; **(C) Types of Basidiospores** - (a) Allantoid, (b) Cylindrical, (c) Ellipsoid, (d) Ovoid (e) Subglobose, (f) Globose, (g) Navicular, (h) Subangular, (i) Fusiform, (j) Sigmoid, (k) Verrucose; **(D) Types of cystidia and other sterile hymenial organs-** (a) Pseudocystidia, (b) Leucocystidia, (c) Encrusted cystidia, (d) Metullo-cystidia, (e) Gloeocystidia, (f) Smooth thin walled cystidia, (g) Dendrohyphidia, (h) Acanthohyphidia, (i) Acanthocystidia, (j) Stephanocystidia, (k) Asterosetae. *(Redrawn from The Corticiaceae of North Europe by Hjortstam, Larsson & Ryvarden 1987)*

The texture may vary with the age of the fungus, and also with the relative rates of tangential and radial growth. Thus the margin of a fungus often differs in texture from the older parts, owing to a difference in clompactness of growth. Purely superficial modifications of texture may also occur in some genera, e. g., *Peniophora* and *Hymenochaete* by late production of cystidia and setae respectively. These often impart a glitter to the surface, and feel scabrid if they are sufficiently robust. The texture may be modified by chemical change. Ceraceous fungi appear waxy. The tissues are usually compacted by a waxy substance which may obliterate the hyphae or render them indistinct, giving the impression of degeneration. The wax often forms an obstructive emulsion in alkaline mounts of sections. Gloecystidia filled with oily globules can also give the fungus a ceraceous appearance. This state, where the waxy or oily substance is intracellular, contrasts with the commoner one where it invests the outside of the cells.

The hyphae of gelatinous species often appear filamentous, only the lumen showing distinctly with stains like phloxine. Gelatinised hyphae are best stained with iodine or with ammoniacal aqueous methylene blue, which reveals that the hyphae are frequently loosely intertexed but held together by a gelatinous substance. This has the power of imbibing water, in contrast with non-gelatinous matrices which absorb water by capillarity (Buller, 1922). The gelatinous modification is characteristic of the majority of common Haterobasidiomycetes, where it would seem to be an ecological adaptation for securing quick imbibition and retention of water. Mucilaginous species have much the same construction a gelatinous ones, but differ in having a matrix of thinner, slimy consistency. A corneous texture usually results from the lox of water from tissues which were previously ceraceous, gelatinous or mucilaginous.

For taxonomic work on the lower Hymenomycetes, the types of texture described below are important:

1. **Membranaceous** : The hymenium is a compact palisade but not otherwise sharply differentiated from the subhymenial and deeper tissues, which are usually also fairly compact. The fructification may be adnate, separable or held to the substratum by superficial intrusive hyphae. This group contains most of the species with thick, moist hymenophores.

2. **Pellicular** : The hymenium is sharply differentiated as a thin crust laying above loosely arranged inferior tissues. The hymenium is continuous and compact at maturity, but at times may be discontinuous in patches thus appearing minutely poroid under the lens. Pellicular species are generally rather thin and dry; the open nature of the tissues is doubtless responsible for quick drying out. The hyphae are usually very distinct. Most of the species are separable from the substratum by snapping of the basal tissues under tension.

3. **Byssoid or Floccose** : Here there are no compact tissues, even the hymenium is irregular, discontinuous and often staged at different levels.

A fairly thick, tufted or woolly structure results, in which the hyphae are loosely arranged throughout. The byphae are easily detached from the substratum by breaking under tension.

4. **Arachnoid** : This type of texture is common in the young stages of growth but may persist into maturity, and then the hymenial elements are quite scattered and discontinuous. The hyphae are usually delicate and more or less appressed to the substratum in an open work structure reminiscent or cobwebs. When the elements of the fungus are so dry and discontinuous as to appear scurfy or mealy, then the terms furfuraceous or farinaceous are applicable; in a finer and more powdery state the structure may be termed pruinose or pulverulent.

5. **Crustose or Arid** : Certain resupinate fungi are thin, dry and compact throughout their tissues, and lack a distinct cortex. They are conveniently termed crustose. Usually they are adnate frequently they are charged with mineral matter. But it is more common for fungi to be arescent, that is becoming crustrose only on drying, a character well shown in the section Coloratae (Bourdot and Galzin, 1928) of the genus *Peniophora.*

II. MARGIN

The type of margin depends typically on the compactness of the hyphae composing it, and the direction in which they are intertwined. It is observed that as a resupinate fructification grows, new hyphae and young hymenial elements are differentiated behind the radially expanding periphery. Radial growth generally precedes the tangential growth which is responsible for knitting the hyphae together into a firm fructification. It is thus probable that the presence or absence of marginal hyphal strands depends largely upon the relative rates of radial and tangential growth. The marginal character is fairly constant, and useful in taxonomy. The more important types of margins are; fibrillose, fimbriate, arachnoid, byssoid, villose or pubescent, farinaceous and prunose. Hyphae may be aggregated into thick mycelial strands which sometimes meander over the surface of the fructification or extend beyond the margin. Certain of the lower Hymenomycetes also develop rhizomorphs.

III. THE FRUITBODY

The fruitbody may vary considerably from one species to another, and it is often difficult in a few words to cover the variation properly. The most commonly used terms are as follows (Fig. 3A).

(a). **Resupinate:** The Resupinate basidiocarp is characterized by the absence of any sterile parts, except for a margin. This is the general fruitbody type among the corticiaceae.

(b). **Effused:** In some species of resupinate Aphyllophorales pileus stretched out flat, ± irregularly, over the substrate.

(c). **Effuso-reflexed:** In some species a pileus develops with age along the upper edge of the basidiocarp. Such basidiocarp is called effused-reflexed. Other species develop a pileate basidiocarp directly without any effused - resupinate part and there are transitions between the two types.

Construction of the fruit body

The fruitbody in the Corticiaceae is rather simple but may, in the most complicated type, consist of the following layers (Fig. 1).

(i) The Hymenium which consists of the basidia eventually mixed with sterile organs such as cystidia, etc.

(ii) The Hymenium which is a layer of hyphae below the basidia usually growing vertically and strongly branched. It is some times very compact and then the individual hyphae may be difficult to observe. In some species with very thin fruitbodies: a subhymenium may be lacking.

(iii) The trama is a layer of hyphae supporting the hymenium and subhymenium, and the term is only used when the hymenophore is poroid, hydnoid etc. not for smooth species. The hyphae of the trama are often wider and of a looser consistency than those of the subhymenium.

(iv) The subiculum is the layer of hyphae next to the substrate. As in the trama, its hyphae are often wider and looser than in the rest of the fruitbody and mostly grow parallel to the substratum.

HYMENOPHORE

The hymenophore consists of the subiculum, trama and sub- hymenium, if all of them are present, i.e.; they are the part of the fruitbody on which the hymenium is placed. The hymenophore can have a very variable configuration and the following terms are used to describe the variation:

Poroid with pores.

*Irpicoid w*ith irregular and flattened teeth.

Reticulate with netlike ridges.

Merulioid with radial or more or less netlike folds.

Grandinoid with small granules.

Odontioid with teeth or small spines.

Hydnoid with prominent spines.

Tuberculate with (usually) spares and irregular warts.

Colliculose with low, rounded hymenium swelling and blisters.

Epitheloid with scattered, sterile pegs.

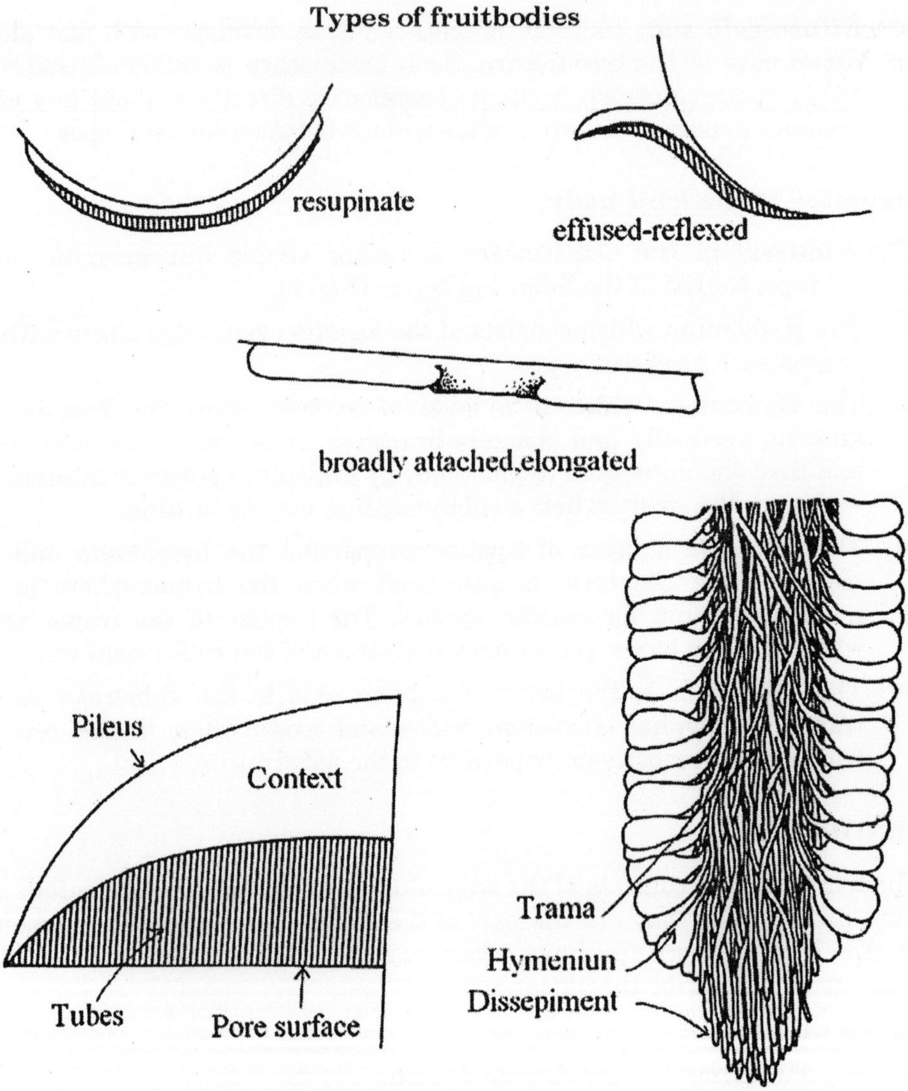

Fig. 3. Section of pileate fruitbody and outer part of a tube *(Redrawn from North American Polypores by Gilbertson & Ryvarden 1986)*

CONSISTENCY

According to structure, the fruitbody can vary from very loose to almost horny hard when dry. The following terms have been used to describe this variation.

Byssoid has a cotton- like or floccose, usually discontinuous surface.

Farinaceous has mealy, minutely granulose, more or less discontinuous surface.

Athelioid has a thin, usually pliable membrane over a loose subiculum, also called *Pellicular.*

Membranaceous like athelioid but thicker.

Ceraceous with a waxy, closely adnate appearance.

Phlebioid very dense and hard. Watery as fresh, horny when dry.

(B) External morphology of Poroid Resupinate Aphyllophorales

The poroid resupinate Aphyllophorales are collected generally in sexual stage, that is the fruitbody stage. The vegetative mycelial stage is nothing but a vast network of hyphae which remain embedded inside the host. The classification and keys used in this investigation are almost exclusively based on the sexual reproductive stage (Gilberton and Ryvarden, 1986).

Macroscopic characters of the reproductive stage (Fig. 3)

1. **Types of Basidiocarps:** For most species the shape of the basidiocarps is distinct and is an important character. In some species there may be a development from resupinate to effuso-reflexed or both resupinate and distinct pileate basidiocarps may occur on the same log or stump. The various types of basidiocarps are resupinate and effuso-reflexed. (Fig. 3).
2. **Attachment of Basidiocarps:** The type of attachment is known broadly attached (elongated) resupinate (Fig. 3)
3. **Consistency:** Perennial fruitbodies with a dominance of skeletal hyphae will normally be woody, while the trimitic ones will be tough and almost impossible to tear apart. The monomitic ones may be soft or very sappy and may shrink considerably during drying, often accompanied by a colour change.

Construction of the basidiocarps

The basidiocarp consists of following parts (Fig. 3).

1. **Pileus Surface :** The pileus may be glabrous or covered by some types of hairs. When glabrous, it may be dull or glossy-to-shiny (Ryvarden and Johansen, 1980). In many species the pileus has a distinct colour which remains reasonably unchanged throughout the life span of the basidiocarp. White or light coloured species frequently become darker with age as the upper layer of hyphae collapse and become darker in ochraceous to brownish shades, resulting in distinct zonation from base to the margin, where new hyphae are under active development. This is especially typical in perennial basidiocarps. In some species with normal light coloured basidiocarps, a thin, black or deep reddish cuticle may develop from the base, making the pileus strikingly bi-coloured (Gilbertson and Ryvarden, 1986).

A glabrous pileus may be smooth without any zonation, radial lines, or raised ridges. The type and consistency of the pileus cover however plays an important taxonomic role and the term used in this connection are explained in the glossary. Basidiocarps of some species with hairs on the pileus are long lived or perennial. Furthermore, in other species the hairs may tear away with age or become agglutinated to the surface as warts or lines of scrupose tufts of hyphae. In some species there is a distinct dark cuticle below a tomentum and with age the hair may fall off zonately and expose the cuticle which then will contract strongly with the lighter coloured tomentum. In many perennial species the pileus may crack up with age, either in radial lines or into an irregular pattern of angular pieces, often so strongly and characteristically that the surface is called rimose.

2. **Margin:** Especially for species with resupinate basidiocarp, the margin and its development may be an important character. Most species have more or less continuous margin several mm. wide, often white or coloured. The hyphae grow at an equal rate and the margin becomes soft and finally tomentose. For other species the margin can end abruptly with an almost vertical slope. Such basidiocarps often have a tendency to curl up and loosen along the margin after drying.

3. **Pore Surface:** The colour of the pore surface is characteristic and diagnostic feature for many species. With age, it has a tendency to become darker and this has to be remembered when using keys. More important than colour is the shape of pores per mm. or cm. is a rather distinct character, although the pores have a tendency to become larger with age and in large specimen. While some species have consistency poroid basidiocarps, there are a few confusing ones where the hymenophore may change with age and development. In these species, poroid specimen have been found side by side with lamellate; ones and single basidiocarp may even be variable in different parts of the hymenophore, indicating that the configuration of the hymenophore may not be so strongly genetically fixed as often assumed. With age, the dissepiments in some species may becomes incised or become dentate as they develop at different rates. The pores may be round, radially elongated, irregular, sinuous, daedeloid, angular, hexagonal, or lamellate (Ryvaden and Johansen, 1980).

4. **Tubes:** In most species the tubes are more or less concolorous with the pore surface or with age become paler than the pore surface. In many species there is no apparent difference either in colour or consistency between the tubes and context or subiculum. However, in other species the tubes may have a different colour, consistency and even hyphal system than the context. These differences may be important taxonomic characters and a specimen should therefore always be sectioned to observe the tubes and context.

5. **Context:** In the resupinate Aphyllophorales the context is the sterile part between the tubes and the substrate is often called the subiculum although there is no principle difference between the two types of tissue. In most species the context is homogeneous with regard to both colour and consistency and will normally have radial structure as the hyphae grow from the base towards the margin.

Micro-morphology of the Resupinate Aphyllophorales (Fig. 4).

The importance of micro-morphology of basidiocarp in taxonomy was introduced by Corner by introducing the concept of "hyphal system". The basidiocarp consists of hyphae for a long time no attention, in particular, was paid to them, except for mentioning of septation, rarely. Corner's fundamental work on the hyphal system brought out the importance of the microstructures of the basidiocarp to the taxonomy of Aphyllophorales. With the advancement in knowledge over a period of time through experience it was analyzed that hyphal is considered more complicated than originally assumed. Hence there has been some modifications in corner's system for hyphal classification.

In a series of publications, Corner described the anatomy of the basidiocarp in the five polyporoid fungi namely *Polystictus xanthocarpus* Fries, *Fomes lavigatus* Corner, *Fomes lamaensis* (Murr). Sace. ex. Trott., *Fomes maxium* Corner, *Fomes pachyphloes* Pat. Cunningham (1946, 1965) realised the significance of Corner's (Corner, 1948, 1950) finding and confirmed his results through a series of publications. Cunningham (1946-1965) further brought out the importance of the hyphal configuration in the understanding of the limits in the Aphyllophorales.

Corner (1932 a) discovered that the context of members of Aphyllophorales may consist of morphologically distinct hyphae which he classified into three basic groups namely; Generative, Skeletal and Binding hyphae. When only generative hyphae are present in a sporophore, the hyphal system is monomitic; when skeletal hyphae are present along with generative hyphae, the hyphal system is called dimitic; when all the three types of hyphae are present in the sporophore it is called trimitic.

Cunningham observed and stressed the presence of clamps as an important character and it is significantly connected with hyphal types and systems.

TYPES OF HYPHAE (Fig. 4 A)

1. **Generative Hyphae:** These are the basic units of structure universally present in all sporophores. They are either hyaline or coloured, usually thin-walled, rarely thick walled, freely branched, septate, often with clamps. Generally they are readily stained. They are produced directly from the secondary mycelium and are of limited growth. They give rise to the skeletal hyphae, binding hyphae and other hyphal structure found in the basidiocarp. When dried, the hyphae collapse and are therefore difficult

to be found in the dried specimen. The types of septation in the generative hyphae are a very important character in the hyphal classification of polypores. The septation between two hyphal cells can either be simple or resulting from the formation of a clamp. The septation of the generative hyphae is fundamental as it is consistent for a species. In Corticiaceae, both clamp and simple septa are present. The wall thickness of generative hyphae may vary considerably although generally they are thin-walled.

2. **Skeletal Hyphae:** They are normally unbranched and thick walled to solid, aseptate, without clamps, may be hyaline or coloured; skeletal hyphae often reacts with Melzer's solution showing strong reddish brown colour. Cunningham distinguishes two types of skeletal hyphae; the 'bovista' and 'long'. The 'bovista' type has a main axis, 3-10 μm. in width, with several lateral branches, aseptate, hyaline or coloured. The 'long' type is slender, 3-7 μm. in diam., septate or aseptate, branched or typically unbranched, usually thick but sometimes thin-walled, hyaline or coloured and loosely interwoven.
3. **Binding Hyphae:** They are branched, solid to very thick- walled; non-septate, which often have a more randomly oriented growth than the skeletal hyphae. In trimitic genera like Trametes and Perenniporia they are intertwined between the radially (in the context) or vertically (in the trama) oriented skeletal hyphae.
4. **Mediate Hyphae:** They are intermediate between generative and skeletal, generative and binding hyphae forming the transition from one to another type of hyphae.
5. **Gloeoporous Hyphae:** In some polypores there may occur rather wide and thin-walled hyphae with a refractive content distinctly different from that of the normally hyaline generative hyphae. They are often yellowish and the content may be homogeneous or grainy. In some cases the content darkens when the hyphae are mounted in Melzer's solution or stains brightly with phloxine. The occurrence of gloeoporous hyphae is rather scattered in the group and they have significance, only on specific level.
6. **Ampoule Hyphae:** In some lower Hymenomycetes, a number of hyphae become swollen into shapes resembling ampoules. These are not always easy to find and appear to be less obvious in material which is not fresh. They are generally associated with clamp connections.
7. **Conducting Hyphae:** In many Hymenomycetes, the context contains, in addition to the usual hyphal system, a network of special hyphae with dense contents which are known as "conducting hyphae" or "laticiferous hyphae". They are of intermediate lengths, much branched, anastomosing, of wide diameter, relatively thin-walled, at first non-septate and coenocyclic but later becoming sparsely septate. The dense species it changes colour on exposure to light and air. Their content stains easily

with Melzer's or chloral hydrates and iodine or with phloxine. The function of the conducting hyphae is distinguished from gloeocystidia by the fact that the later are unbranched or limited in length.

8. **Vesicles:** Vesicles are pyriform or sub-globose, hyaline, thin- walled swelling, usually borne terminally but occasionally in an intercalary position in the context hyphae. In section, they may appear devoid of contents or filled with a homogenous or globular granular matter which deeply stains with phloxine. They occur typically, in the trama and are conspicuous objects when numerous or large.

Other structures (Fig. 4C)

(A) **Paraphysoid Structures:** A number of sterile accessory hymenial structures are often grouped for convenience under the general name of "paraphyses". They are a heterogenous collection and not homologous with the paraphyses found in Ascomycetes. For this reason the various types are sometimes distinguished and sub-divided according to their form as follows:

(i) **Simple Paraphyses:** These are colourless, cylindrical or filiform, elongated, unbranched, rarely septate, smooth, thin-walled, straight, flexous or spirally twisted elements which are prominent in the hymenium. In form, there is often little to distinguish them from hyphae or immature basidia or cystidioles.

(ii) **Pseudophyses:** These structures are thin- walled, non-septate, hyaline, unbranched or occasionally forking slightly, somewhat clavate and constricted especially near the apex into a number of bead like parts. The cells are homologous and deeply stain with phloxine.

(iii) **Acanthophyses:** Pilat (1926) introduced this term for the peculiar form of dendrophysis, which is clavate or cylindrical or decorticated with a number of short, pin-like outgrowths chiefly near the apex. They are hyaline cells with hyaline or occasionally dilutely coloured, pin-like processes. They occur in hymenium. They may be thick-walled and cylindrical or clavate, resembling some from of cystidia or thin-walled with a few processes.

(iv) **Dichophysis:** This term was introduced by Pilat (1926) for a special form of dendrophysis, distinguished by its more or less regular dichotomous branching. They are undoubtely a form of hyphae having rather thick walls, narrow lumen, and subulate apices to the dichotomous branches. They are usually coloured and may occur in the hymenium or in the tramal tissue or in both. When present only in the hymenium, they should be considered as a type of paraphyses.

(v) **Dendrophysis:** These are thin-walled, usually colourless but not dichotomously branched hyphae which occur intermingled with other hymenial elements. Pilat (1926) and Wakefield (1931) restricted this term to paraphyses of tree-like form with irregular branching.

(B) **Hyphal Pegs:** These are erect, wart like, papillate, tuberculate or spiny structure projecting from the hymenium and often big enough to see with the naked eye. They are composed of sterile hyphae arising in coherent fasicles from the subhymenial or deeper tramal tissue. The term "hyphal pegs" has been commonly applied only to the polyporaceae, the group in which sterile tufts of hyphae project into the lumen of the tubes. Bose (1944) noted that hyphal pegs common in certain species at high altitudes, may be scarce or absent in the same species at low altitude. Sometimes the presence of hyphal pegs has been the basis for splitting of new genera eg; *Dentrothele* separated from *Aleurodiscus* or some times they are regarded only as confirmatory specific characters. Generic distinctions involving hyphal pegs have been based on their hyphal composition and their arrangement either as discrete spines or united into sterile linear ridges or shallow poroid network.

(C) **Cystidioles:** A cystidioles is defined by Ainsworth and Bisby (1945) as a "sterile structure in the hymenium at the same level as a basidia, wider than paraphyses, thin-walled and little differentiated". They are usually inconspicuous, colourless, subulate, bottle-shaped, clavate or fusoid often with a sharply pointed apex. They are thin-walled, sometimes finely encrusted or smooth, sometimes projecting a short distance above the basidia, usually have no septa. In *Peniophora cremea* the cystidioles are large and can be called "cystidia". They are only diagnostic on the specific level. They can be distinguished from immature basidia by their pointed and tapering apex, while immature basidia are normally widened slightly towards the apex and with rounded tip.

(D) **Cystidia:** These are a rather heterogenous assemblage of sterile bodies, commonly occurring in the members of Aphyllophorales and Agaricales. Certain generic segregates have been based on the presence of cystidia. The term 'cystidium' was used in the Agaricaceae by Montague (1842). They are sterile, hyaline, unicellular, ovoid, cylindrical or clavate, blunt or pointed, smooth or ornamented and encrusted, or originating as the terminal cells of branched hyphae and projecting from the hymenium. Buller (1924), classified them according to their position on the fructification as Pilocystidia, on the surface of pileus; Cheilocystidia, on the edge of the gills; Pleurocystidia, on the sides of the gills; Cantocystidia, on the stipe; Dermatocystidiam, on the cuticle. Cystidia may be divided into two groups according to their origin; Hymenial cystidia arises in the subhymenium and normally are of same size or slightly larger than the basidia. Tramal cystidia arises in the trama; may or may not bend into the hymenium and are often far beyond it. Cystidia have been thought to function as excretory hydathodes. In Aphyllophorales, the pointed, encrusted 'Metuloides' were suggested to have protective function against small animals feeding on the hymenium. The cystidia are very heterogenous even in the single genus, e. g. *Peniophora*. They may be conical or cylindrical in shape or sometimes narrow and flexous. The walls are sometimes thick and encrusted but some species show smooth, thin-walled cystidia. Encrustation of the wall may take the form of very fine amorphous to large crystalline deposits of mineral matter. Cystidia are usually hyaline, but if dilutely coloured, then the colour residues occurs in their walls and not in their contents.

Gloeocystidia are sterile organs mainly distinguished by their thin- walled, smooth, dense and deeply staining contents; They are often tubular and slightly sinuous or constructed, without encrustation. The contents may be hyaline or lightly coloured, yellow to brownish; they are highly refractile, homogenous or granular, globular of an oily nature. They arise from the hyphae situated in the sub-hymenium or deeper tissues and may be completely embedded or may extend as far as the level of the basidia. Only in few species they are found projecting above the hymenium, with colourless walls. Gloeocystidia stain deeply with phloxine or eosine in KOH mounts and one which are normally coloured, take on a deeper, brownish colour with iodine solution. Presence of Gloeocystidia is confirmed with sulpho-benzaldehyde test (Rattan, 1977). Gloeocystidia are concerned with nutrition. Overholts (1929) suggested that they might have protective function. In some species of Aphyllophorales they turn blackish in sulpho-benzaldehyde.

(E) Setae: They may be defined as sterile, rigid, dark coloured, thick-walled, spine like organs, usually having more or less pointed apex and possessing the property of darkening in alkali. They are yellowish or reddish brown in colour, the pigment mainly located in the walls. The lumen is very narrow, simple, unbranched, vary in size, shape and in the character of their apices. Cunningham (1946) has studied the origin of setae in some species of polypores. They usually develop below the hymenium and project beyond the basidium. Such setae are called hymenial setae. Tramal setae are mostly embedded in the trama and run parallel to the tube walls and may project obliquely into hymenium. They are distinctly conical and are widest in the middle part and taper slightly towards the apex (Fig. 4 D).

It is supposed that setae may provide rigidity to the fructification and protect the tissue from damage by small animals (Overholts, 1929). The setae have been used as a characteristic in the erection of new genera. They are common in *Phellinus, Hymenochaete, Tubulicrinis*. For the diagnosis of species, some of the differences seen in setae are the shape, size, colour, apex, position, the type of hypha from which the setae arise, branching etc.

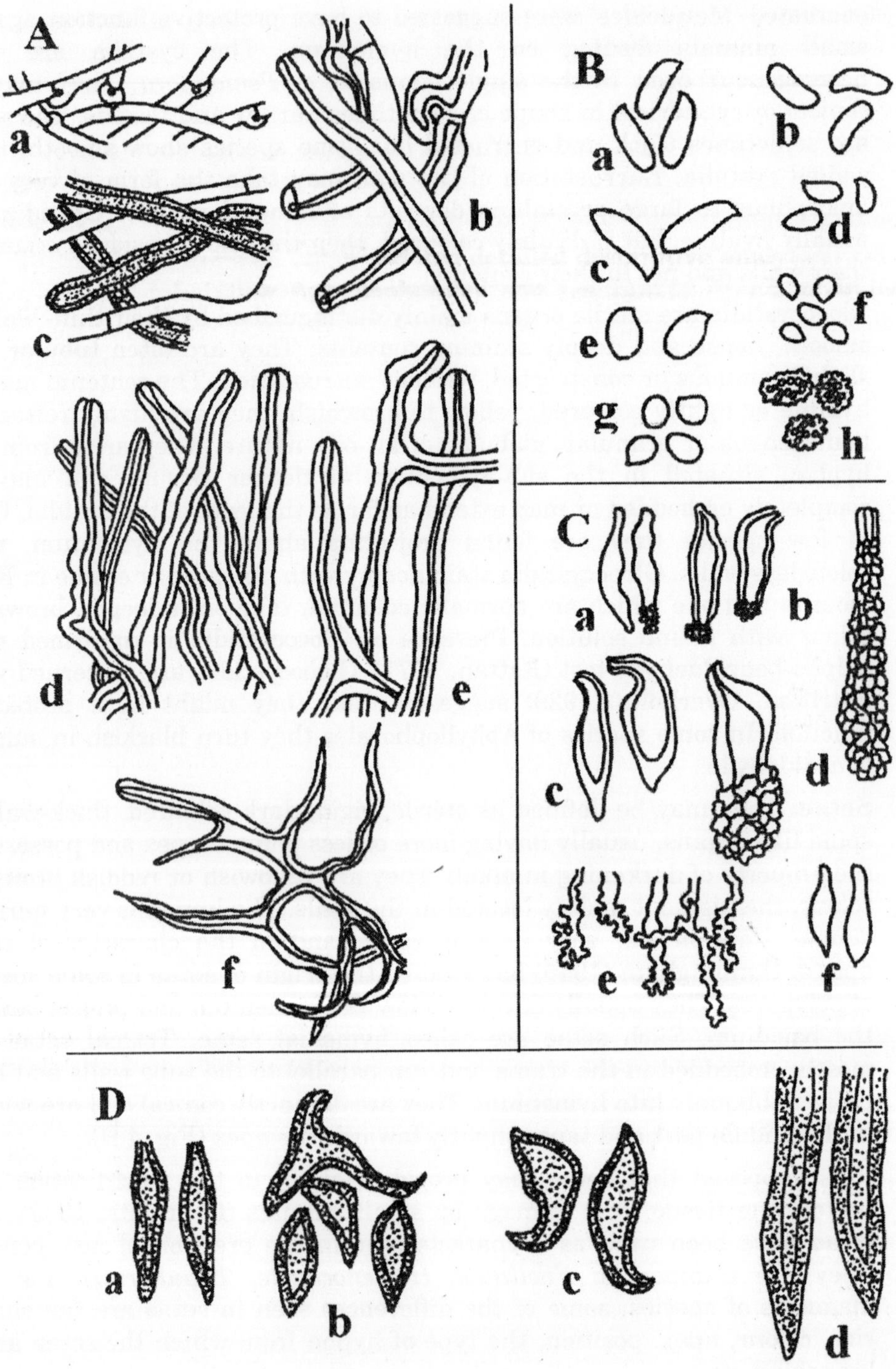

Description of Figure is on P. 25.

Fig. 4. Micromorphology of Poroids : (A) Types of hyphae and septation - (a) Thin-walled generative hyphae with clamps, (b) Thick-walled generative hyphae with clamps, (c) Thin to thick-walled generative hyphae with simple septa, (d) Thick-walled, unbranched and flexuous skeletal hyphae, (e) Thick-walled, sparingly branched and flexuous skeletal hyphae, (f) Binding hyphae; **(B) Type of Basidiospores** - (a) Cylindrical, (b) Allantooid, (c) Navicular, (d) Oblond-ellipsoid, (e) Broadly ellipsoid, (f) Subglobose, (g) Globose, (h) Warted; **(C) Types of cystidia and other sterile hymenial elements** - (a) Thin-walled, apically encrusted cystidia, (b) Thick-walled, apically encrusted cystidia, (c) Ventricose cystidia, (d) Apically and completely, coarsely encrusted cystidia, (e) Dendrohyphidia, (f) cystidioles; **(D) Setae and setal hyphae** - (a) Straight setae, (b) Ventricose setae, (c) Hooked setae, (d) Straight setae hyphae *(Redrawn from North American Polypores by Gilbertson & Ryvarden 1986)*

(F) Basidia: Basidia are the hymenial cells in which fusion of nuclei or Karyogamy and meiosis occur. They bear basidiospores on terminal projections called sterigmata. The term basidia includes pro-basidia, meta-basidia and sterigmata as parts of the whole basidium (Talbot, 1954).

(i) Probasidium: It is primary basidial cell, the part or stage of the basidium in which Karyogamy occurs. The term is intended to denote the "first stage of the basidium" rather that which precedes the basidium.

(ii) Metabasidium: It is that part or stage of the basidium in which meiosis of the diploid nucleus occurs. In many basidia it replaces the probasidium. This term is intended to denote the final stage of the basidium.

(iii) Sterigmata: This is apart of the basidium that comes between the metabasidium and the basidiospores or the elongation of the metabasidium through which the nuclei migrates to the spores which are borne terminally. The sterigmata is composed of a basal, filamentous or inflated part called the Prosterigmata and apical point called the apiculum on which the spore is borne.

SIZE AND SHAPE OF BASIDIA

In higher Hymenomycetes the hymenium is compact and resist lateral expansions of the basidia, with the result that they tend to be clavate or cylindrical. Basidia may arise singly at the apices of subhymenial hyphae or in botryose clusters. Basidia ranging in size from 3 x 7 μm (*Corticium galzini*) to 25 x 210 μm (*Aleurodiscus amorphous*) have been noted.

(G) Basidiospores: They are generally accepted as very important character for classification of Basidiomycetes. Basidiospores are borne singly at the apex of each sterigmata. There are usually two to four sterigmata per basidium their shape may be subglobose, globose, oblong, cylindrical, elliptical, navicular, reinform, alloantoid (Fig. 4B). They frequently have one side depressed and their shape may be modified by an apiculus. They may be hyaline or coloured or light brown or they are bright coloured in a mass. Such colours usually being rosy, violaceous or blue green, often the spores are guttulate. The majority of spores are smooth walled but in some, they may be minutely sculptured or in others they bear long spines, warts or reticulations. Sculpturing is frequently accompanied by a slight thickening of the spore wall and colouration. The range of the spore size is large. The basidiospores are amyloid or non-amyloid. The amyloid spores can be detected with Melzer's reagent (chloral hydrate and iodine solution) which turns blue (Talbot, 1954).

METHODOLOGY

Spore prints were obtained from fresh materials. The spore prints were maintained in air-tight plastic envelopes (Ryvarden and Johansen, 1980; Gilbertson and Ryvarden, 1986). The fungi were then properly dried and preserved in the dried condition. The dried specimen were wrapped in butter paper and then in brown paper folders, properly numbered and finally registered. The herbarium specimen, after well furnishing have been deposited in the Herbarium, Department of Botany, University of Pune, Pune and are indicated in the text as VH (Vaidya-Hakimi) number.

The specimens were examined for external and internal morphology. For external morphology the material was observed for colour, texture, type of attachment to host, pore morphology, dissepiments character; margin, hymenial and pileal surface of basidiocarp.

For internal morphology, thin hand sections were taken from fruiting body passing through hymenium, which was done by chopping method. To loosen the hyphae the sectioned material was treated with 10% KOH, washed with water and stained with 1% phloxine. The sections were again washed with water and finally stained with cotton blue. The sections were then teased with glass rod in lactoglycerine with l% cotton blue. Semi-pemanent slides prepared in lactoglycerine were maintained by sealing with nail polish (Beneke, 1958). For amyloid and non-amyloid reaction, spores were treated with Melzer's reagent (chloral iodine solution), which turns blue. For Xanthocronic reaction 10% KOH solution was used. The slides were observed under Bausch and Lomb compound microscope having a combination of 10×, 45× and oil immersion (i.e. 100 ×), objectives. Photographs were taken using digital camera. Measurements of hyphae, basidia cystidia, setae, spores, cutis element etc. were taken using objective micrometer or calibrated ocular Dimension of microscopic characters are given in micrometer (μm) eyepiece.

Standard chemical tests were done such as spraying of 1% benzidin in 90% ethanol on decaying wood samples to confirm the type of rot.

For culture dry material (the decaying wood or basidiocarps) was used after rewetting in absolute alcohol for few minutes, and surface sterilization in sodium hypochlorite. The treated material was then placed in petridishes containing 2% potato dextrose agar (PDA) medium at equidistance. The Petri plates were incubated in an incubator at 25°C under dark. Pure culture was obtained by transferring of hyphal tip from dog culture to a to freshly poured and sterilized Petri plates and slants for further studies such as characterizing the culture, using standard procedure as described by Stalpers (1978).

To detect the various extracellular enzymes, few chemical tests were performed on growing mycelia in pure culture. These enzymes are of several types such as phenol oxidases, laccases, lignases, peroxidases and tyrosinases. The type of rot could be identified by detecting these enzymes. The type of rot is considered to be an important taxonomic character (Davidson *et al.,* 1938; Nobles, 1948; Boidin, 1951; Domanski *et al.,* 1965). Also, the possibility of detecting the enzymes responsible for the decomposition of lignin in pure cultures is important.

IDENTIFICATION

In the next phase of investigation, fungi were identified by studying their macro and micro morphology. In external morphological study, the material was observed for its colour, texture, type of attachment with wood, its margin etc., while for internal morphological examination, the material was cut in dorsiventral manner by chopping method to get free hand sections along with context and basidial (hymenial) layer. For obtaining a section, the dry material was rewetted in 70% absolute alcohol for reviving, followed by 2-3 washings with water. The sections were then transferred in 5% KOH solution for about 5 to 10 minutes. In KOH, hyphae and other elements get swelled, loosened and then become easy to identify. The sections were then transferred to water and given a wash 2-3 times to remove the traces of KOH. The sections were then teased in lactoglycerine with 1% cotton blue. For the staining of basidia, 1% Phloxine was used. Semi-permanent slides prepared in lactoglycerine were maintained after sealing with nail polish (Beneke, 1958). The permanent slides were prepared by mounting the sections in P.V.A. (Poly Vinyl Alcohol) medium. P.V.A. is a plastic resin medium, recently used for preparations of permanent slides. This medium is compatible with acidic stains like safranin (Omar *et al.,* 1979).

The advantage of this medium is due to its viscous nature, excess of blue stain is removed from the section, giving the section a virtually colourless back ground. The slides were then left for 24-36 hours at 40ºC to harden completely. The permanent slides were examined for identification of the fungus. The taxonomic identity was assigned with the help of standard identification keys and descriptions (Talbot, 1973; Ryvarden and Johansen, 1980; Natarajan and Kolanduvelu, 1985, 1998; Gilberston and Ryvarden, 1986; Rajchenberg, 1987, 1989; Rajchenberg and Wright, 1987; Larsen and Cobb-Poulle, 1990; Ryvarden, 2000; CBS and MycoBank (http://www. mycobank. org/MycoTaxo. aspx?link=Tand.) and http://www. punjabenvironment. com/bd_listhtm.

Few standard chemical tests such as spraying of 1% Benzidine in 90% Ethanol (Hintikka, 1970) on decaying wood sample were taken to confirm the type of rot. For amyloid and non-amyloid reaction, basidiocarps were treated with Melzer's reagent (chloral iodine solution) (Hawksworth *et al.*, 1983). The presence of gloeocystidia was confirmed by using sulpho-benzaldehyde reagent (Rattam, 1977). For xanthochroic reaction in case of poroid Aphyllophorales, 5% KOH solution was used (Hawksworth *et al.*, 1983). 'Camera Lucida' (Melzer s, India Optical Instrument Co.) drawings were made from temporary slides under oil emulsion. Measurement of hyphae, basidia, cystidia and spores were taken with "Ochnar micrometer", or "Calibrated ocular dimension of microscopic characters which were given in micrometer (μm) eye piece".

To detect the various extracellular enzymes, few chemical tests were performed on growing mycelia in pure culture. These enzymes are of several types such as phenol oxidases, laccases, lignases, peroxidases and tyrosinases. The type of rot could be identified by detecting these enzymes. The type of rot is considered to be an important taxonomic character (Davidson *et al.*, 1938; Nobles, 1948; Boidin, 1951; Domanski *et al.*, 1965). Also, the possibility of detecting the enzymes responsible for the decomposition of lignin tin pure cultures is important.

The following **enzymatic tests** were performed on mycelium for extra cellular oxidase enzymes:

1. A drop of 0.1% α -napthol in 96% ethanol when added to pure culture, a purple colour was obtained (Kaarik, 1965). This showed the presence of enzyme laccase.
2. Equal parts of 0.4% hydrogen peroxide and 1% pyrogallol in water, freshly prepared, when added to pure culture, gave a yellowish brown colour, indicating the presence of peroxidase.
3. 0.1 M p-cresol in 96% ethanol when added to pure culture gave an orange brown colour. This showed the presence of enzyme tyrosinase.
4. 0.1% syringaldehyde (Harkin and Obst; 1973) in 96% ethanol when added marginally to growing culture gave a purple violet colour. This showed the presence of enzyme laccase.

All reactions were performed on actively growing marginal hyphae (mycelium at the edge of the colony). It is possible to perform almost all the tests on growing (marginal) mycelium in the same plate, because most of the reagents do not interfere with each other.

For cultural studies, detailed observation of hyphal characterization, colony colour, characters of mat, cystidial and chlamidospore structures and other macroscopic characters such as odour, reverse side of the culture plate along with enzyme tests were carried out according to Stalpers (1978), The species were described with species code as proposed by Stalpers (1978).

The present work has mycological significance and will update the knowledge of resupinate Aphyllophorales of Maharashtra, which presently lacks in the scenario of Indian mycology. This will be of help for further studies and investigations. Taking into consideration the taxonomic study of Wood Rotting Fungi, this kind of approach is studied for the first time for understanding thoroughly the compilation of the resupinate Aphyllophorales from a particular region.

Attention has been paid towards poroid and non-poroid resupinate Aphyllophorales by Bagchee *et al.* (1950, 1953 and 1954), Bakshi (1958, 1971), Bakshi *et al.* (1970), Thind *et al.* (1956 and 1978), Rehill and Bakshi (1965, 1966). Thind (1973, 1975), Sathe and Rahalker (1976), Rattan (1977), Thind and Dhanda (1978); Anjali Roy and others (1979, 1987), Harsh (1982); Natarajan *et al.* (1980, 1985, 1998); Ganesh and Leelavathy (1986), Vaidya and his colleagues (1987, 1990, 1991, 1993 and 1994), Rabba (1994), Sharma (1995) and Nanda (1996) pertaining mainly to Himalayas and some other parts of Central Peninsula. Besides bracket Aphyllophorales and reresupinate Aphyllophorasles, the other members of Basidiomycotina, namely an Agaricales and jelly fungus has also received scanty attention. Agaricales occur in rainy season and jelly fungi of Auriculariales and the Tremellales continue their appearance almost even in winters as per the studies carried out by Cook (1891); Llyod (1898-1925); Theissen (1913 a,b); Butler and Bisby (1931); Thite and Kulkarni (1976); Patil and Thite (1979, 1980); Thite and Patil (1983); Khundalkar (1984); Khundalkar and Patil (1985).

The present investigation was therefore, undertaken to reveal the floristic affinities of the resupinate Aphyllophorales from WG with that of tropical Africa, South-East Asia and tropical America. The investigation was undertaken along with a thorough understanding of resupinate Aphyllophorales with respect to its taxonomic identity, floristic affinities, distribution patters, detailed macro and micro morphological characters, enzymatic patterns, along with their culture characters. In the present study a total of 110 fungal species from the 46 respective hosts were identified out of 355 collected specimens.

Out of which *Aleurodiscus aberrans, A. cremeus, Amylocorticium cebennense, A. olivaceoalbum, Amylocystis sericeomollis, Amylosporus bracei, Amylostereum laevigatum, Anomoporia dumontii, Auriporian aurulenta, Botryohypochnus anomalus, Candelabrochaete verruculosa, Cerrena unicolor, Cystostereum murraii, Diplomitoporus hondurensis, Earliella scabrosa, Flavodon flavous, Gloeocystidiellum fistulatum, G. flammeum, G. furfuraceum, G. insidiosum, G. irpiscescens, G. kenyense, G. lacticolor, G. luridum, G. luteocystidiatum, G. percuriosum, G. turpe, Grammothelopsis puiggarii, Hymenogramme javensis, Hyphodontia alienata, Hypochnicicum cymosum, H. eichlerie, H. globosum, Kavinia himantia, Laeticorticium simplicibasidium, Laxitextum lutescens, Lepidomyces subcalceus, Leucogyrophana mollis, Metulodontia flavidoalba, Oxyporus latamarginatus, Peniophora aurantiaca, P. gladiola, P. laurentii, P. pithya, P. gomezii, Phanerochaete calotricha, P. joseferroirae, P. tuberculata, Phlebiopsis galochroa, P. peniophoroides, Phlebia rufa, P. subcretacea, Phlebiopsis galochroa, P.*

peniophoroides, Phellinus cesatii, P. luctuosus, P. purpureogilva, P. punctatus, P. sonorae, Pseduoxnasma verrucisporum, Pycnoporellus alboluteus, Schizopora carneolutea, S. trichiliae, Scopuloides hydnoides, Scytinostromella cerina, Steccherinum setulosum, Steccherinum setulosum, Tubulicrinis ellipsoideus, Tyromyces undosus, Vuilleminia acerina, Wolfiporia cocos, Wrightoporia Africana, W. avellanea, W. cremea, W. iobaphus, Xylobolus apricans and *X. frustulatus* are reported for the first time in India and to Maharashtra as well, in this investigation.

It was observed that out of 110 reported species, 73 species were non-poroid and 37 species were poroid resupinate Aphyllophorales. As comparison, *Mangifera indica* Linn have demonstrated to be the most susceptible host, which harbors the highest number fungi. Where as the least number of fungi were found on *Adhatoda vasica* Nees. and *Albizzia lebbek* (L.) Benth.

The mixed vegetation of *Artocarpus heterophyllus* Lam, *Azadirachta indica* Juss, *Bauhinia purpurea* L., Ficus benganina L., *Mangifera indica* Linn. and *Memecylon umbellatum* Burm. shows 46 species of investigation areas in which non-poroid members are dominant (27 non-poroid and 11 poroid). Resupinate Aphyllophorales occur almost in equal proportion; non-poroid members are more common on the plants which show either well developed bark or hard wood or under second layer canopy trees. It is also confirmative that most of the hosts which is decaying or wounded in lower side of a standing tree shows luxuriant growth of non-poroid resupinate Aphyllophorales, mainly the species of *Hyphoderma, Phanerochaete, Laeticorticium* and *Gloeocystidiellum,* which shows white to pinkish-white resupinate basidiocarps. The observations made also conclude that species of *Hymenochaete, Peniophora, Phanerochaete* and *Phlebiopsis* occur more frequently on the fallen branches of second layer trees of bushes or climbers. The habitat of fungi can be grouped in three major categories namely live standing tree, dead standing tree and fallen branches. Species of *Amylocorticium, Amylocystis, Anomoporia, Oxyporus, Trichaptum* and *Vararia* exhibit brown rot type decay. The proportion of white rot fungi to brown rot fungi is not in concurrence with that of North American polypore worked out by Gilbertson and his co-workers (1979), as the area under investigation calls under typical southern tropical vegetation.

To confirm the rot, Benzidine test was carried out and it was revealed that 15 species belongs to brown rot and 95 to white rot. Amongst white rot, majority of that species belonging to *Gloeocystidiellum, Hymenochaete, Peniophora, Phanerochaete, Phellinus* and *Wrightoporia*.

Among the members of resupinate Aphyllophorales 74 are non-amyloid (species of *Hymenochaete, Hypochnicium, Oxyporus, Peniophora, Phanerochaete, Phellinus, Rigidopus, Schizopora*, etc.), 33 are amyloid (species of *Amylocorticium, Amylostereum, Gloeocystidiellum, Leucogyrophana, Wrightoporia, Xylobolus*, etc.) and 3 are dextronoid (species of *Perenniporia* and *Grammothelopsis*).

It was observed that 81 identified species belong to resupinate basidiocarps (species of *Aleurodiscus, Amylocorticium, Amylostereum, Cejpomyces, Gloeocystidie-*

llum, Hymenochaetae, Peniophora, Phanerochaete, Phellinus, Vararia, Wrightoporia, etc.) 16 species belong to effused (species *of Diplomitoporus, Kavinia, Laeticorticium, Perenniporia, Phanerochaeta, Phellinus, Phlebiopsis & Scopuloides*) and 13 species belong to effuso-reflexed.

The presence of white rot humus is another very important and significant feature in the investigation areas. The decomposed wood finally lying on the vegetation floor gets heavily infested with the termite or they lead to the development of so called white humus in the need special attention with respect to floristic composition of white rot humus, moisture regime, gaseous exchange, nitrogen metabolism and geochemical cycles. These could be some of the important topics for further research, to be worked in greater details.

TAXONOMY

In order to study the detailed Mycobiota of resupinate Aphyllophorales, a total of 110 fungal species from the 46 respective hosts were identified out of 355 collected specimens (Host-Fungus Index).

TAXONOMY OF RESUPINATE APHYLLOPHORALES

Taxonomy of identified species and distribution of the species in the investigation areas is as follows:

Family and Generic name	Areas
1. Amylostereaceae Boidin, Mugnier & Canales	
a. ***Amylostereum*** Boidin	
i. *A. laevigatum* (Fr.) Boidin	**SA**
2. Atheliaceae Julich	
a. ***Amylocorticium*** Pouzar	
i. *A. cebennense* (Bourdot) Pouzar	**PU**
ii. *A. olivaceoalbum* (Bourdot & Galzin) Boidin, Lanq. & Gilles	**RT**
3. Bondarzewiaceae Kolt. and Pouzar	
a. ***Amylosporus*** Ryvarden	
i. *A. bracei* (Murrill) David & Rajchenb.	**SA**
b. ***Wrightoporia*** Pouzar	
i. *W. avellanea* (Bresadola) Pouzar	**PU**
ii. *W. cremea* Ryvarden	**PU**
iii. *W. iobaphus* (Patouillard) Ryvarden	
iv. *W. africana* Johansen & Ryvarden	**PU**

Family and Generic name	Areas
4. Botryobasidiaceae (Parmasto) Julich	
a. ***Botryohypochnus*** Donk	
i. *B. anomalus* Hjortstam	**PU**
5. Ceratobasidiaceae Martin	
a. ***Cejpomyces*** Svrcek & Pouzar	
i. *C. terrigenus* (Bresadola) Svrcek & Pouzar	**PU**
6. Corticiaceae Herter	
a. ***Laeticorticium*** Donk	
i. *L. simplicibasidium* Lindsey & Gilbertson	**PU**
b. ***Vuilleminia*** Maire	
i. *V. acerina* (Persoon) Parmasto	**SA**
7. Cystostereaceae Julich	
a. ***Cystostereum*** Pouzar	
i. *C. murrayi* (Berkeley & Curtis) Pouzar	**PU**
8. Fomitopsidaceae (Polyporaceae) Julich	
a. ***Amylocystis*** Bondartsev & Singer ex Singer	
i. *A. sericeomollis* (Romell) Teixeira	**PU**
b. ***Auriporia*** Ryvarden	
i. *A. aurulenta* David, Tortic & Jelic	**SA**
c. ***Wolfiporia*** Ryvarden & Gilbertson	
i. *W. cocos* (Wolf) Ryvarden & Gilbertson	**RT**
d. ***Anomoporia*** Pouzar	
i. *A. dumontii* Hjortstam & Ryvarden	**PU**
9. Gloeocystidiellaceae (Parmasto) Julich	
a. ***Gloeocystidiellum*** Donk	
i. *G. citrinum* (Persoon) Donk	**PU, SA, RT**
ii. *G. convolvens* (Karsten) Donk	**PU, SA, RT**
iii. *G. fistulatum* (Cunningham) Boidin	**PU**
iv. *G. flammeum* Boidin	**RT**
v. *G. furfuraceum* (Bresadola) Donk	**PU, RT**

Family and Generic name	Areas
vi. *G. insidiosum* (Bourdot & Galzin) Donk	**SA**
vii. *G. irpiscescens* Boidin	**SA**
viii. *G. kenyense* Hjortstam	**PU**
ix. *G. lactescens* (Berkeley) Boidin	**RT**
x. *G. lacticolor* (Bresadola) Stalpers & Hjortstam	**PU**
xi. *G. leucoxanthum* (Bresadola) Boidin	**ST**
xii. *G. luridum* (Bresadola) Boidin	**PU**
xiii. *G. luteocystidiatum* (Talbot) Boidin	**SA**
xiv. *G. percuriosum* Parmasto	**RT**
xv. *G. sulcatum* (Rehill & Bakshi) Boidin	**PU**
xvi. *G. turpe* Freeman	**PU**
b. ***Pseudoxenasma*** Larsson & Hjortstam	
i. *P. verrucisporum* Larsson & Hjortstam	**SA**
10. Grammotheleaceae Julich	
a. ***Hymenogramme*** Montagne & Berkeley	
i. *H. javensis* Montagne & Berkeley	**SA**
11. Hericiaceae Donk	
a. ***Laxitextum*** Lentz	
i. *L. lutescens* Hjortstam & Ryvarden	**PU**
12. Hygrophoropsidaceae Kuhner	
a. ***Leucogyrophana*** Pouz.	
i. *L. mollis* (Fr.) Parm.	**PU**
13. Hymenochaeteceae Imazeki	
a. ***Hymenochaete*** Lév.	
i. *H. attenuata* (Lév.) Lév.	**SA, RT**
ii. *H. cinnamomea* (Pers.) Bres.	**SA**
iii. *H. rubiginosa* (Dicks.) Lév.	**SA**
iv. *H. tabacina* (Sowerby) Lév.	**PU, RT**
b. ***Phellinus*** Quélet	
i. *P. chryseus* (Lév.) Ryvarden	**PU**

Family and Generic name	Areas
ii. *P. luctuosus* (Cesati) Ryvarden	PU, SA
iii. *P. carterii* (Berkeley ex Cooke) Ryvarden	PU, SA, RT
iv. *P. purpureogilvus* (Petch) Ryvarden	PU, SA, RT
v. *P. cesatii* (Bresadola) Ryvarden	PU, SA
vi. *P. inermis* (Ellis & Everhart) Cunningham	PU
vii. *P. lamaensis* (Murrill) Patouillard	PU
viii. *P. punctatus* (Fries ex Karsten) Pilát	PU
ix. *P. sonorae* Gilbertson	PU
14. Hyphodermataceae Julich	
a. ***Hyphoderma*** Wallroth	
i. *H. radula* Fries	PU
b. ***Hypochnicium*** Eriksson	
i. *H. cymosum* (Rogers & Jackson) Larsson & Hjortstam	SA
ii. *H. eichlerie* (Bresadola ex Saccardo) Eriksson & Ryvarden	PU
iii. *Hypochnicium globosum* Wu	SA
15. Lachnocladiaceae Reid	
a. ***Vararia*** Karsten	
i. *V. sphearicospora* Gilbertson	PU
16. Meripilaceae Julich	
a. ***Rigidoporus*** Murrill	
i. *R. ulmarius* (Sowerby) Imazeki	RT
17. Meruliaceae Karst	
a. ***Phlebia*** Fries	
i. *P. rufa* (Persoon) Christiansen	PU, SA
ii. *P. subcretacea* (Litschauer) Christiansen	PU
b. ***Scopuloides*** Höhnel & Litschauer	
i. *S. hydnoides* (Cooke & Massee) Hjortstam & Ryvarden	RT
ii. *S. rimosa* (Cooke) Jülich	PU
18. Peniophoraceae Lotsy	
a. ***Peniophora*** Cook	

Family and Generic name	Areas
i. *P. aurantiaca* (Bresadola) Höhnel & Litschauer	SA
ii. *P. farinose* (Bresadola) Höhnel & Litschauer	SA
iii. *P. gladiola* Cunningham	RT
iv. *P. laurentii* Lundell	SA
v. *P. nuda* (Fries) Bresadola	PU, RT
vi. *P. pithya* (Persoon) Eriksson	PU
vii. *P. violaceolivida* (Sommerfelt) Massee	SA
19. Phanerochaetaceae Julich	
a. ***Candelabrochaete*** Boidin	
i. *C. verruculosa* Hjortstam	SA
b. ***Lopharia*** Kalchbrenner & MacOwan	
i. *L. fulva* (Lév.) Boidin	PU
c. ***Metulodontia*** Parmasto	
i. *M. flavidoalba* (Cooke) Malençon & Bertault	SA
d. ***Phanerochaete*** P. Karsten	
i. *P. cacaina* (Bourdot & Galzin) Burdsall & Gilbertson	SA
ii. *P. calotricha* (Karsten) Eriksson & Ryvarden	SA
iii. *P. joseferroirae* (Reid) Reid	SA
iv. *P. laevis* (Fries) Eriksson & Ryvarden	RT
v. *P. sanguinea* (Fries) Pouzar	PU, SA, RT
vi. *P. sordida* (Karsten) Eriksson & Ryvarden	PU, RT
vii. *P. tuberculata* (Karsten) Parmasto	PU
e. ***Phlebiopsis*** Jülich	
i. *P. galochroa* (Bresadola) Hjortstam & Ryvarden	SA
ii. *P. gigantea* (Fries) Jülich	SA
iii. *P. peniophoroides* Gilbertson & Adaskaveg	PU
iv. *P. roumeguerei* (Bresadola) Jülich & Stalpers	RT
20. Polyporaceae Fr. ex Corda	
a. ***Cerrena*** Gray	
i. *C. unicolor* (Bulliard) Murrill	PU, RT
b. ***Earliella*** Murrill	

Family and Generic name	Areas
i. *E. scabrosa* (Persoon) Gilbertson & Ryvarden	SA
c. ***Grammothelopsis*** Jülich	
i. *G. puiggarii* (Spegazzini) Rajchenberg & J. E. Wright	PU
d. ***Perenniporia*** Murrill	
i. *P. gomezii* Rajchenberg & J. E.	PU
ii. *P. medulla-panis* (Jacquin) Donk	PU, SA, RT
e. ***Pycnoporellus*** Murrill	
i. *P. alboluteus* (Ellis & Everhart) Kotlaba & Pouzar	PU
f. ***Tyromyces*** Karsten	
i. *T. undosus* (Peck) Murrill	SA
g. ***Trichaptum*** Murrill	
i. *T. versatile* (Berk.) Cunn	SA
21. Ramariaceae Corner	
a. ***Kavinia*** Pilát	
i. *K. himantia* (Schweinitz) Eriksson	SA, RT
b. ***Ramaricium*** J. Eriksson	
i. *R. polyporoideum* (Berkeley & Curtis) Ginns	SA
22. Schizoporaceae Julich	
a. ***Hyphodontia*** J. Eriksson	
i. *H. alienata* (Lundell) Eriksson	SA
b. ***Oxyporus*** (Bourdot & Galzin) Donk	
i. *O. cervino-gilvus* (Junghuhn) Ryvarden	PU, RT
ii. *O. latamarginatus (Durieu & Montagne) Donk*	SA, RT
c. ***Schizopora*** Velen.	
i. *S. carneolutea* (Rodway & Cleland) Kotlaba & Pouzar	SA
ii. *S. paradoxa* (Schrader) Donk	SA
iii. *S. roseotigens* Hjortstam & Ryvarden	SA
iv. *S. trichiliae* (Van der Byl) Ryvarden	SA
23. Sistotremataceae Julich	
a. ***Fibriciellum*** Eriksson & Ryvarden	
i. *F. silvae-ryae* Eriksson & Ryvarden	SA
24. Steccherinaceae Parmasto	
a. ***Diplomitoporus*** Domanski	

Family and Generic name	Areas
i. *D. hondurensis* (Murrill) Ryvarden	**SA**
b. ***Flavodon*** Ryvarden	
i. *F. flavous* (Klotzsch) Ryvarden	**PU, SA, RT**
c. ***Junghuhnia*** Corda	
i. *J. crustacea* (Junghuhn) Ryvarden	**PU**
d. ***Steccherinum*** Gray	
i. *S. setulosum* (Berkeley & Curtis) Miller	**PU**
e. ***Aleurodiscus*** Rabnh ex. Schrot.	
i. *A. aberrans* Cunningham	**SA**
ii. *A. cremeus* Patouillard	**PU**
25. Stereaceae Pilat	
a. ***Scytinostromella*** Parmasto	
i. *S. cerina* (Bresadola) Hjortstam & Ryvarden	**ST**
b. ***Xylobolus*** P. Karsten	
i. *X. apricans* (Bourdot) Wu, Boidin & Chien	**PU**
ii. *X. frustulatus* (Persoon) Boidin	**SA**
26. Tubulicrinaceae Julich	
a. ***Tubulicrinis*** Donk	
i. *T. ellipsoideus* Rajchenberg	**PU, SA**
27. Xenasmataceae Oberw	
a. ***Lepidomyces*** Jülich	
i. *L. subcalceus* (Litschauer) Jülich	**PU**

SA: Satara **PU**: Pune **RT**: Ratnagiri

A. KEYS TO THE GENERA OF NON-POROID APHYLLOPHORALES

1. Spores amyloid 2
1. Spores non-amyloid 11
2. Clamps lacking 3
2´. Clamps present 4

3. With thin-walled pseudocystidia, other cystidial elements Lacking... .... ***Gloeocystidiellum*** (9)

3´. With thick-walled pseudocystidia, acuto-or/and acanthocystidia Present .... ***Xylobolus*** (33)

4. Cystidial elements present 5

4´. Cystidial elements lacking 8

5. Acanthocystidia present .... ***Aleurodiscus*** (1)

5´. Acanthocystidia lacking 6

6. Spores thich-walled, grayish in Melzer's reagent, cystidia sparse .... ***Leucogyrophana*** (18)

6´. Spores thin-walled, not grayish in Melzer's reagent, cystidia mostly numerous 7

7. Cystidia strongly encrusted of metuloid appearance .... ***Amylostereum*** (3)

7´. Cystidia different shaped .... ***Amylocorticium*** (2)

8. Dimitic (or trimitic) species .... ***Scytinostromella*** (28)

8´. Monomitic species 9

9. Spores 8-10 × 7-8 µm .... ***Pseudoxenasma*** (25)

9´. Spores 4.5-7 × 2.5-3. 5µm 10

10. Fruitbodies resupinate to reflexed, tramal hyphae light brown, cystidia without sulphoaldehyde reaction .... ***Laxitextum*** (16)

10´. Fruitbodies resupinate, tramal lacking, subicular hyphae hyaline, cystidia with sulphoaldehyde reaction .... ***Gloeocystidiellum*** (9)

11. Basidia with a basal clamp 12

11´. Basidia without a basal clamp 28

12. Dichohyphidia present .... ***Vararia*** (31)

12´. Dichohyphidia lacking 13

13. Spores ornamented 14

13´. Spores smooth 15

14. Cystidial elements present .... ***Hypochnicium*** (13)

14´. Cystidial elements lacking .... ***Ramaricium*** (26)

15. Cystidial elements present 16

15´. Cystidial elements lacking .... ***Kavinia*** (14)

16. With lyocystidia two-rooted .... ***Tubulicrinis*** (30)

16´. With other kinds of cystidia 17

17. Basidia and spores very long, normally 80-100 and 20 µm respectively, common on still attached branches of deciduous trees .... ***Vulleminia*** (32)

17´. Basidia and spores smaller 18

18. Dendrohyphidia present 19

18´. Dendrohyphidia lacking 20

19. Cystidia protruding, often more than 30µm above the basidia, without Sulphoaldehyde Reaction .... ***Laeticorticium*** (15)

19´. Cystidia not or very slightly protruding with sulphoaldehyde reaction.. ***Peniophora*** (21)

20. Hyphale system dimitic 21

20´. Hyphale system monomitic 24

21. Fruitbodies odontoid, hydnoid or subporoid.. ***Steccherinum*** (29)

21´. Fruitbodies smooth or tuberculate 22

22. With brown pseudocystidia, strongly encrusted .... ***Lopharia*** (19)

22´. Without such cystidia 23

23. Skeletal hyphae strongly cyanophilous, cystidia lacking, but with spares hyphoids between the Basidia .... ***Fibriciellum*** (8)

23´. Skeletal hyphae not cyanophilous, cystidia well differentiated ...... ***Cystostereum*** (7)

24. Fruitbodies reticulately plicate, poroid, grandinoid, odontoid or hydnoid 25

24´. Fruitbodies smooth but some species pilose by protruding cystidia of fruitbodies Colliculose 26

25. Cystidia of two kinds .... ***Metulodontia*** (20)

25. Cystidia of one kind .... ***Hyphoderma*** (11)

26. Lagenocystidia present, other kind of cystidia smooth .... ***Hyphodontia*** (12)

26. Lagenocystidia lacking 27

27. Spores subglobose to ellipsoide, up to 6-7 µm. long.. ***Phlebia*** (23)

27´. Spores allantoid, sigmoid, sub-cylindrical, longer than 7µm........... ***Lepidomyces*** (17)

28. Basidiocarp xanthochroic .... ***Hymenochaete*** (10)

28´. Basidiocarp non-xanthochroic.. 29

29. Spores ornamented, yellowish brown .... ***Botryohypochuns*** (4)

29´. Spores smooth, as a rule hyaline 30

30. Dimitic species .... ***Steccherinum*** (29)

30´. Monomitic species.. 31

31. Hymenophore grandinoid, odontoid or hydnoid.. 32

31´. Hymenophore smooth or meruloid 34

32. Cystidia strongly encrusted, of metuloid appearance, spores small, 3.5-4.5 × 1.8- 2.2 µm .... ***Scopuloides*** (27)

32´. Not with this combination of characteristics 33

33. Cystidia septate .... ***Candelabrochaete*** (5)

33′. Cystidia not septate.. ***Phanerochaete*** (22)

34. Fruitbidies hard and parchment-like with distinct or indistinct subicular tissue, subhymenial tissue dense, cystidia metuloid .... ***Phlebiopsis*** (24)

34′. Not with this combination of characteristic, spores 12-17×6-7μm...... ***Cejpomyces*** (6)

DESCRIPTION OF NON-POROID GENERA AND SPECIES

(1). *ALEURODISCUS* (Rabenhorst) ex. Schroeter. Cohn, Kryptogamen-flora Schlesiens **3:** 429, 1888.

The genus *Aleurodiscus* was proposed by Rabenhorst in 1888, with type species, *Aleurodiscus amorphous* (Pers.) Rabh. The name *Aleurodiscus* was first used by Rabenhorst in 1874 for specimen issued in the exsiccatum fungi Eur., Cent. XIX, 1824. No validating description was published, either with specimen or in his explanatory note in Hedwigia **13:** 184, 1874. Although Cooke (1875, p. 137) recognised the genus it was not until 1888 that it was validly published by Schroeter.

The genus is cosmopolitan in distribution with 141 species recorded. At first *Aleurodiscus* was limited to species with cupulate or pezizae for fructification and large basidia, features which equally well apply to many species of cystidia. Later workers added species possessing various types of ancillary or accessory organs, such as pseudophyses, acanthophyses and dendrophyses; other species again placed under this genus, have or small basidia together with one or more of these organs, and fructification either pezizaeform, disciform or resupinate and effused. The genus contains basidiocarp formed from radiate hyphae parallel with the hymenial surface and bearing large or small basidia, associated with pseudophyses, acanthophyses and gloeocystidia. From the investigation areas two species collected. The specimen with characteristic white hymenial surface, without clamp was identified as *A. aberrans.* The specimen with characteristic cream-buff coloured hymenial surface, with clamp was identified as *A. cremeus*.

KEY TO THE SPECIES OF *ALEURODISCUS*

1. Hymenial surface white, remaining so or becoming cream, or occasionally sulphur yellow, generative hyphae clampless.. ***A. aberrans***

1′. Hymenial surface cream-buff colored, rimose, generative hyphae with clamp...***A. cremeus***

(1-1). *Aleurodiscus aberrans* Cunningham. Trans. Roy. Soc. New Zealand **84 (2) :** 257, 1956.

= *Acanthophysium aberrans* (Cunn.) Cunn., *Bull. N. Z. Dept. Sci. Industr. Res., Pl. Dis. Div.* **145:** 154, 1963. (Plate 1, 8; Fig. 5).

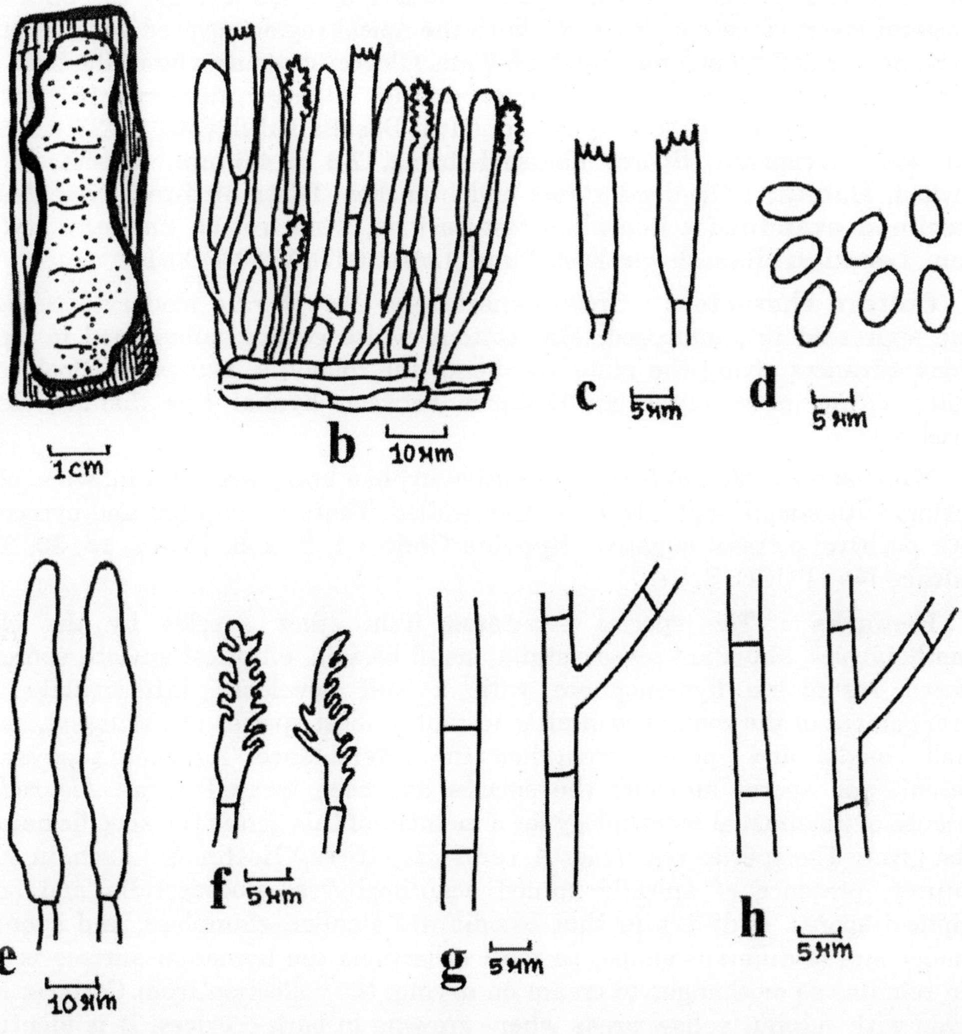

Fig. 5- *Aleurodiscus aberrans* : a. Habit b. T.S. hymenium, c. Basidia. d. Basidiospores. e. Gloeocystidia. f. Acanthophyses. g. Generative hyphae. h. Generative hyphae in culture.

Morphological Characters: Hymenophore annual, resupinate, adnate, ceraceous-cretaceous, 5-6 × 1-2 cms., hymenial surface white, remaining so or becoming cream, or occasionally sulphur yellow, pruinose, at length deeply areolately creviced; margin thinning out, fibrillose, white, adnate. Context white, 120-160 µm thick, basal layer narrow, of mainly parallel hyphae, intermediate layer occupying the greater part of the context, of closely arranged upright hyphae soon cemented by their walls to form a pseudoparenchyma, embedding masses of crystals. Hyphal system monomitic, generative hyphae to 4-6 µm wide, hyaline, branched, septate, without clamp. Hymenial layer to 50 µm deep, a loose palisade of basidia, acanthophyses and gloeocystidia. Acanthophyses forming the bulk of the hymenial layer, clavate or fusiform, with the apical region covered with several (5-14) blunt spines 2-3 μm long, 15-20 × 5-7 μm. Gloeocystidia arising in the base of the context and subhymenium, some projecting slightly, fusiform or flexuous-cylindrical, 50-80 × 8-10 μm, apices rounded. Basidia subclavate, 20-25 × 5-6 μm., with four sterigmata. Spores ellipsoid-oblong, 7-9 × 4-5 μm, smooth, hyaline, amyloid. **Habitat :** On dead stems and branches. **Distribution :** New Zealand. **Specimen examined :** Benzidine test positive, white rot on *Capparis zeylanica* Linn., **Location :** Baneshwar; Dist: Pune (13/01/2007), VH-NPO-611.

Culture characters : Growth characters: Growth rate moderate; advancing zone appressed to submerged; Mat cottony white getting submerged in agar as colony advances giving the plate and hence the colony, a transparent look; colony white; outline of colony even; Distance between hyphal tips distant; Reverse bleached.

Microscopic characters: Generative hyphae branched, 5-6 μm, wide, clamps lacking, with simple septa, hyaline, thin-walled. **Tests :** α-napthal and pyrogallol + H_2O_2 positive; p-cresol negative. **Species Code :** 1, 2, 3, 6, 13, 14, 15, 30, 37, 54. **Culture No :** PUCC. B: 01.

Remarks : The species separated from other species by the clavate acanthophyses, abundant gloeocystidia, small basidia, elliptical smooth spores and effused resupinate hymenophore with a well developed intermediate layer. Arrangement of the context is similar to that of most species of *Corticium*, and the small basidia and spores strengthen the resemblance. As acanthophyses are present, and spores amyloid, the species has been treated as an *Aleurodiscus*. Because of its unusual morphology for a member of this genus the specific name has been given. The species resembles *A. cerussatus* (Bres.) Hoehn. & Litsch. in surface features, presence of apically spined acanthophyses, gloeocystidia and smooth elliptical spores. It differs in that basidia are smaller, clampless, and spores are smaller and of different shape. In most collections the hymenial surface is white, and remains so or changes to cream on drying; the collection from Coriaria is rich cream with sulphur yellow areas where growing in bark crevices. It is identical in microfeatures. This and the following species are resupinate and effused, with the context composed of both basal and intermediate tissues. *A. candidus*, though colonies are usually pulvinate, is treated under the resupinate section since it possesses a similar type of context and colonies are sometimes coalesced and form resupinate areas., Species may be divided into four sections according to the types of acanthophyses and dendrophyses. Acanthophyses of *A. aberrans* are clavate with apices crowned with digitate processes or. spines, coralloid in *A. coralloides, A. coronatus and A. sbarsus*, and botryose in *A. botryosus*. Dendrophyses of somewhat simple type are present in *A. acerinus, A. candidus, A. nivosus* and *A. aurantius*. The species is a new record from India.

(1-2). *Aleurodiscus cremeus* Patouillard. *Bull. trimest. Soc. Mycol. Fr.* **31:** 73, 1915 (Plate: 1, 8; Fig. 6).

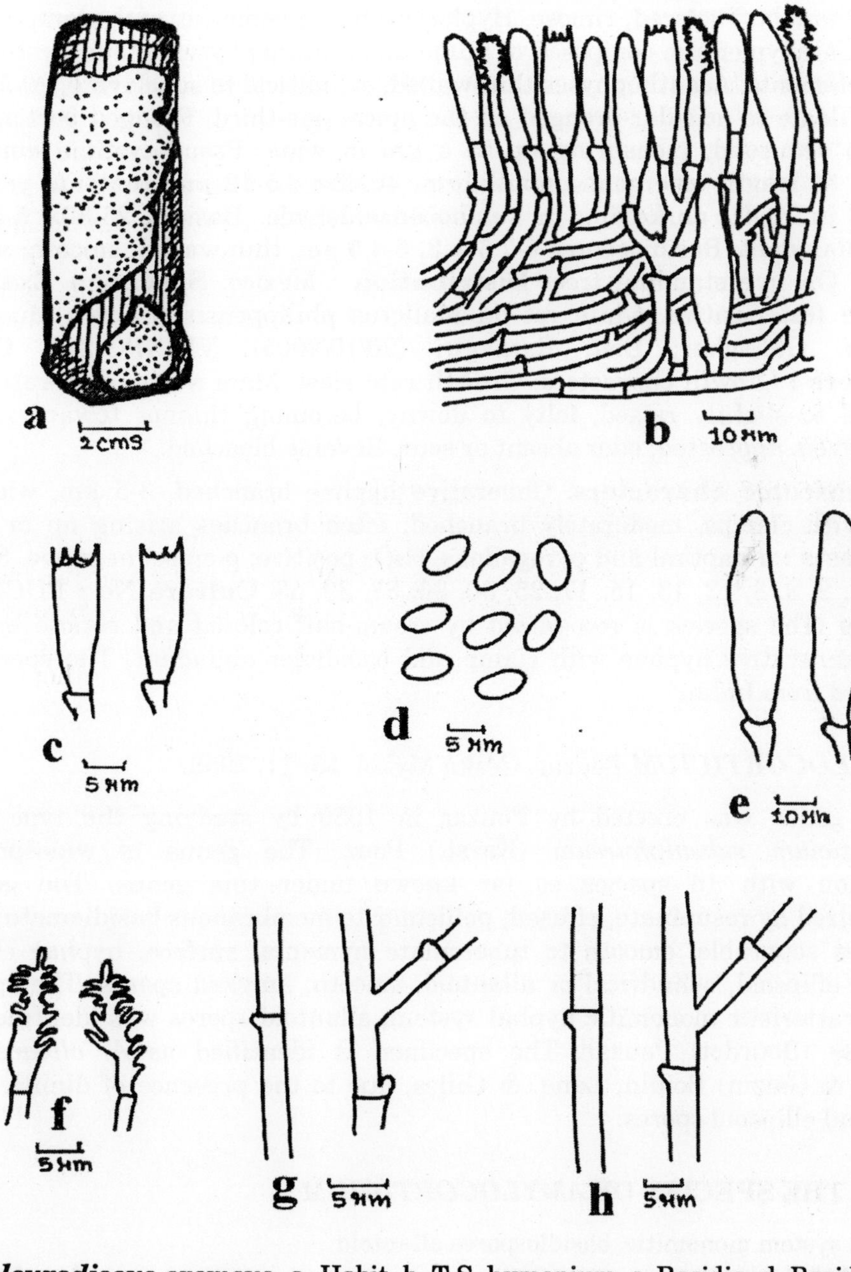

Fig. 6 *Aleurodiscus cremeus,* a. Habit. b. T.S. hymenium. c. Basidia. d. Basidiospores. e. Pseudocystidia, f. Acanthocystidia. g. Generative hyphae. h. Generative hyphae in culture.

Morphological Characters : Basidiocarp initially pulvinate, readily confluent, effused; margin determinate, adnate, 5-7×3-4 cms. Fructification 500-700 μm thick in cross section; texture compact-farinose to subceraceous. Hymenial surface cream-buff colored, rimose. Hyphal system monomitic, with clamps, 3-4 μm in wide. Catahymenium composed of numerous acanthophyses, pseudocystidia, and basidial elements. Acanthophyses thin-walled, cylindrical to subclavate, 20-30 × 4.5-6 μm; aculeate to acicular-pronged on the apical one-third. Pronged portion thick-walled to semisolid, measuring up to 4 μm in wide. Pseudocystidia embedded-cylindric, at length emergent-ampulliform, 40-60× 8.5-12 μm; contents yellowish-refractile in KOH, darkening in sulphobenzaldehyde. Basidia 25-30× 5-5.5 μm, clavate, abundant. Basidiospores 6-7.5 × 3. 5-4.5 μm, thin-walled, smooth, amyloid. **Habitat:** On live standing tree. **Distribution** : Mexico. **Specimen Examined**: Benzidine test positive, white rot on Mallotus philippensis (Lamk.) Muell. Arg. **Location**: Guhaghar; Dist: Ratnagiri (20/10/2005), VH-NPO-273. **Culture Characters :** Growth characters: Growth rate slow. Mats white, moderately thin, appressed to slightly raised, felty to downy, becoming thinner toward margins, margins even, appressed; odor absent or sour. Reverse bleached.

Microscopic characters: Generative hyphae branched, 3-5 μm, wide, thin walled, with clamps, moderately branched, often branches arising up or from a clamp. **Tests :** α-napthal and pyrogallol + H_2O_2 positive; p-cresol negative. **Species Code :** 1, 2, 3, 8, 12, 13, 15, 17, 25, 30, 36, 37, 39, 53. **Culture No :** PUCC. B:02. **Remarks :**The species is recognized by cream-buff colored and rimose hymenial surface; generative hyphae with clamp and basidioles abundant. The species is a new record from India.

(2). *AMYLOCORTICIUM* Pouzar. Ceská Mykol. **13:** 11, 1959.

The genus was erected by Pouzar in 1959 by studying the type species *Amylocorticium subsulphureum* (Karst.) Pouz. The genus is widespread in distribution with 15 species so far known under this genus. The genus is characterized by resupinate, effused, pellicular to membranous basidiomata;adnate, sometimes separable; smooth to tuberculate hymenial surface; hyphae clamped; narrowly ellipsoid, cylindrical or allantoid, smooth, amyloid spores. The specimen with characteristic monomitic hyphal system, allantoid spores was identified as *A. cebennense* (Bourdot) Pouzar. The specimen is identified as *A. olivaceoalbum* (Bourdot & Galzin) Boidin, Lanq. & Gilles, due to the presence of dimitic hyphal system and ellipsoid spores.

KEY TO THE SPECIES OF *AMYLOCORTICIUM*

1. Hyphal system monomitic, basidiospores allantoid
   6-8 × 1. 5-2 μm .... ***A. cebennense***

1´. Hyphal system dimitic, basidiospores ellipsoid,
   4. 0-5. 0 × 2- 2.5 μm .... ***A. olivaceoalbum***

(2-1). *Amylocorticium cebennense* (Bourdot) Pouzar. *Ceská Mykol.* **13 (1)** : 11, 1959.

= *Corticium cebennense* Bourdot, *Rev. Sci. Bourb. Centr. Fr.* **23:** 7, 1910. (Plate 1, Fig. 7).

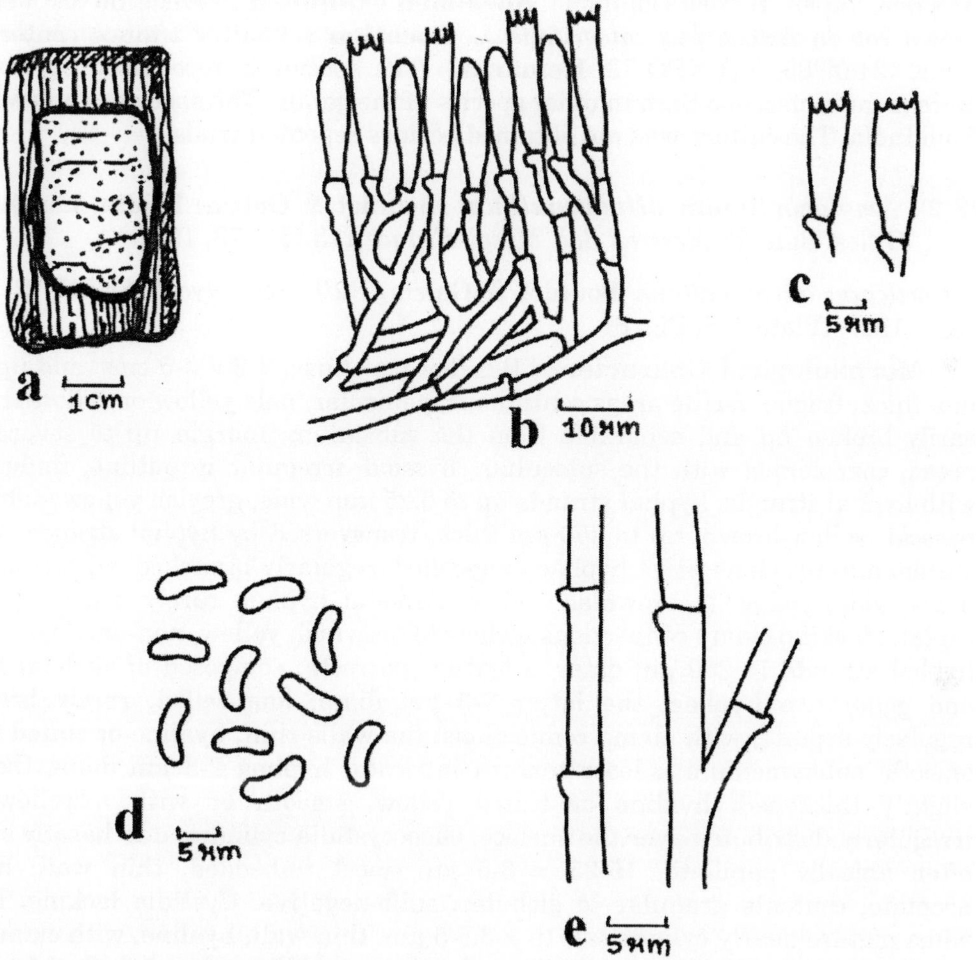

Fig. 7 *Amylocorticium cebennense* a. Habit, b. T.S. hymenium, c. Basidia, d. Basidiospores, e. Generative hyphae.

Morphological Characters : Basidiomata resupinate, loosely adnate, effused, 3 × 2 cms., hymenial surface white to pale cream, smooth; margin white, thinning out arachnoid to fimbriate. In section subhyaline, pellicular to membranous. Hyphae hyaline, 4-6 μm wide, thin-walled, with clamps, branches frequent and at right angles. Subhymenium dense; hyphae hyaline, more closely branched and narrower than subicular hyphae, wall thin, septa clamped throughout. Hymenium compact and homogeneous, single depth along with closely intertwined subhymenium 40-60 μm sometimes thickening, with zonate deposition of spores. Cystidia none. Basidia clavate, 20-30 × 4-6 μm, with 4 sterigmata. Basidiospores hyaline, thin-walled, amyloid, narrowly cylindrical to vaguely ovoid, 6-8 ×1. 5-2.5 μm. **Habitat :** On live standing tree. **Distribution :** Thailand, Europe, North America, Japan, British Columbia. **Specimen examined :** Benzidine test negative, brown rot on *Artocarpus integrifolia* L. **Location :** Poultry trainee center; Dist: Pune, (24/09/05) VH-NPO-73. **Remarks :** The species is recognized by frutibody more membranaceous than in other species of the genus. The species is a new record from India. The culture was not obtained besides repeated trials.

(2-2). *Amylocorticium olivaceoalbum* (Bourdot & Galzin) Boidin, Lanquetin & Gilles. *Bull. Trimestriel Soc. Mycol. France* **113 (1) :** 76, 1997.

= *Corticium olivaceoalbum* Bourdot & Galzin, *Bull. Soc. mycol. Fr.* **27 (2) :** 239, 1911 (Plate 1, 8, Fig. 8)

Morphological Characters : Basidiocarp effuse, 7-8 × 4-5 cms. and up to 0.3 mm thick, fragile, fertile areas continuous, pellicular, pale yellow or yellowish white easily broken up and separable from the subiculum; margin up to several mm. broad, concolorous with the subiculum, byssoid, irregular in outline, fimbriate or with hyphal strands; hyphal strands up to 0.25 mm wide, greyish yellow; subiculum byssoid, yellow-brown, up to 250 μm thick, transversed by hyphal strands. Hyphal system dimitic. Generative hyphae long-celled, regularly branched, with clamp, 3-4. 5 μm wide, yellow to brownish yellow. Skeletal hyphae rarely branched, rarely septate, lacking clamp connections, yellow to brownish yellow, non-amyloid, smooth; hyphal strands 10-250 μm diam, a textura porrecta, composed of skeletal hyphae and generative hyphae, the latter 2-3 μm diam, long-celled, rarely branched, regularly septate, with clamp connections, the walls thin, hyaline or tinted yellow, smooth; subhymenium a loose textura intricata, hyphae 2-3 μm diam, the walls slightly thickened, hyaline or tinted yellow, smooth or with a yellow layer irregularly distributed over the surface. Gloeocystidia cylindrical to basally swollen, often apically papillate, 15-25 × 3-5 μm, most imbedded, thin wall, hyaline, aseptate; contents granular to globular, sulfo-negative. Cystidia lacking. Basidia when mature nearly cylindrical, 15 × 3.5-5 μm, thin wall, hyaline, with clamp, with four sterigmata. Basidiospores ellipsoid to narrow ellipsoid, 4-5 × 2- 2.5 μm, the walls weakly amyloid, smooth, thin-walled, acyanophilous. **Habitat :** On live standing tree. **Distribution :** United States. **Specimen examined :** Benzidine test negative, brown rot on *Areca catechu.* **Location:** Guhaghar; Dist: Ratnagiri (20/10/2005), VH-NPO-278. **Culture characters:** Growth characters: Growth rate slow; Mats white to pale yellow-orange, moderately thick, raised, cottony throughout or limited around inocula, then becoming white to light buff, moderately thin, arachnoid to cottony toward margins; margins even, raised, cottony; odor absent or sour; Reverse bleached. Microscopic characters: Marginal hyphae 2-4 μm diam thin walled, with clamps, sparsely branched, branches arising below and occasionally from clamps. Submerged hyphae 1.5-5 μm diam, thin walled, nodose septate, sparsely to moderately branched. Aerial hyphae similar to submerged hyphae. Arthroconidia cylindrical with rounded ends, 6-15 × 1.5-2 μm, hyaline, thin walled. **Tests :** α-napthal and pyrogallol + H_2O_2 positive; p-cresol negative. **Species Code :** 1, 2, 3, 8, 12, 15, 21, 35, 36, 38, 39, 52, 53, 84. **Culture No :** PUCC. B:03. **Remarks :** The species is recognized by dimitic hyphal system. Generative hyphae yellow to brownish yellow. Basidiospores ellipsoid to narrow ellipsoid, the walls weakly amyloid. The species is a new record from India.

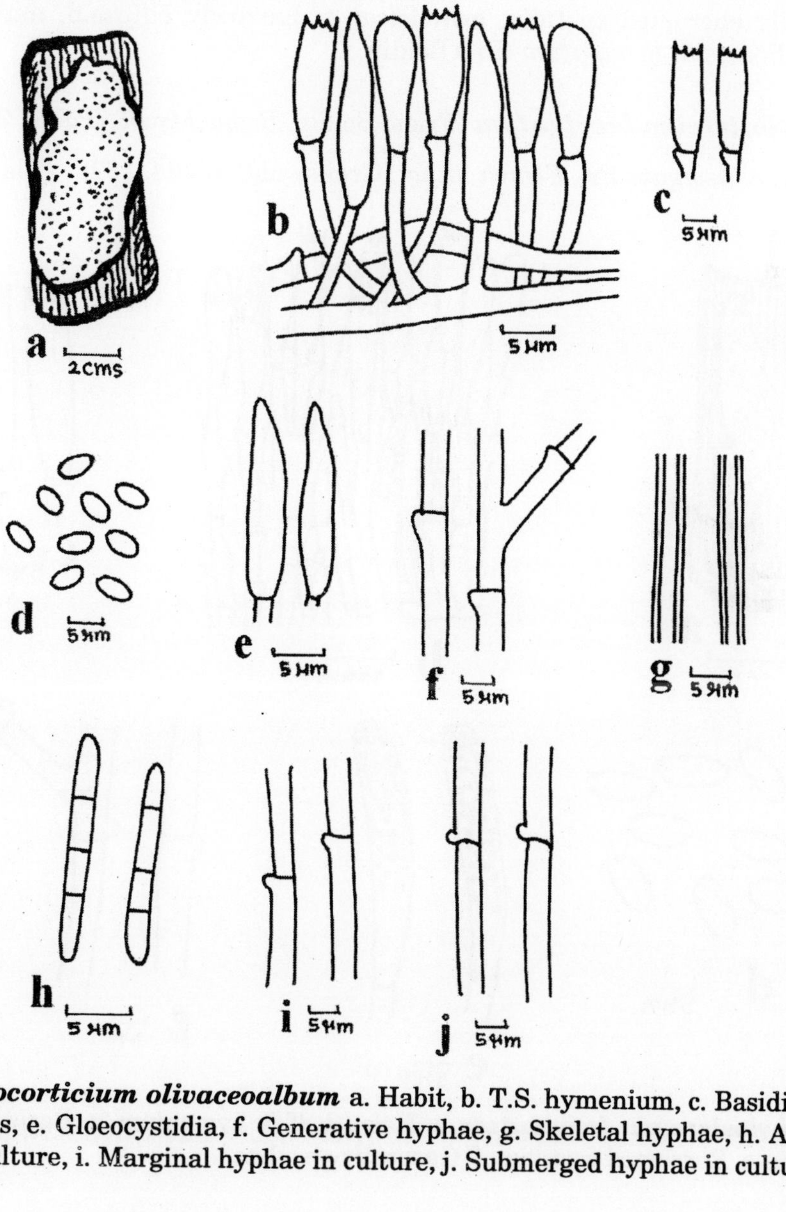

Fig. 8. ***Amylocorticium olivaceoalbum*** a. Habit, b. T.S. hymenium, c. Basidia, d. Basidiospores, e. Gloeocystidia, f. Generative hyphae, g. Skeletal hyphae, h. Arthroconidia in culture, i. Marginal hyphae in culture, j. Submerged hyphae in culture.

(3). *AMYLOSTEREUM* Boidin. Revue Mycol. **23:** 345, 1958.

This genus erected by Boidin in 1958, by studying the type of species *Thelephora chailletii* Pers. The genus has five species reported under this genus. The genus is characterized by usually perennial; resupinate, effuso-reflexed or sessile; coriaceous to corky basidiomata; even, yellowish brown to brown hymenial surface; with clamps hyphae; acute or somewhat rounded, brown; thick-walled; encrusted, contents weakly sulfo-positive cystidia; ellipsoid to cylindrical; and amyloid spores. The specimen with characteristic numerous, thick-walled brown or yellowish brown and apically encrusted cystidia; cylindrical or narrowly ellipsoid, amyloid spores was identified as *A. laevigatum* (Fr.) Boidin.

(3-1). *Amylostereum laevigatum* (Fries) Boidin. *Revue Mycol. (Paris)* **23:**345, 1958

= *Thelephora laevigata* Fr., *Elench. fung.* (Greifswald) **1:** 224, 1828. (Plate 1, Fig. 9)

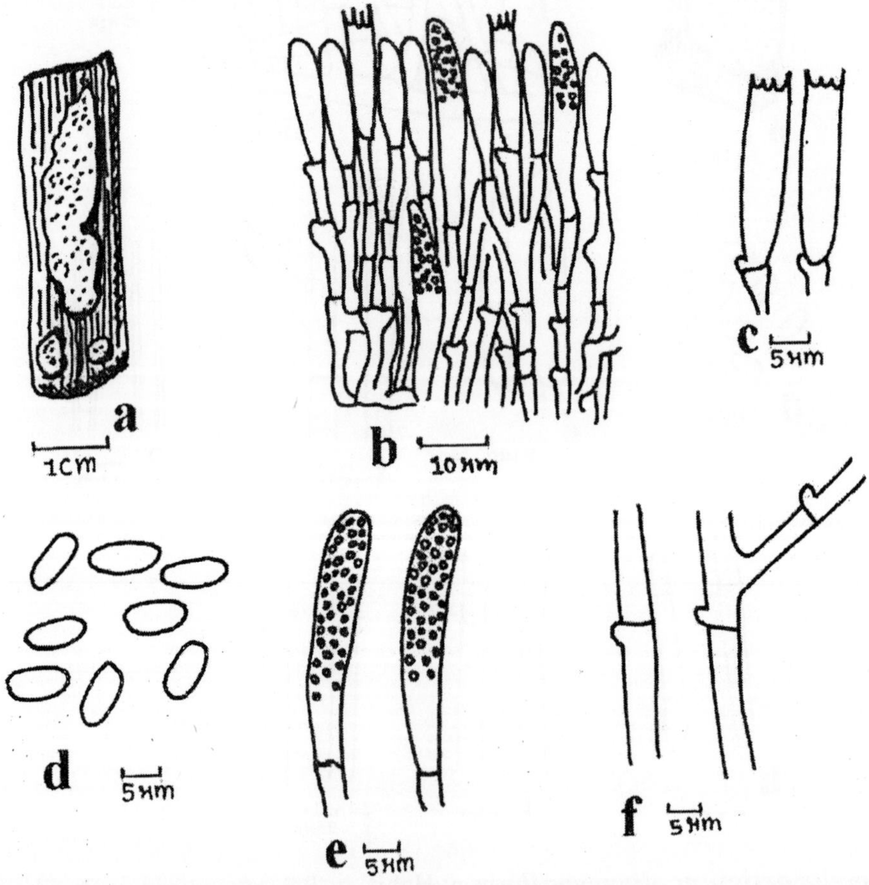

Fig. 9 *Amylostereum laevigatum* a. Habit, b. T.S. hymenium, c. Basidia, d. Basidiospores, e. Encrusted cystidia, f. Generative hyphae.

Morphological Characters : Fruitbody resupinate, adnate, but the margin together with the outer layer of the bark may loosen from the substrate, 3-4 × 1-1.5 cms. Hymenium smooth, when dry often finely cracked, light brownish to ochraceous or even greyish. Hyphal system monomitic, consisting of richly branched, thin to thick-walled hyphae with clamps, 3-4 μm wide. Basidia narrowly clavate, 25-30 × 4-6 μm., with 4 sterigmata and a basal clamp. Cystidia numerous, thick-walled, brown or yellowish brown and apically encrusted, 30-50 × 7-10 μm. Basidia clavate, thin wall, hyaline, with clamp, with four sterigmata, 20-30 × 5-7 μm. Spores cylindrical or narrowly ellipsoid, 7-12 × 3-4 μm, smooth and amyloid. **Habitat :** On dead standing tree. **Distribution :** Sweden. **Specimen examined :** Benzidine test negative, brown rot on *Albizzia lebbek* (L.) Benth. **Location :** Baneshwar; Dist: Pune (13/01/2007), VH-NPO-584. **Remarks :** The species is easily to recognize by light brownish to ochraceous or even greyish hymenium surface. Cystidia numerous, thick-walled, brown or yellowish brown and apically encrusted. The species is a new record from India. The cultures could not obtain from this specimen.

(4) *BOTRYOHYPOCHNUS* Donk. *Meded. Nederl. Mycol. Ver.* **18-20:** 118, 1931.

The genus was erected by Donk in 1931 by studying the type species *Botryohypochnus isabellinus* (Fr.) Erikss. The genus is widespread in distribution with 11 species so far known under this genus. The genus is characterized by loosely adnate, reticulate to hypochnoid basidiomata. In section sub-hyaline to pale yellowish brown, arachnoid to byssoid; monomitic hyphal system; obovate to short-cylindrical Basidia; globose, verrucose to echinulate, thick-walled spores. The specimen is identified as *B. anomalus* Hjortstam, due to the presence of hyphae with simpl septa; echinulate to aculeate, subglobose and pale brown spores.

(4-1) *Botryohypochnus anomalus* Hjortstam. *Mycotaxon* 17: 562, 1983. (Plate 1, 8, Fig. 10)

Morphological Characters : Fruitbody resupinate, loosely adnate, 10-12 × 3-4 cms., hypochnoid to slightly pellicular, pale ochraceous. Hyphal system monomitic, 5-6 μm wide, thin-walled, without clamps. Basidia stout, mostly obovate, 18-25 ×10-12 μm, with four sterigmata. Spores more or less distinctly thick-walled, at least when fully mature, echinulate to aculeate, subglobose and normally 8-9 μm. wide, pale brown. **Habitat :** On live standing tree. **Distribution :** Kenya, North America. **Specimen examined :** Benzidine test positive, white rot on *Swietenia mahogany* (L.) Jaca. Chun. **Location :** Sambhaji park; Dist: Pune (21/10/2005), VH-NPO-293. **Culture characters :** Growth characters: Growth rate moderate; Mats cottony white, moderately thin, raised, cottony, then becoming thinner, slightly raised toward margins, outline of colony even; Reverse unchanged. Microscopic characters: Generative hyphae 4-6 μm wide, thin-walled, simple septate, sparsely branched; branches formed below septa, constricted at base. **Tests :** α-napthal and pyrogallol + H_2O_2 positive; p-cresol negative. **Species Code :** 1, 2, 3, 7, 12, 15, 21,

30, 37, 53, 54. **Culture No :** PUCC. B: 04. **Remarks :** The species is somewhat intermediate between *B. bondarcevii* Parm. and *B. isabellinus* (Fr.) John Erikss. The first one has decidedly smaller spores and thicker walls, while the latter has globose spores which are distinctly echinulate. The species is a new record from India.

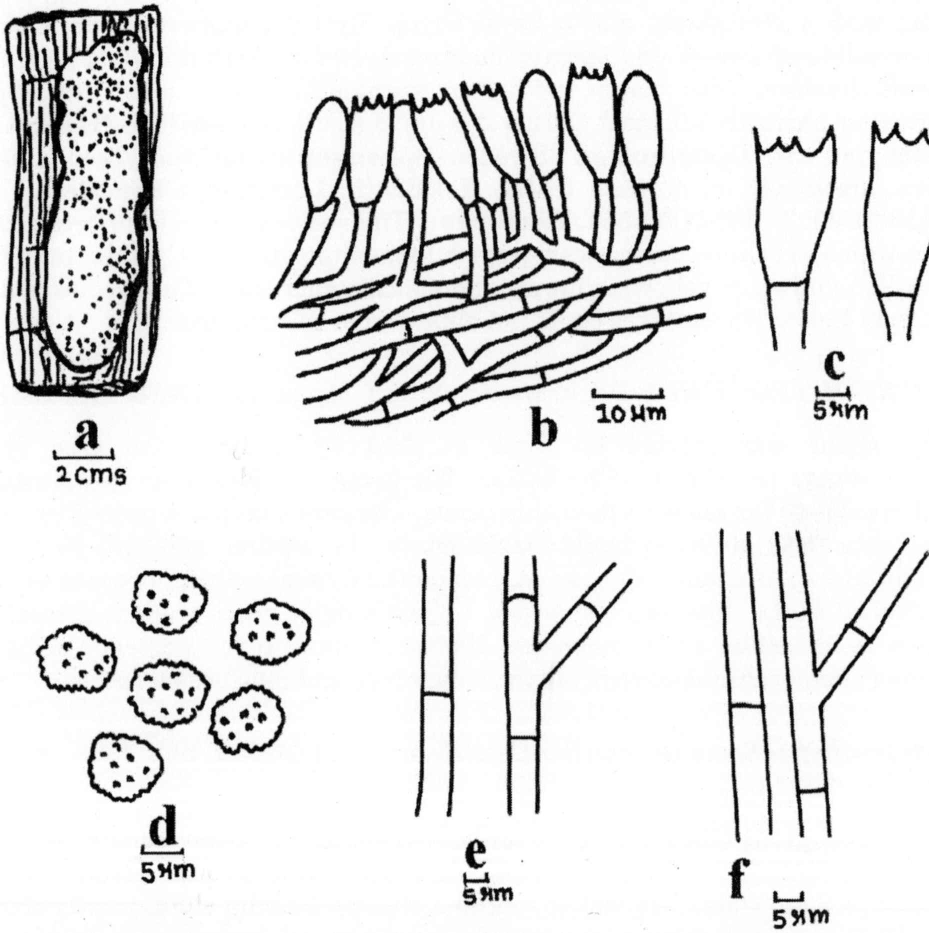

Fig. 10. ***Botryohypochnus anomalus*** a. Habit, b. T.S. hymenium, c. Basidia, d. Basidiospores, e. Generative hyphae, f. Generative hyphae in culture.

(5). ***CANDELABROCHAETE*** Boidin. Cah. Maboké **8:** 24, 1970.

The genus was erected by Boidin in 1970 with type species *Candelabrochaete africana* Boidin. The genus is widespread in distribution with 13 species known under the genus. The genus is characterized by resupinate, effused, byssoid to submembranous basidiomata; smooth to odontoid hymenial surface; monomitic hyphal system, clampless; cylindrical cystidia; ellipsoid to allantoid, smooth, thin-walled, non-amyloid basidiospores. The specimen with characteristic buff, vinaceous-buff to ochreous, grandinioid to odontoid hymenial surface;narrowly clavate to cylindrical cystidia was identified as *C. verruculosa* Hjortstam.

(5-1) *Candelabrochaete verruculosa* Hjortstam. *Mycotaxon* 17: 566, 1983.

=*Odonticium verruculosum* (Hjortstam) Zmitr. & Spirin, *Mycena* **6:** 40 (2006) (Plate 1, 8, Fig. 11)

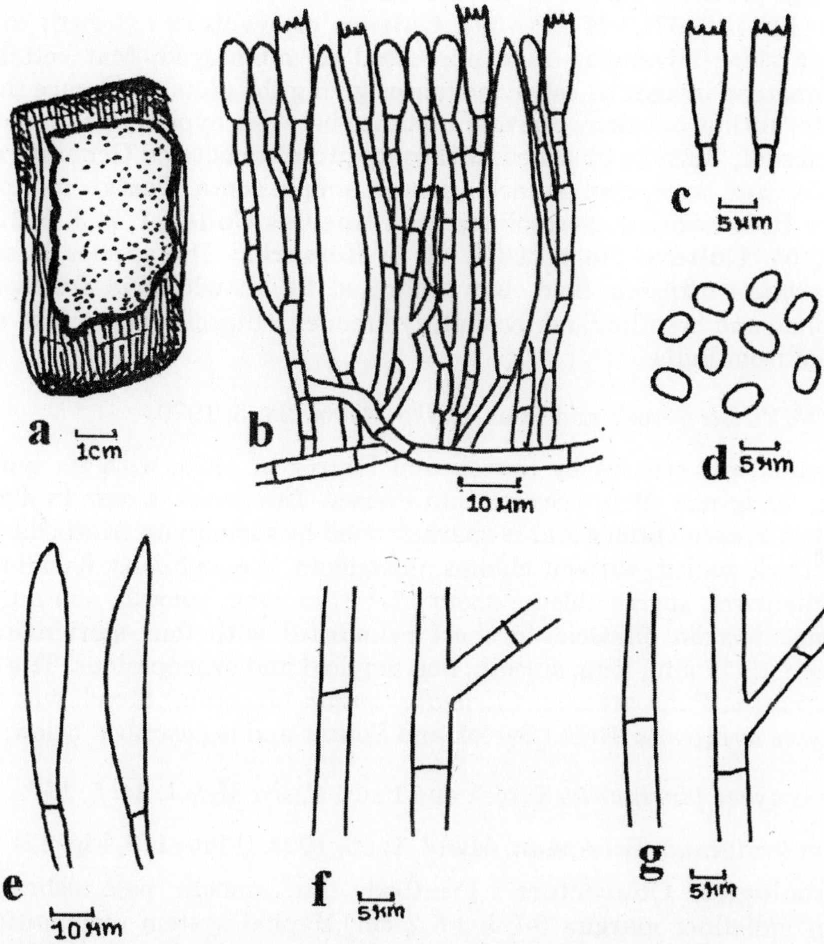

Fig. 11 ***Candelabrochaete verruculosa*** a. Habit. b. T.S. hymenium. c. Basidia. d. Basidiospores. e. Cystidia. f. Generative hyphae. g. Generative hyphae in culture.

Morphological Characters : Basidiomata resupinate, adnate, effused, 4 × 3 cms; hymenial surface buff, vinaceous buff to ochreous, grandinioid to odontoid; aculei 15-25 per mm², conical, up to 400 µm. in length; margin concolorous with the hymenial surface, thinning out, indeterminate. In section subhyaline, submembranous, containing crystalloid materials in the subhymenium, the subiculum, and especially in the aculei. Hyphal system monomitic, hyphae 3-6 μm. wide, smooth, thin- to slightly thick-walled, clampless. Cystidia narrowly clavate to cylindrical, up to 60-90 × 8-12 μm, smooth, thin-walled, but often thickened at the base. Basidia clavate to cylindrical, sometimes slightly constricted, 10-20 × 4.5-5 μm, without a basal clamp, producing 4 sterigmata; basidiospores ellipsoid, 5-6.5 × 3-3.5 μm, smooth, thin-walled, non-amyloid. **Habitat :** On fallen branch. **Distribution :** Africa, Europe, Japan. Norwegian, Center Africana, Brazil. **Specimen examined :** Benzidine test positive, white rot on Tamarindus indica Linn. **Location :** Panshet; Dist: Pune (13/01/ 2007), VH-NPO-597. **Culture characters :** Growth characters: Growth rate slow; advancing zone appressed to submerged; Mat cottony white getting submerged in agar as colony advances giving the plate and hence the colony, colony white; outline of colony uneven; Distance between hyphal tips distant; Aerial mycelium absent; Reverse bleached. **Microscopic characters:** Generative hyphae branched, 5-6 μm, wide, clamps lacking, with simple septa. **Tests :** α-napthal and pyrogallol + H_2O_2 positive; p-cresol negative. **Species Code :** 1, 2, 3, 6, 13, 14, 15, 16, 30, 37, 54. **Culture No :** PUCC. B: 5. **Remarks:** The species is similar to *Candelabrochaete africana* Boid. but separated by an odontioid fruitbody, thin-walled hyphae and cystidia, and by slightly broader, ellipsoid spores. The species is a new record from India.

(6) *CEJPOMYCES* Svrcek and Pouz. *Ceska Mykol.* **24:** 5, 1970.

The genus was created by Svrcek and Pouzer in 1970, with its type species *Cejpomyces terrigenus* (Bres.) Svrcek and Pouzer. The genus is rare in distribution with only two known species and is characterized by resupinate fruitbodies; hyphae somewhat thick-walled, without clamps, dextrinoid, cyanophilous; basidia ellipsoid to short cylindrical; spores oblong, about 15-17 μm. long, smooth, non amyloid and cyanophilous. basidia ellipsoid to short cylindrical with four sterigmata; spores oblong, about 15-17 μm., long, smooth, non amyloid and cyanophilous. The specimen collected from the Pune area, on *Gliricidia sepium* (Jacquin.) Steud. was identified as *Cejpomyces terrigenus* (Bres.) Svrcek and Pouzer and is described below.

(6-1). *Cejpomyces terrigenus* Svrcek and Pouz. *Ceska Mykol*. **24:** 5, 1970.

= *Corticium terrigenum* Bres., *Ann. Mycol.* **1:** 99, 1903. (Plate 1, 8, Fig. 12)

Morphological Characters : Fruitbody thin, smooth, pale, ochraceous and with an in indistinct margin, 5-7 × 1.5-2 cm. Hyphal system monomitic, hyphae dextrinoid in Melzer's reagent, cyanophilous and without clamps, 8-10 μm., wide; basal hyphae parallel to substrate and with a bit thickened walls, hymenial hyphae thin-walled and dextrinoid. Basidia ellipsoid to short cylindrical, 20-25 × 8-10 μm., with four sterigmata; sterigmata long, 10-17 μm., bearing four spores on them. Spores oblong, narrowly ovate to subcylindrical, apically obtuse, 12-17 × 5-7 μm., smooth, non-amyloid, cyanophilous. **Habitat :** On live standing tree. **Distribution :** Poland, England, Czechoslovakia, Sweden, Gabon. **Specimen examined:** Benzidine test positive, white rot on *Gliricidia speium.* **Location :** Panchwatii; Dist:

Pune (24/09/2005), VH-NPO-61. **Culture characters :** Growth characters: Growth rate moderate; advancing zone appressed to submerged; Mat cottony white getting submerged in agar as colony advances giving the plate and hence the colony, a transparent look; colony white; outline of colony even; Distance between hyphal tips distant; Aerial mycelium absent; Reverse bleached. Microscopic characters: Generative hyphae branched, 7-9 μm, wide, clamps lacking, with simple septa. **Tests :** α-napthal and pyrogallol + H_2O_2 positive; p-cresol negative. **Species Code :** 1, 2, 3, 6, 13, 14, 15, 16, 30, 37, 54, 55. **Culture No :** PUCC. B: 06. **Remarks :** The species seems to be close to *Ceratobasidium,* but differs from it in the absence of repetitive spores. It also seems to be close to *Botryobasidium,* but is distinguished from it by having much larger and longer sterigmata; also it doesn't has more than four spores *Cejpomyces* has a dense subhymenium formed by wide, shortcelled and frequently branched hyphae. Basidia are also arranged in a dense palisade, quite different from those of *Botryobasidium.* This species has been earlier reported from India (The Western Ghats of Maharashtra) by Nanda (1996).

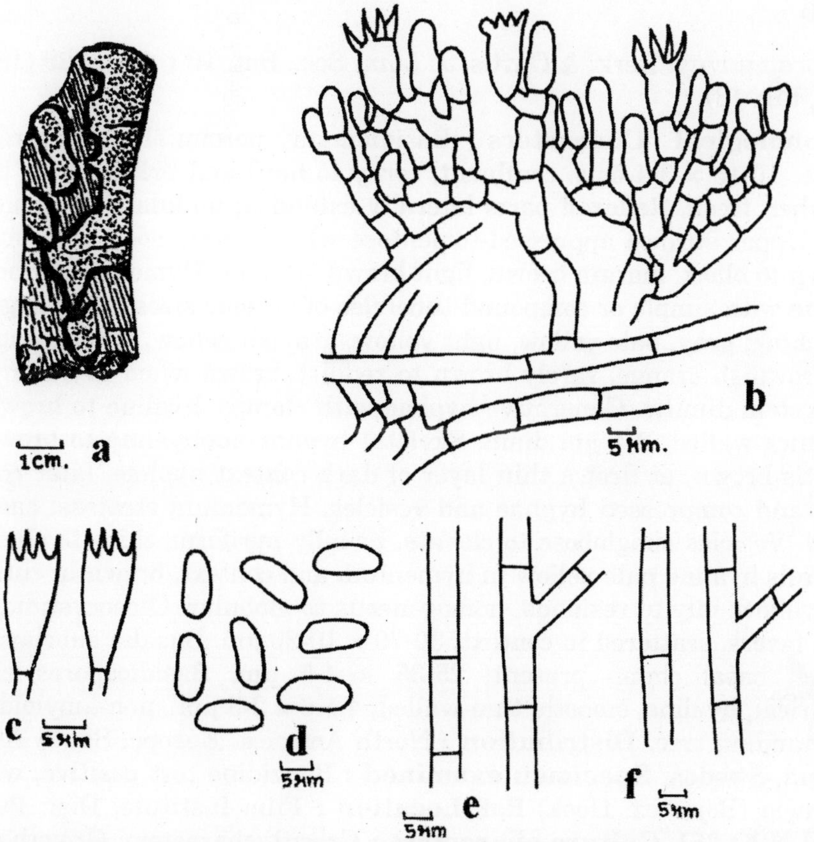

Fig. 12 ***Cejpomyces terrigenus*** a. Habit, b. T.S. hymenium, c. Basidia, d. Basidiospores, e. Generative hyphae, f. Generative hyphae in culture.

(7). ***CYSTOSTEREUM*** Pouzar. Ceská Mykol. **13 (1)** : 18, 1959.

The genus was erected by Pouzer in 1959 by studying the type species *Cystostereum murrayi* (Berkeley & Curtis) Pouzar. The genus is widespread with 11 species known. The genus is characterized by perennial, pileate or mostly resupinate fruitbodies; hymenium odontioid or tuberculate; dimitic hyphal system; with clamps; fruitbodies consisting of two layers, a subiculum of mainly horizonyal hyphae and a subhymenial trama of vertical hyphae. Gloeocystidioid vesicles in the hymenium and in the subhymenium, but can also occur in the subiculum. The vesicles are developed in the hymenium but become enclosed in the subhymenium as the fruit body increase in thickness; narrowly ellipsoid or subcylindrical, thin walled, smooth, stained in cotton-blue spores. The specimen with characteristic hard texture, coarsely tuberculate and cracked hymenial surface, and by the presence of numerous vesicles was identified as *C. murrayi* (Berkeley & Curtis) Pouzar.

(7-1). ***Cystostereum murrayi*** (Berkeley & Curtis) Pouzar. *Ceská Mykol.* **13(1)**:18, 1959

=*Thelephora murrayi* Berk. & Curtis, *J. Linn*. Soc., Bot. **10 (46)** : 329 (1868) (Plate 1, 8, Fig. 13)

Morphological Characters :Basidiomata perennial; effuso-reflexed to resupinate, 10-13 × 3-4 cms. confluent; corky to hard and brittle; with the odor of coconut when fresh. Reflexed parts laterally extended; undulate to plicate; up to 1 cm wide. Upper surface appressed-tomentose when young, soon glabrous; sulcate; dark brown to black, margin brown, light brown or paler. Hymenial surface coarsely tuberculate with simple or compound tubercles of various sizes, often rugose, rarely even; cracking; gray, pale yellow, light yellow, grayish yellow, light orange, grayish orange, brownish orange, rarely brown to reddish brown when dry; margin white. Hyphal system dimitic. Generative hyphae with clamps, hyaline to brown, thin- to slightly thick-walled, 2-4 μm diam. Skeletal hyphae subhyaline to brown, 1-3 μm diam. Cutis brown, at first a thin layer of dark context hyphae, later composed of darkened and compressed hyphae and vesicles. Hymenium stratose, each stratum ill-defined. Vesicles subglobose to clavate, usually pyriform; thin- to slightly thick-walled; walls hyaline pale yellow in hymenium and context, brown in cutis, smooth, contents yellow, oily to resinous, homogeneous to globular. Gloeocystidia formed in hymenial layers, scattered in context; 30-70 × 10-20 μm. Basidia subclavate, with 4 sterigmate, basal clamp present; 25-35 × 4-5 µm. Basidiospores ellipsoid to subcylindrical, hyaline, smooth, thin-walled, 4.5-6 × 2-3 µm., non-amyloid. **Habitat:** On live standing tree. **Distribution** : North America, Europe, South Africa, Asia, Australasia, Sweden. **Specimen examined** : Benzidine test positive, white rot on *Delonix regia* (Bojer ex. Hook) Raf. **Location** : Film Institute; Dist: Pune (20/10/2005), VH-NPO-261. **Culture characters** : Growth characters: Growth rate moderate; Mats white, thin to moderately thick, appressed or raised, velvety to downy; outline of colony even, appressed or raised; odor absent; then becoming faintly to

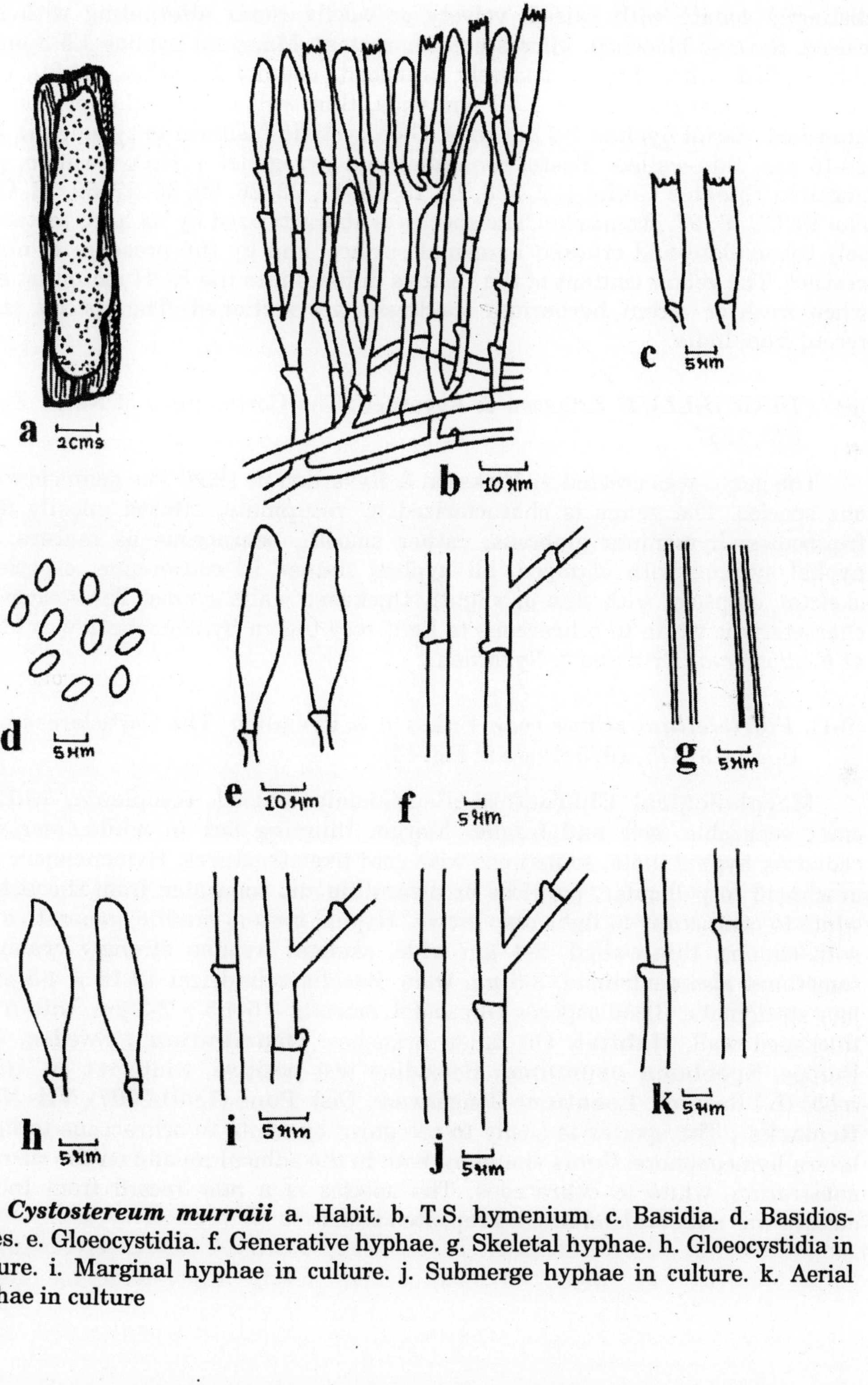

Fig. 13. ***Cystostereum murraii*** a. Habit. b. T.S. hymenium. c. Basidia. d. Basidiospores. e. Gloeocystidia. f. Generative hyphae. g. Skeletal hyphae. h. Gloeocystidia in culture. i. Marginal hyphae in culture. j. Submerge hyphae in culture. k. Aerial hyphae in culture

distinctly zonate with raised, velvety or woolly zones alternating with slightly raised, Reverse bleached. Microscopic characters: Marginal hyphae 1.5-3 μm diam, thin walled, with clamps, sparsely branched, usually branches arising opposite clamps. Submerged hyphae 1.5-3 μm wide, thin walled, with clamps, moderately branched. Aerial hyphae 1-2 μm wide. Gloeocystidia fusiform to cylindrical, 30-40 × 10-15 μm, thin-walled. **Tests:** α-napthal and pyrogallol + H_2O_2 positive; p-cresol negative. **Species Code:** 1, 2, 3, 8, 12, 13, 15, 17, 22, 26, 30, 36, 37, 52, 73. **Culture No:** PUCC. B: 07. **Remarks:** The species is characterized by its hard texture, coarsely tuberculate and cracked hymenial surface, and by the presence of numerous vesicles. The yellow content of the vesicles diffuses into the KOH mounting medium when fresh or recent herbarium specimens are sectioned. The species is a new record from India.

(8). *FIBRICIELLUM* Eriksson & Ryvarden. The Corticiaceae of North Europe 3: 373, 1957

The genus was erected by Eriksson & Ryvarden in 1957. The genus is with only one species. The genus is characterized by resupinate, effused, mostly fibrillose fruitbodies; hymenium glabrous, rather smooth, ochraceous as mature; dimitic hyphal system; with clamped, all hyphae stained in cotton-blue, especially the skeletal; ellipsoid, with thin or slightly thickened walls spores. The specimen with characteristic white to ochraceous to light rosy brown hymenophore was identified as *F. silvae-ryae* Eriksson & Ryvarden.

(8-1). *Fibriciellum silvae-ryae* Eriksson & Ryvarden. The Corticiaceae of North Europe 3: 375, 1975 (Plate 2; Fig. 14)

Morphological Characters : Basidiomata effused, resupinate, 10-12 × 3-5 cms., separable, soft and fragile. Margin thinning out in white, sterile, often radiating hyphal mats, sometimes with cord-like structures. Hymenophore smooth, arachnoid to pellicular, porulose or discontinuous, separable from the subiculum, white to ochraceous to light rosy brown. Hyphal system dimitic, generative hyphae with clamps, thin-walled, 2-4 μm wide, skeletal hyphae strongly cyanophilous, sometimes also dextrinoid, 3-5 μm wide. Basidia cylindrical 12-15 × 4-5 μm, with four sterigmata. Basidiospores ellipsoidal, smooth, 3.5-4.5 × 2-3 μm. with a slightly thickened wall. **Habitat:** On fallen branches. **Distribution :** Sweden, Canada, Europe. **Specimen examined:** Benzidine test positive, white rot on *Allophylus cobbe* (L.) Raeusch. **Location:** Dongerwadi, Dist. Pune (13/01/2007), VH- NPO-647. **Remarks :** The species is easily to recognize by white to ochraceous to light rosy brown hymenophore. Cords abundant both in the subiculum and on the surrounding substratum, white to ochraceous. The species is a new record from India. The culture was not obtained besides repeated trials.

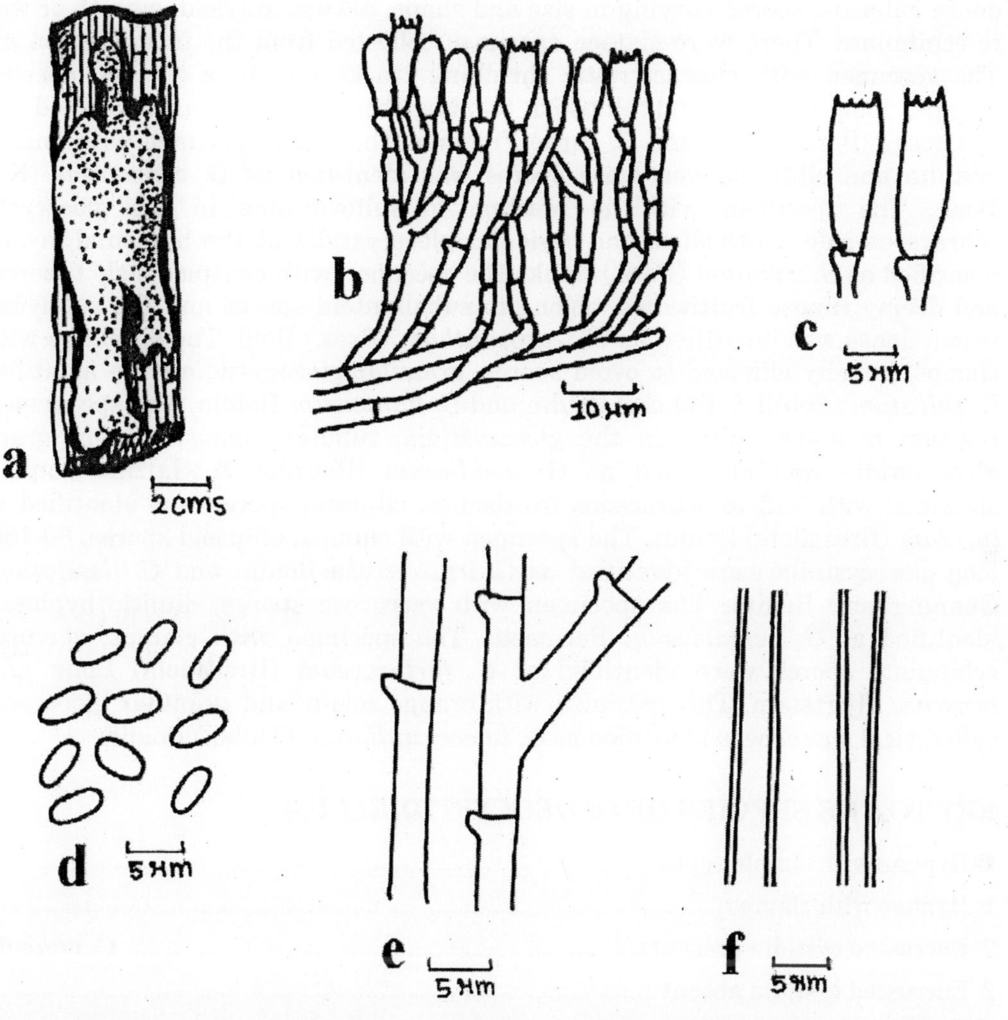

Fig. 14 ***Fibriciellum silvae-ryae*** a. Habit. b. T.S. hymenium. c. Basidia. d. Basidiospores. e. Generative hyphae. f. Skeletal hyphae

(9). *GLOEOCYSTIDIELLUM* Donk. *Meded. Ned. Mycol. Ver.* **18-20:** 156, 1931.

The genus was created by Donk in 1931 with type species *Gloeocystidiellum porosum* (Berk. & Curt.) Donk. The genus is wide spread in distribution with 74 known species.

The genus is characterized by the presence of gloeocystidia and amyloid spores, as dominant characters. It has resupinate, effused, thin to considerably thick fruitbodies; hyphal system monomitic, with or without clamps; gloeocystidia are always present; thin-walled, tubular, sinouse, reacting positively to sulfovaniline, encrusted cystidia may or may not be present; basidia clavate, sterigmata four, in a dense palisade; spores varying in size and shape, always amyloid, smooth or warted to echinulate. There were sixteen specimen collected from the investigation areas. The specimen with characteristic shrinking cracks and in a section, absence of clamps, long cystidia and ellipsoid to subglobose spores was identified as *G. lactescens* (Berk.) Boid. and *G. turpe* Freeman The other specimen with encrusted cystidia and ellipsoid, verrucose spores was identified as *G. convolvens* (Karst.) Donk. The specimen with no reaction to sulfovaniline in the gloeocystidia, ochraceous-pale fruitbodies, and vesicular gloeocystidia at the hymenial layer was identified as *G. citrinum* (Pers.) Donk. The specimen with conspicuously tuberculate and deeply rimose fruitbodies, oblong to suballantoid spores and the subhymenial trama dense was identified as *G. leucoxanthum* (Bres.) Boid. The specimen without clamps, broadly ellipsoid to ovoid spores, granular gloeocystidia were identified as *G. sulcatum* (Rehill & Bakshi) Boidin and *G. flammeum* Boidin. The specimen with reaction to sulfovaniline in the gloeocystidia, tubular sinous, content granular gloeocystidia was identified as *G. insidiosum* (Bourdot & Galzin) Donk. The specimen with buff to ochraceous fruitbodies, ellipsoid spores was identified as *G. luridum* (Bresadola) Boidin. The specimen with clamps, ellipsoid spores, 80-160 µm long gloeocystidia were identified as *G. irpiscescens* Boidin and *G. fistulatum* (G. Cunningham) Boidin. The specimen with verrucose spores, dimitic hyphae was identified as *G. percuriosum* Parmasto. The specimen with clamps, verrucos or echinulate spores were identified as *G. furfuraceum* (Bresadola) Donk and *G. kenyense* Hjortstam. The specimen with orange colour and granular gloeocystidia, cylindirical spores was identified as *G. luteocystidiatum* (Talbot) Boidin.

KEY TO THE SPECIES OF *GLOEOCYSTIDIELLUM*

1. Hyphae with simple septa 2
1. Hyphae with clamps 9
2. Encrusted cystidia present .... ***G. convolvens***
2. Encrusted cystidia absent 3
3. Spores globose to subglobose 4
3. Spores ellipsoid 5
4. Spores smooth .... ***G. citrinum***
4. Spores ornamentation .... ***G. lacticolor***
5. Spores 3-5 µm long 7
5. Spores 6-10 µm long 8
6. Gloeocystidia 80-90 × 9-14 µm., spore oblong to ovoid, 3-3. 5 µm. wide .... ***G. flameum***
6. Gloeocystidia 30-80 × 4-9 µm., spore sub-cylindrical to ellipsoid, 2-3 µm. wide 7
7. Gloeocystidia tubular, sinuous, content granular with positive reaction to sulfovanilline .... ***G. insidiosum***

7. Gloeocystidia clavate-cylindrical, content granular staining deeply with phloxine .... ***G. sulcatum***
8. Gloeocystidia up to 150 µm. long, spores 6-7µm long........ ***G. turpe***
8. Gloeocystidia up to 180-270 µm. long, spores 6.5-10 µm. Long .... ***G. lactescens***
9. Spores verrucose or echinulate 10
9. Spores smooth 12
10. Spores globose, echinulate........ ***G. furfuraceum***
10. Spores subglobose-ellipsoid, verrucose........ 11
11. Gloeocystidia up to 30 µm. Long, cylindrical to fusoid or clavate ...... ***G. percuriosum***
11. Gloeocystidia up to 50-100 µm. Long, subfusiform........ ***G. kenyense***
12. Spores allantoid, 12-20 µm. Long .... ***G. leucoxanthum***
12. Spores ellipsoid or cylindrical, up to 10 µm. Long........ 13
13. Spores cylindrical........ ***G. luteocystidiatum***
13. Spores ellipsoid 14
14. Spores 5-9 × 4-5.5 µm., gloeocystidia 80-160 µm. Long........ 15
14´. Spores 4-6 × 3-3.5 µm., gloeocystidia 80-400 µm. Long .... ***G. fistulatum***
15. Spores 5-7 × 4-4.5 µm., gloeocystidia 80-160 × 11-16 µm .... ***G. irpiscescens***
15´. Spores 6-9 × 4-4.5 µm., gloeocystidia 50-150 × 6-11 µm .... ***G. luridum***

DESCRIPTION OF SPECIES OF *GLOEOCYSTIDIELLUM*

(9-1). ***Gloeocystidiellum citrinum*** (Persoon) Donk. *Fungus* 26: 9, 1956.

=*Thelephora citrina*, Pers., *Mycol. europ.* 1p. 136, 1882

=*Thelephora radiosa* Fr., ex. Pers., *Mycol. europ*. 1p. 130, 1882.

=*Thelephora radiosa* Fr., *Elench. fung.* 1p. 206, 1828.

= *Gloeocystidiellum radiosum* (Fr.) Boid. *Pulb. Mus. Nat. Hist. Nat.* 17p. 122, 1957 (Plate 2, 8, Fig. 15)

Morphological Characters : Basidiomes effuse, 8-10 × 3-4 cms. and 100-500 µm thick, membranous; color ivory yellow, cartridge buff, cream-buff, light buff, or pale yellow-orange when dry; surface smooth to slightly tuberculate, margin usually distinct, white, fibrillose to fimbriate, sometimes farinaceous or abrupt. Hyphal system monomitic, generative hyphae 4-5 µm. wide, clampless. Gloeocystidia numerous, ventricose, ventricose-rostrate, and lageniform, thin-walled, 50-70 × 7-15 µm. Among the hymenial layer, we also come across few thin-walled encrusted cystidia, hyaline in texture, 22-25 × 4.5-6 µm., few hymenial cystidia without any encrustation or protoplasmic granular content, hyaline, cylindrical, 22-25 × 4-6 µm. Basidia 35-50 × 5-7.5 µm, clavate or cylindrical, with four sterigmata. Basidiospores 5.5-6.5 µm diam, globose to subglobose, thin wall, smooth, amyloid. **Habitat :** On

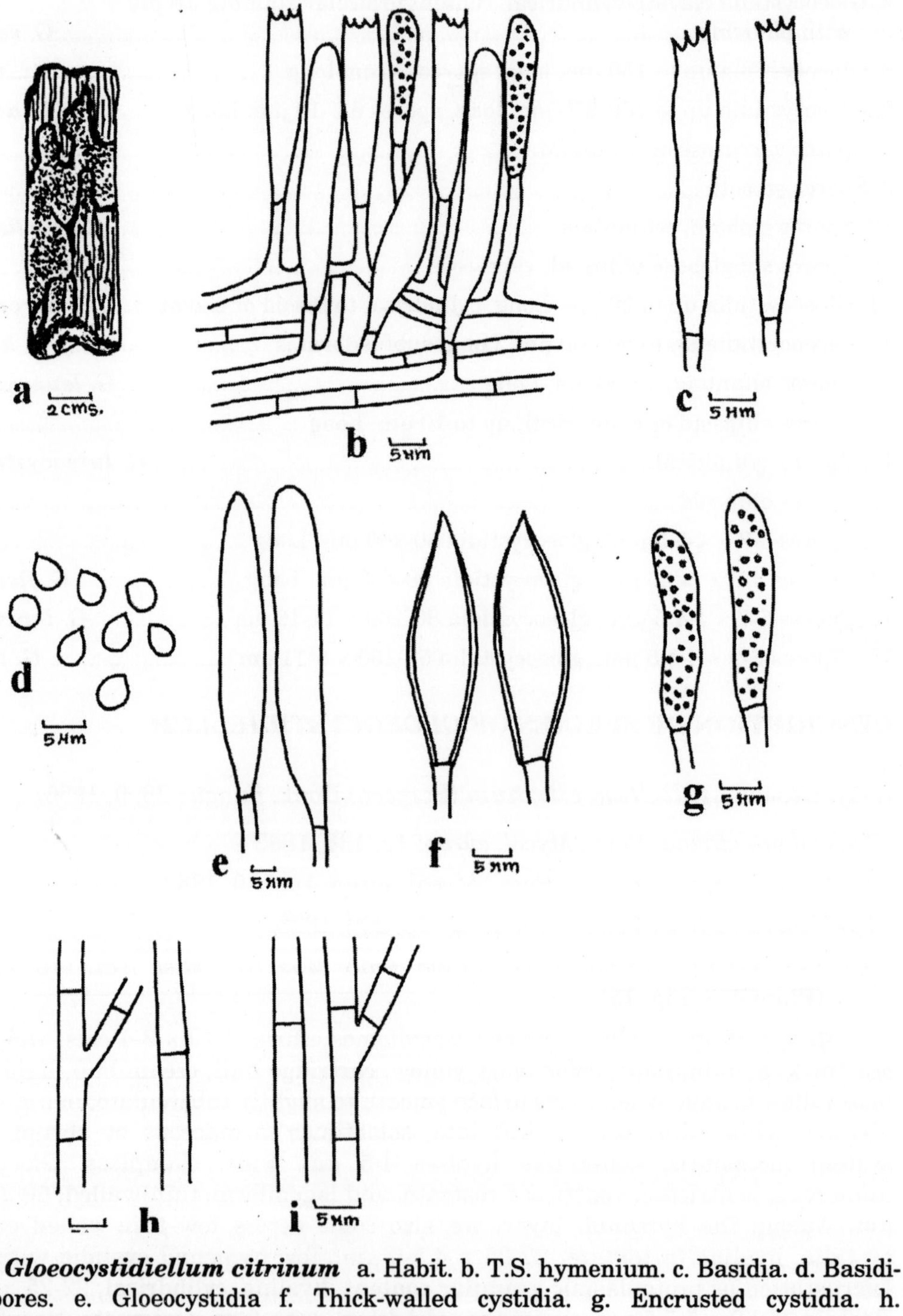

Fig. 15 ***Gloeocystidiellum citrinum*** a. Habit. b. T.S. hymenium. c. Basidia. d. Basidiospores. e. Gloeocystidia f. Thick-walled cystidia. g. Encrusted cystidia. h. Generative hyphae. i. Generative hyphae in culture

live standing tree. **Distribution :** North Europe, Scandinavia, India. **Specimen examined :** Benzidine test positive, white rot on: *Gliricidia speium* (Jacquni.) Stend. **Location :** Panshet (06/10/2005), VH-NPO-69. **Specimen examined :** Benzidine test positive, white rot on: *Azadriachata indica* Juss., NCL Park; Dist: Pune (10/09/2005), VH-NPO-32. **Specimen examined :** Benzidine test positive, white rot on: *Bauhinia purpurea* L. **Location :** Poultry trainee center; Dist: Pune (24/09/2005), VH-NPO-48. **Culture characters :** Growth characters: Growth rate moderate; advancing zone raised; Mat cottony white initially, turning ash coloured or watery grey, latter on. Initially it is highly fluffy, cushoinly in appearance, later settling down and getting dense; Distance between hyphal tips dense; Colony ash coloured; outline of colony even; Aerial mycelium absent; Reverse completely darkened latter on after getting bleached initially.

Microscopic characters: Generative hyphae branched, 3-4 μm., wide, hyaline, thin-walled without clamps, with simple septa. **Tests :** α-napthal and pyrogallol + H_2O_2 positive; p-cresol negative. **Species Code:** 1, 2, 3, 8, 12, 14, 15, 16, 33, 38, 52, 53. **Culture No :** PUCC. B: 08. **Remarks :** This species differs from all other by having no reaction to sulfovaniline in the gloeocystidia. It is close to *G. lactescens* but differs from it in ochraceous-pale fruitbody and vesicular gloeocystidia at the base of hymenial layer.

(9-2). *Gloeocystidiellum convolvens* (P. Karsten) Donk. *Fungus* 26: 9, 1956.

= *Corticium convolvens,* Karst., *Bidr. Kann. Finl. Nat. Folk.* 37 p. 148, 1882.

= *Peniophora lundellii* Litsch,. Lund., Nannf., *Fungi exs. Suec. n.* 573, Sv. Bot. Tidskr. 32p. 288, 1938. (Plate 2, Fig. 16)

Morphological Characters : Basidiomes effuse, 3-6 × 2-3 cms. and 100-500 μm thick, membranous to ceraceous, the hymenial surface smooth to tuberculate, not cracked or extensively fissured, light buff to cream color, light ochraceous-buff or ochraceous buff when dry; margin pruinose to fimbriate, white or concolorous. Hyphal system monomitic, generative hyphae clampless, 2-4 μm diam, the walls thin to slightly thickened, nonamyloid. Gloeocystidia 70-100 × 7-14 μm, numerous, fusoid to cylindric, and with a ventricose base and a rounded acuminate tip, thin-walled, the contents hyaline to yellowish in KOH under bright field, more or less homogeneous, sulfo-positive. Cystidia (lamprocystidia) numerous, conical, 30-65 × 7-10 μm, heavily encrusted, thick-walled, imbedded or projecting to 20 μm, arising throughout the thickness of the subhymenium, few hymenial cystidia hyaline without any granularcontent, cylindrical, 22-25 × 4-6 μm. Basidia 25-35 × 5-7 μm, clavate, lacking a basal clampless, with four sterigmata. Basidiospores 5-6 × 3.5-4.5 μm, ellipsoid to broadly ellipsoid, the wall verrucose to spiny, amyloid, thin, with a small apiculus. **Habitat**: On live standing tree. **Distribution:** Scandinavia, Sweden, Finland, Norway, India **Specimen Examined:** Benzidine test positive, white rot on *Artocarpus heterophyllus* Lam. **Location :** Film Institute; Dist: Pune (13/01/2006), VH-NPO-263. **Specimen examined :** Benzidine test positive, white

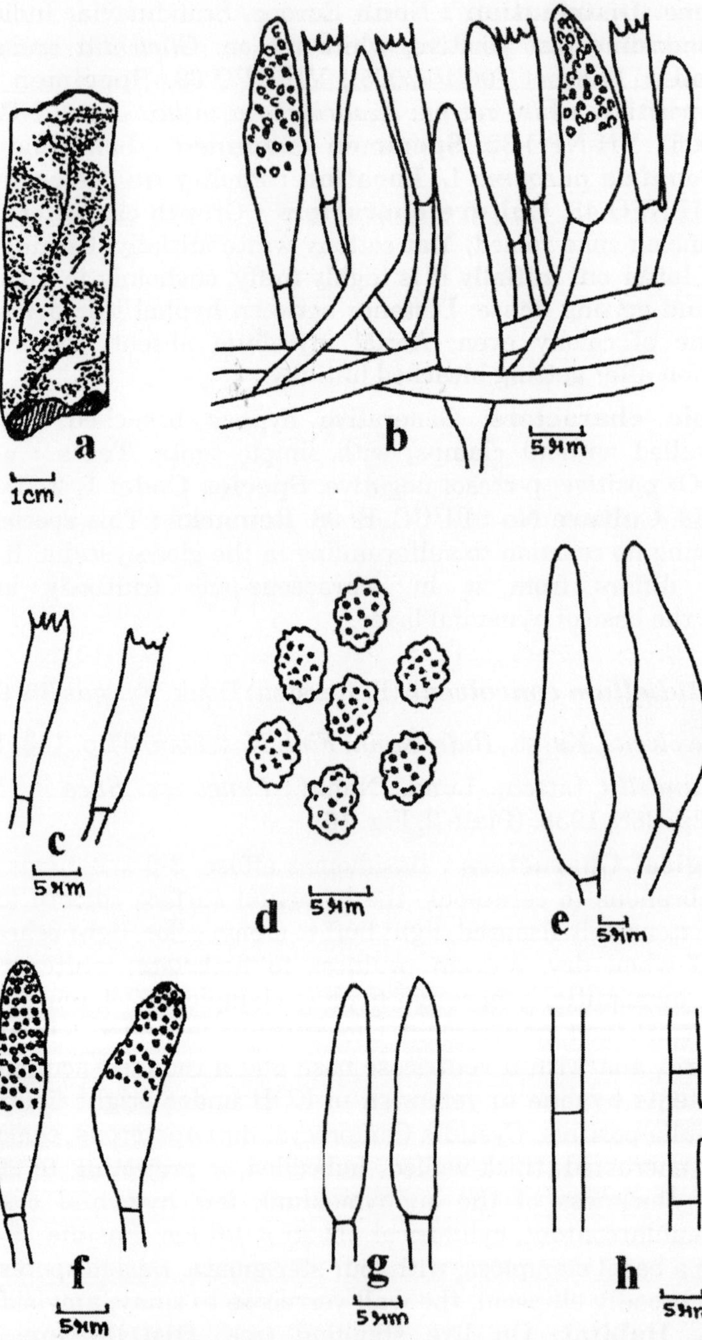

Fig. 16. ***Gloeocystidiellum convolvens*** a. Habit. b. T.S. of hymenium. c. Basidia. d. Basidiospores. e. Gloeocystidia. f. Encrusted cystidia. g. Cystidia. h. Generative hyphae

rot on *Ficus bengiamina*. **Location:** Pune university; Dist: Pune (09/10/2005), VH-NPO-188. **Remarks :** This is the only species of the genus with encrusted cystidia, hence is recognised easily. The spores sculpture is sometimes inconspicuous and difficult to observe, but with Melzer's reagent can be studied easily. The culture was not obtained besides repeated trials.

(9-3). *Gloeocystidiellum fistulatum* (Cunningham) Boidin. *Cah. Maboké* **4(1):** 13, 1966.

= *Corticium fistulatum* Cunn., *Trans. Roy. Soc. New Zealand* **82:** 292, 1954. (Plate 2, Fig. 17)

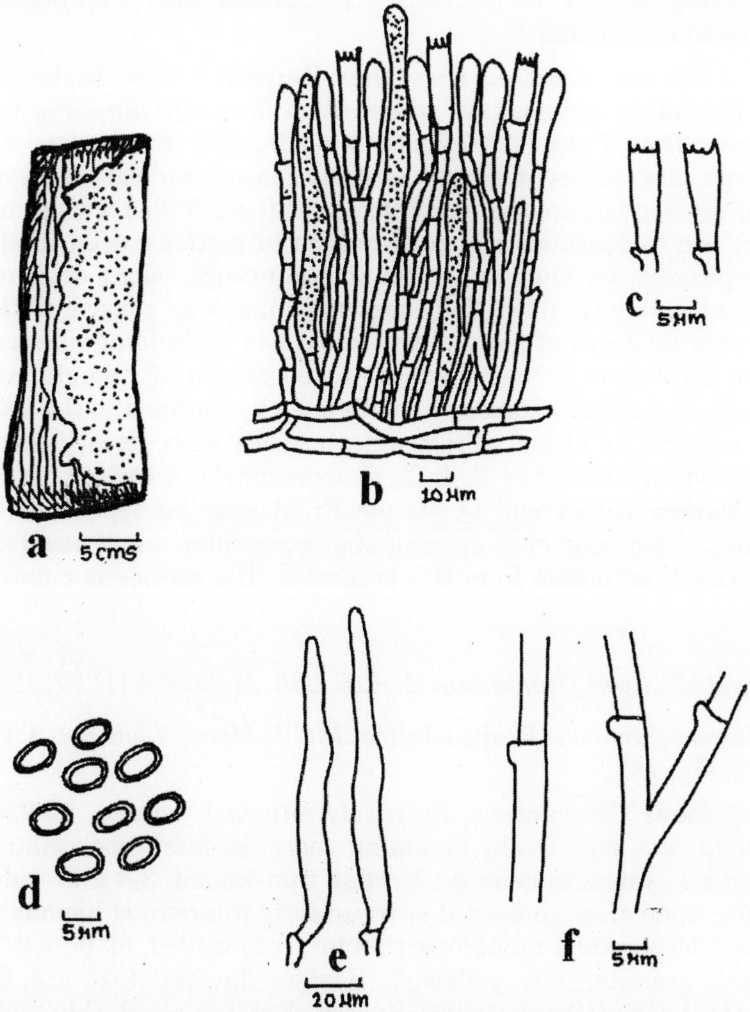

Fig. 17. *Gloeocystidiellum fistulatum* a. Habit. b. T.S. of hymenium. c. Basidia. d. Basidiospores. e. Gloeocystidia. f. Generative hyphae

Morphological Characters : Hymenophore annual or perennial, adnate, ceraceous-coriaceous, effused 25-30 × 3-5 cm., surface white, becoming cream or alutaceous, sometimes tinted reddish-buff, laterally closely creviced; margin thinning out, white or tinted brown, abrupt, sometimes receding, membranous, adnate. Context white, 200-400 μm thick, of 1-3 obscure layers, defined partly by masses of crystals and tinted hyphae, mainly by rows of gloeocystidia, each layer composed of a narrow base of parallel hyphae, and an intermediate layer of compact vertical hyphae; generative hyphae 4-5 μm wide, hyaline, branched, septate, with clamp. Hymenial layer to 50 μm deep, of basidia and gloeocystidia. Gloeocystidia extending from the base of each context layer, projecting to 30 μm, or not, cylindrical or flexuous-cylindrical, 80-120 × 5-8 μm, contents densely granular and orange. Basidia subclavate, 12-16 × 4-5 μm, with 4 sterigmata. Spores elliptical, 5-6 × 3-3.5 μm, smooth, hyaline, amyloid.

Habitat : On live standing tree. **Distribution :** New Zealand. **Specimen examined :** Benzidine test positive, white rot on *Bauhinia purpurea* L. **Location :** Pune university; Dist: Pune (12/07/2006), VH-NPO-496. **Remarks :** Gloeocystidia are long, narrow, flexuous-cylindrical, regularly spaced, and extend from the base to the surface of the hymenium, some projecting to 30 μm. Filled with granular orange contents, they are conspicuous and readily seen in sections; even in old specimens they are conspicuous, for they either retain the orange colour or contents appear dark-brown and resinous, When folly grown plants may consist of three definite layers, separated by zones of parallel tinted hyphae embedding masses of crystals; but as several specimens in the collections listed consist of a single layer this does not appear to be a constant feature. Crystals may be confined to the base, scattered through the context, or absent. From other species with gloeocystidia containing bright orange contents *G. debile* B. & C., *G. leucoxanthum* Bres., *G. luridum*, Bres., *G. leueocystidiatum* Talbot and *G. seriale* Fr. it may be separated by the often stratose context, singularly even spacing of gloeocystidia, small basidia and spores. The cultures could not obtain from this specimen. The species is a new record from India.

(9-4) *Gloeocystidiellum flammeum* Boidin. *Cah. Maboké* **4 (1) :** 7, 1966.

=*Gloeocystidiopsis flammea* (Boidin) Jülich, *Int. J. Mycol. Lichenol.* **1 (1) :** 28, 1982 (Plate 2, Fig. 18)

Morphological Characters: Fruitbody effusued, adnate, 20-25×10-15 cms., ceraceous when wet and fresh, becoming more or less tuberculate, withish to creamish. Hyphal system monomitic, hyphae thin-walled, 3-4 μm. wide, clampless, basal subicular layer thin, composed of irregularly interwined hyphae;subhymenial trama denser. Gloeocystidia numerous, tubular, thin-walled, 80-90 × 9-14 μm, protoplasmic content granular, oily, yellowish. Basidia clavate, 20-25 × 3, 5-4 μm., with four sterigmata, basal without clamps. Spores oblong to ovoid, thin-walled, hyaline, smooth, amyloid, 4-5.5 × 3-4 μm. **Habitat :** On live standing tree. **Distribution :** Africa. **Specimen examined :** Benzidine test positive, white rot on *Drypetes* *venusta* (Wright) Pax. & Hoffm. **Location :** Khed; Dist:Ratnagiri (12/10/2005), VH-NPO-230. **Remarks:** Fruitbody becoming more or less tuberculate, at first withish and then creamish, gloeocystidia numerous, tubular, protoplasmic content granular, oily and yellowish. Spores oblong to ovoid. The cultures could not be obtained, besides repeated trials. The species is a new record from India.

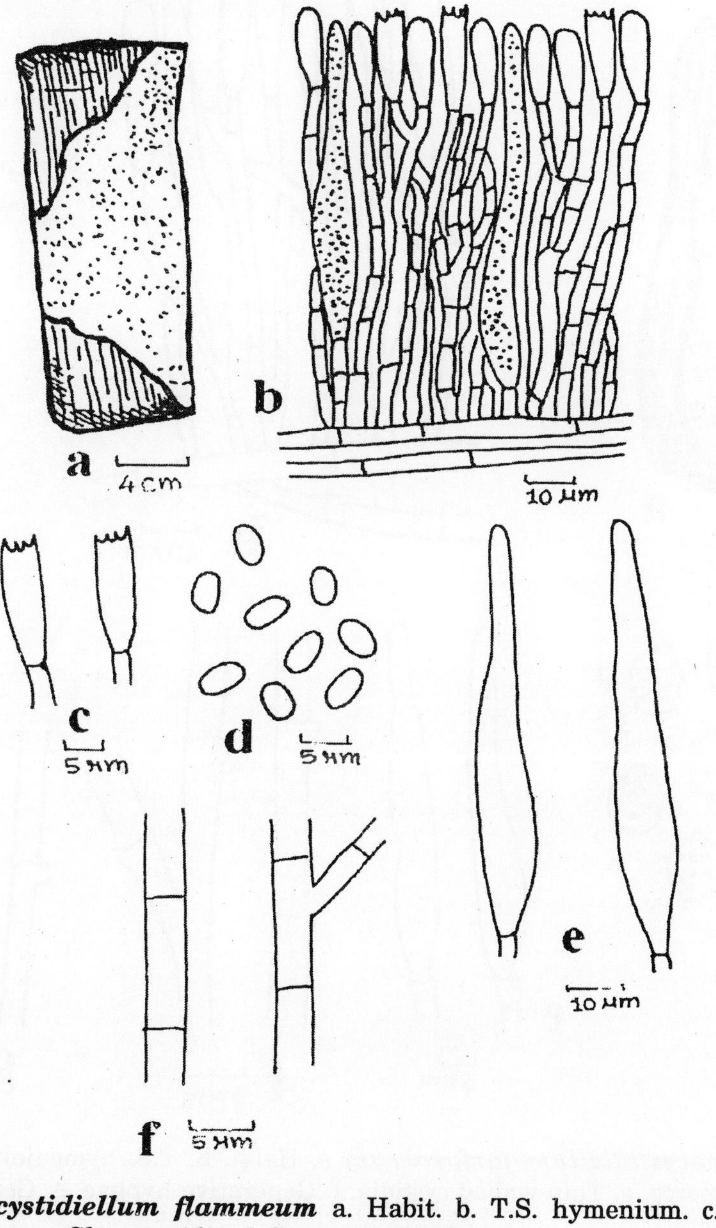

Fig. 18. ***Gloeocystidiellum flammeum*** a. Habit. b. T.S. hymenium. c. Basidia. d. Basidiospores. e. Gloeocystidia. f. Generative hyphae

(9-5) *Gloeocystidiellum furfuraceum* (Bresadola) Donk. *Fungus* **26:** 9, 1956.

=*Hypochnus furfuraceus* Bres., *Fungi trid*. 2p. 97, 1900. (Plate 2, Fig. 19)

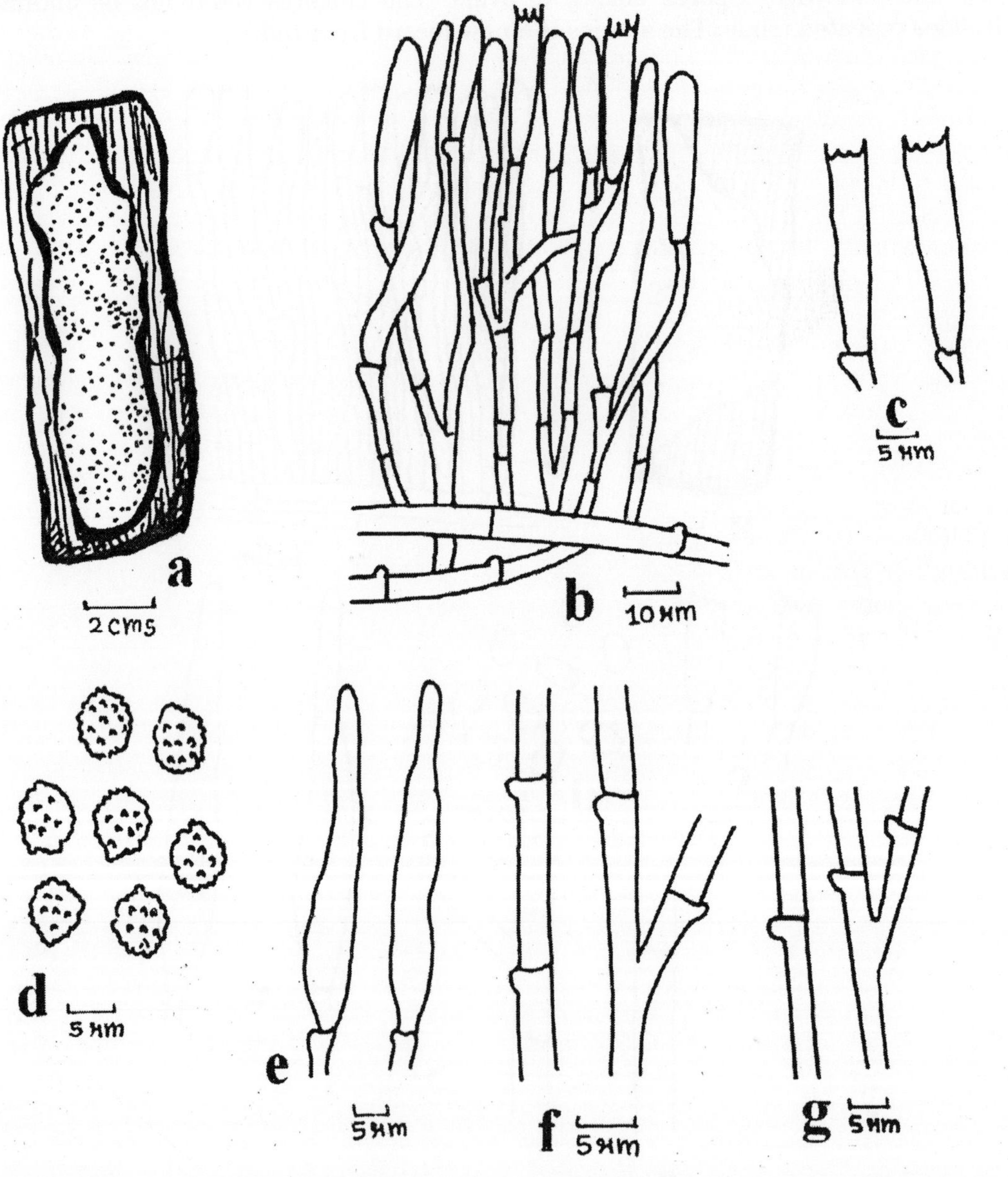

Fig. 19 *Gloeocystidiellum furfuraceum* a. Habit. b. T.S. hymenium. c. Basidia. d. Basidiospores. e. Thin-walled cystidia. f. Generative hyphae. g. Generative hyphae in culture

Morphological Characters : Basidiocarp resupinate, effused, 8-10 × 3-4 cms., arachnoid-pellicular to soft membranaceous, often rimose, separable. Hymenial surface even, somewhat farinaceous, whitish to cream-coloured when fresh, becoming pale ochraceous with age. Margin indistinct. Hyphal system monomitic, generative hyphae hyaline, thin-kwalled, 2-3.5 μm. wide, with clamps. Gloeocystidia cylindrical, often flexuous or somewhat swollen at the base, thin-walled, typically with yellowish contents, 50-90 × 6-10 μm, sulpho-positive. Basidia subcylindrical and often flexuous to clavate, 25-35 × 5-6 μm, typically with 4 sterigmata. Spores hyaline, thin- to somewhat thick-walled, globose to subglobose, echinulate, 5-6.5 μm diam, strongly amyloid. **Habitat :** On live standing tree. **Distribution :** Europe, North, South America, Japan. **Specimen examined :** Benzidine test positive, white rot on: *Adhatoda vasica*. **Location :** Pune University; Dist: Pune (12/09/2005), VH-NPO-19. **Specimen examined :** Benzidine test positive, white rot on: *Samonia samon*. **Location :** Kamal Nehrue Park, Dist. Pune (21/10/2006), VH-NPO-310. **Culture characters :** Growth characters: Growth rate slow; advancing zone appressed to submerged; Mats white, thin, appressed, subfelty; outline of colony even; no odor, reverse unchanged. **Microscopic characters:** Generative hyphae branched, 2-3.5 μm, wide, thin walled, with clamps, sparsely branched, branches usually arising opposite clamps. **Tests :** α-napthal and pyrogallol + H_2O_2 positive; p-cresol negative. **Species Code :** 1, 2, 3, 7, 13, 14, 15, 30, 37, 39, 52, 53. **Culture No :** PUCC. B: 09. **Remarks :** This is the only species of the genus with globose, echinulate spores, it is easily recognized. *Gloeocystidiellum sibiricum* Parm., has similar spores but has thicker fruitbodies, larger gloeocystidia and clampless. The species is a new record from India.

(9-6). *Gloeocystidiellum insidiosum* (Bourdot & Galzin) Donk. *Fungus* **26:** 9, 1956.

=*Gloeocystidium insidiosum* Bourdot & Galzin, *Bull. Soc. mycol. Fr.* **28 (4)** : 370, 1913. (Plate 2, Fig. 20)

Morphological Characters : Fruitbody resupinate, effused, adnate, 7-10 × 2-4 cms. and up to 500 μm thick; hard-crustaceous when dry, stratified, hymenophore smooth - tuberculate, sometimes cracked, cream-coloured - light ochraceous; margin more or less abrupt. Hyphal system monomitic, hyphae without clamps, 3-4 μm wide, densely united; hyphal direction vertical; subiculum inconspicuous or absent; hyphal walls stained by cotton blue. Cystidia tubular, sinuose, thin- thick-walled, 40-70 × 4-7 μm, embedded, often growing through the whole fructification; content granular with positive reaction to sulfovanilline. Basidia clavate, thin-walled, 25-30 × 4-5 μm, with 4 sterigmata and without basal clamp. Spores subcylindrical - ellipsoid, smooth, thin-walled, 4-5 × 2-2.5 μm, amyloid. **Habitat :** On fallen logs and trunks. **Distribution :** Iran. **Specimen Examined** : Benzidine test positive, white rot on *Memecylon umbellatum* Burm. **Location :** Ambaghat; Dist. Satara (19/02/2007), VH-NPO-782. **Remarks:** The specimen is recognized by cream-coloured to light ochraceous hymenophore; tubular, sinuose, thin to thick-walled cystidia, the cell walls sporulating basidial layer are strongly cyanophilous. The species is a new record from India. The cultures could not be obtained, besides repeated trials.

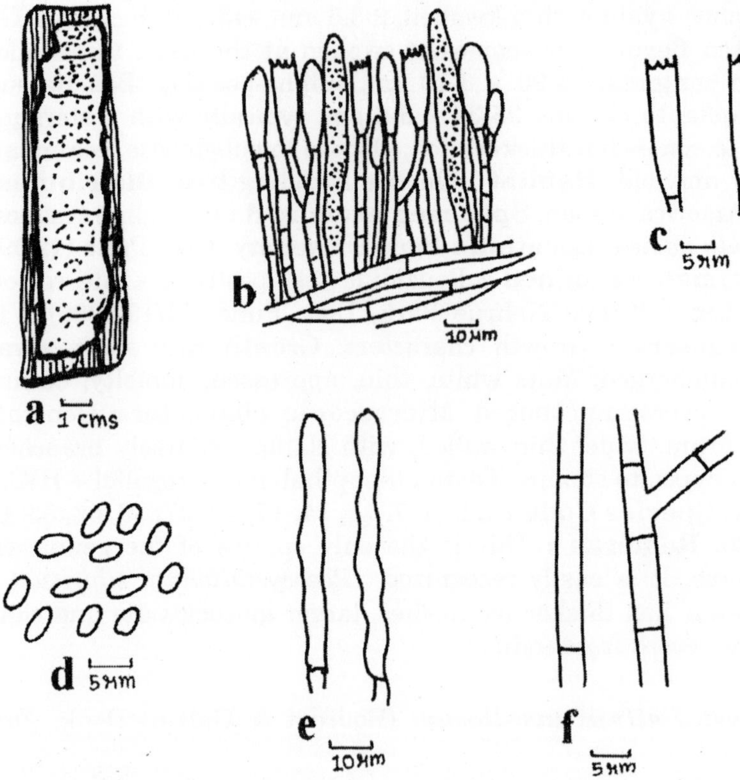

Fig. 20 ***Gloeocystidiellum insidiosum*** a. Habit. b. T.S. hymenium. c. Basidia. d. Basidiospores. e. Gloeocystidia. f. Generative hyphae

(9-7) ***Gloeocystidiellum irpiscenns*** Boidin. *Cah. Maboke* **4(1)**:13, 1966. (Plate 2, Fig. 21)

Morphological Characters : Fruibody effused, membransceous, adnate, 5-7×4-4. 5 µm, even to conspicuously tuberculate, whitish. Hyphal system monomitic, with clamps, hyphae 3 - 4 µm. wide, thin walled, basal hyphae forming a layer of irregularly intertwined hyphal branches in a loose tissue. Gloeocystidia 80-100 × 10-15 µm, thin-walled, with positive reaction to sulfoaldehyde. Basidia clavate, with basal clams, with four sterigmata, 35-45 × 6-7 µm. Spore oblong, smooth, thin-walled, hyaline, 5-7 × 3-4 µm., amyloid. **Habitat:** On live standing tree. **Distribution :** Africa. **Specimen examined :** Benzidine test positive, white rot on *Memecylon umbellatum* Burm. **Location :** Baneshwar; Dist: Pune (13/01/2007), VH-NPO-602. **Remarks:** The specimen is recognized by even to conspicuously tuberculate hymenophor. The species is a new record from India. The culture was not obtained besides repeated trials.

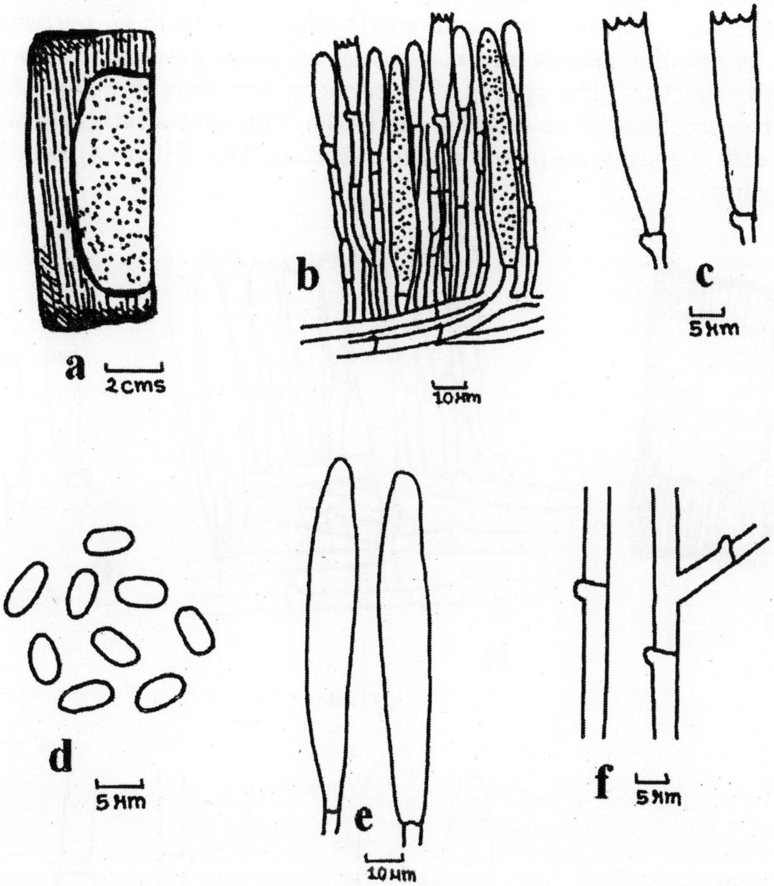

Fig. 21. ***Gloeocystidiellum irpiscescens*** a. Habit. b. T.S. hymenium. c. Basidia. d. Basidiospores. e. Cystidia. f. Generative hyphae

(9-8). ***Gloeocystidiellum kenyense*** Hjortstam. *Mycotaxon* **28(1)** : 29, 1987. (Plate 2, Fig. 22)

Morphological Characters : Fruitbody resupinate, 3 × 3 cms. and about 0.1-0.2 mm thick cream-coloured with yellowish tint, pale ochraceous, smooth or indistinctly rimose, margin not differentiated. Hyhal system monomitic, hyphae thin-walled, with clamps, 2-3 μm wide, forming a fairly dense and somewhat agglutinated tissue. Gloeocystidia numerous, subfusiform and often with one schizopapilla, thin-walled and with oily content, varying in size but commonly 50-80 × 10 μm. Basidia narrowly clavate, 15-20 × 4-5 μm, with 4 sterigmata and a basal clamp. Spores subglobose to ellipsoid, finely verrucose, warts visible in KOH, 4-4.5 ×

3-3.5 μm, strongly amyloid. **Habitat :** On live standing tree. **Distribution :** Kenya, Tanzania **Specimen examined:** Benzidine test positive, white rot on *Ficus benganina* L. **Location:** Pune university; Dist: Pune (25/08/2006), VH-52. **Remarks :** The species is very characteristic and easily separated from *G. porosum* (Berk. & Curt.) Donk by smaller spores and commonly terminal gloeocystidia which lack a sulphoaldehyde reaction. The spores of *G. porosum* are usually up to 5.5-6 μm long, and the warts are hardly observable in KOH. The gloeocystidia are commonly lateral and with strongly sulphoaldehyde reaction. The cultures could not obtain from this specimen.

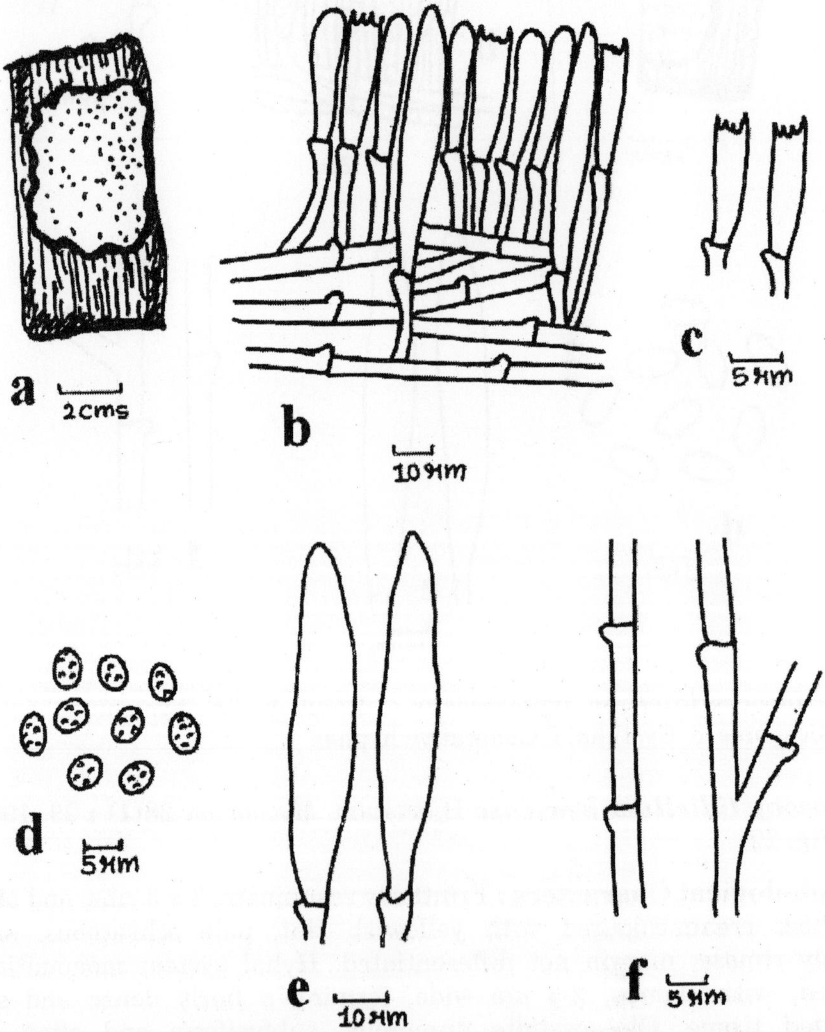

Fig. 22 ***Gloeocystidiellum kenyense*** a. Habit. b. T.S. hymenium. c. Basidia. d. Basidiospores. e. Gloeocystidia. f. Generative hyphae

(9-9). *Gloeocystidiellum lactescens* (Berkeley) Boidin. *C. R. Acad. Sci. Paris* **233:** 1668, 1951.

= *Thelephora lactescens*, Berk., Smith, *Eng. F1.* **5:** 2 p. 169,1836. (Plate 2, 8, Fig. 23)

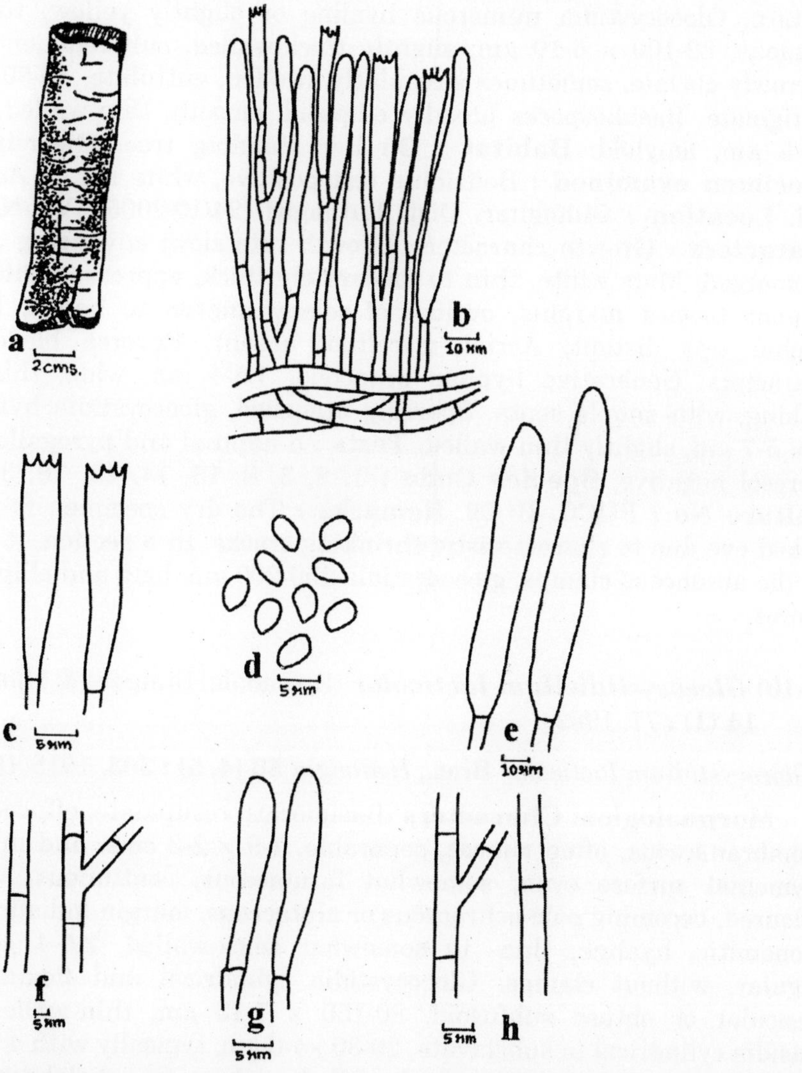

Fig. 23 *Gloeocystidiellum lactescens* a. Habit. b. T.S. hymenium. c. Basidia. d. Basidiospores. e. Gloeocystidia. f. Generative hyphae. g. Gloeocystidia in culture. h. Generative hyphae in culture

Morphological Characters: Fruitbody resupinate, effuse, adnate, subceraceous or ceraceous, 13-15 × 2-3 cms. and 100-300 μm thick in section. Hymenial surface light yellow to pale yellow, smooth, occasionally cracked, margin thinning, paler or white, pruinose. Hyphal system monomitic; hyphae simple-septate.

Subiculum indistinctly bilayered, composed of basal layer and medullary layer; hyphae hyaline, 1.5-3 μm diam, thin-walled. Basal layer with compact texture; hyphae mainly horizontal, usually agglutinated. Medullary layer with somewhat loose tcxture; hyphae variously oriented. Subhymenium thickening, with dense texture. Gloeocystidia numerous hyaline or slightly yellow, tubular, somewhat flexuous, 80-100 × 6-10 μm, slightly thick-walled, sulforeaction positive. Basidia narrowly clavate, sometimes irregularly swollen, guttulate, 40-50 × 6-7 μm, with 4 sterigmate. Basidiospores broadly ellipsoid, smooth, thin-walled, guttulate, 6-7 × 4.5-5 μm, amyloid. **Habitat :** On live standing tree. **Distribution :** Taiwan. **Specimen examined :** Benzidine test positive, white rot on *Accacia nilotica* (L) Del. **Location :** Guhaghar, Dist. Ratnagiri (24/10/2005), VH-NPO-333. **Culture characters :** Growth characters: Growth rate slow; advancing zone appressed to submerged; Mats white, thin to moderately thick, appressed, silky to subfelty but thinner toward margins, outline of colony uneven to bayed; Distance between hyphal tips distant; Aerial mycelium absent; Reverse bleached. Microscopic characters: Generative hyphae branched, 1.5-4 μm. wide, thin walled, clamps lacking, with simple septa. Sparsely branched, gloeocystidia hyaline, tubular, 40-60× 5-7 μm, slightly thin-walled. **Tests :** α-napthal and pyrogallol + H_2O_2 positive; p-cresol negative. **Species Code :** 1, 2, 3, 8, 13, 14, 15, 16, 30, 37, 52, 53, 73. **Culture No :** PUCC. B: 09. **Remarks :** The dry specimen is recognized by the naked eye due to characteristic shrinking cracks. In a section, it can be recognised by the absence of clamps, gloeocystidia 150-200 μm. long and ellipsoid to subglobose spores.

(9-10) ***Gloeocystidiellum lacticolor*** (Bresadola) Stalpers & Hjortstam. *Mycotaxon* **14 (1) :** 77, 1982.

=*Gloeocystidium lacticolor* Bres., *Hedwigia* **56 (4, 5) :** 303, 1915. (Plate 2, 8, Fig. 24)

Morphological Characters : Basidiocarp resupinate, effused, pellicular to soft membranaceous, often rimose, separable, 7-9 × 2-3 cms. and up to 200 μm thick. Hymenial surface even, somewhat farinaceous, continuous, whitish to cream-coloured, becoming pale ochraceous or alutaceous, margin indistinct. Hyphal system monomitic, hyaline, thin- to somewhat thick-walled, 2.5-4 μm wide, typically regular, without clamps. Gloeocystidia cylindrical and flexuous to subclavate, vesicular or obtuse subfusoid, 80-150 × 7-10 μm, thin-walled, sulpho-positive. Basidia cylindrical to subclavate, 20-30 ×4-6 μm, typically with 4 sterigmata. Spores hyaline, thin- to somewhat thick-walled, globose to subglobose, 4. 5-5 μm diam, strongly amyloid. Ornamentation immediately disappearing in KOH 6%. **Habitat :** On dead standing tree. **Distribution :** Philippines. **Specimen examined :** Benzidine test positive, white rot on *Ficus bengalensis* Linn. **Location :** Pune university, Dist. Pune (27/08/2006), VH-NPO-529. **Culture characters :** Growth characters: Growth rate slow; advancing zone appressed to submerged; Mat cottony white getting submerged in agar as colony advances giving the plate and hence the colony, a transparent look; colony white; outline of colony even; Distance between hyphal tips distant; Aerial mycelium absent; Reverse bleached. Microscopic characters: Generative hyphae branched, 4-5 μm, wide, clamps lacking, with simple septa. **Tests :** α-napthal and pyrogallol positive; p-cresol negative. **Species Code :** 1, 2, 3, 8, 13, 14, 15, 16, 30, 37, 53. **Culture No :** PUCC B: 10. **Remarks :** The specimen are recognized by hymenial surface even, whitish to cream-coloured when fresh, becoming pale ochraceous or alutaceous when old. The species can be recognised by the absence of clamps, gloeocystidia 50-200 μm., long and globose to subglobose, thin- to thick-walled spores. The species is a new record from India.

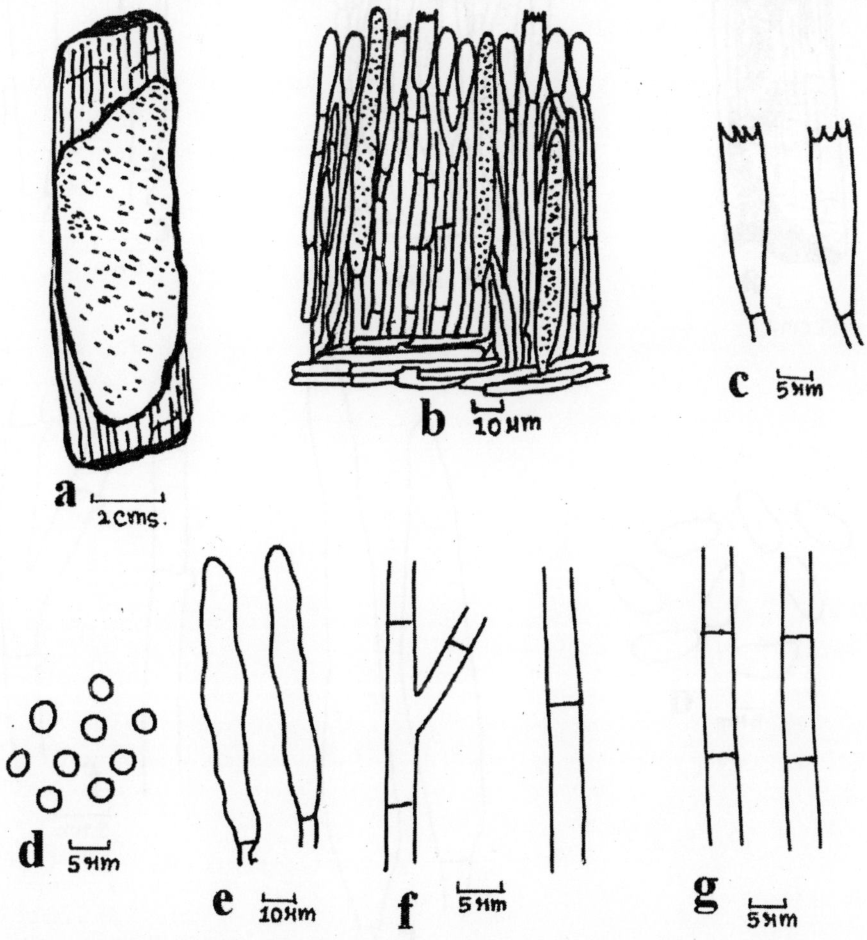

Fig. 24 ***Gloeocystidiellum lacticolor*** a. Habit. b. T.S. hymenium. c. Basidia. d. Basidiospores. e. Gloeocystidia. f. Generative hyphae. g. Generative hyphae in culture.

(9-11) ***Gloeocystidiellum leucoxanthum*** (Bresadola) Boidin. *Publs Mus. natn. Hist. nat.* **17:** 122, 1957.

= *Corticium leucoxanthum* Bres., *Fungi. trid.* 2 p. 57 pl. 166. Fig. 3, 1898. (Plate 2, Fig. 25)

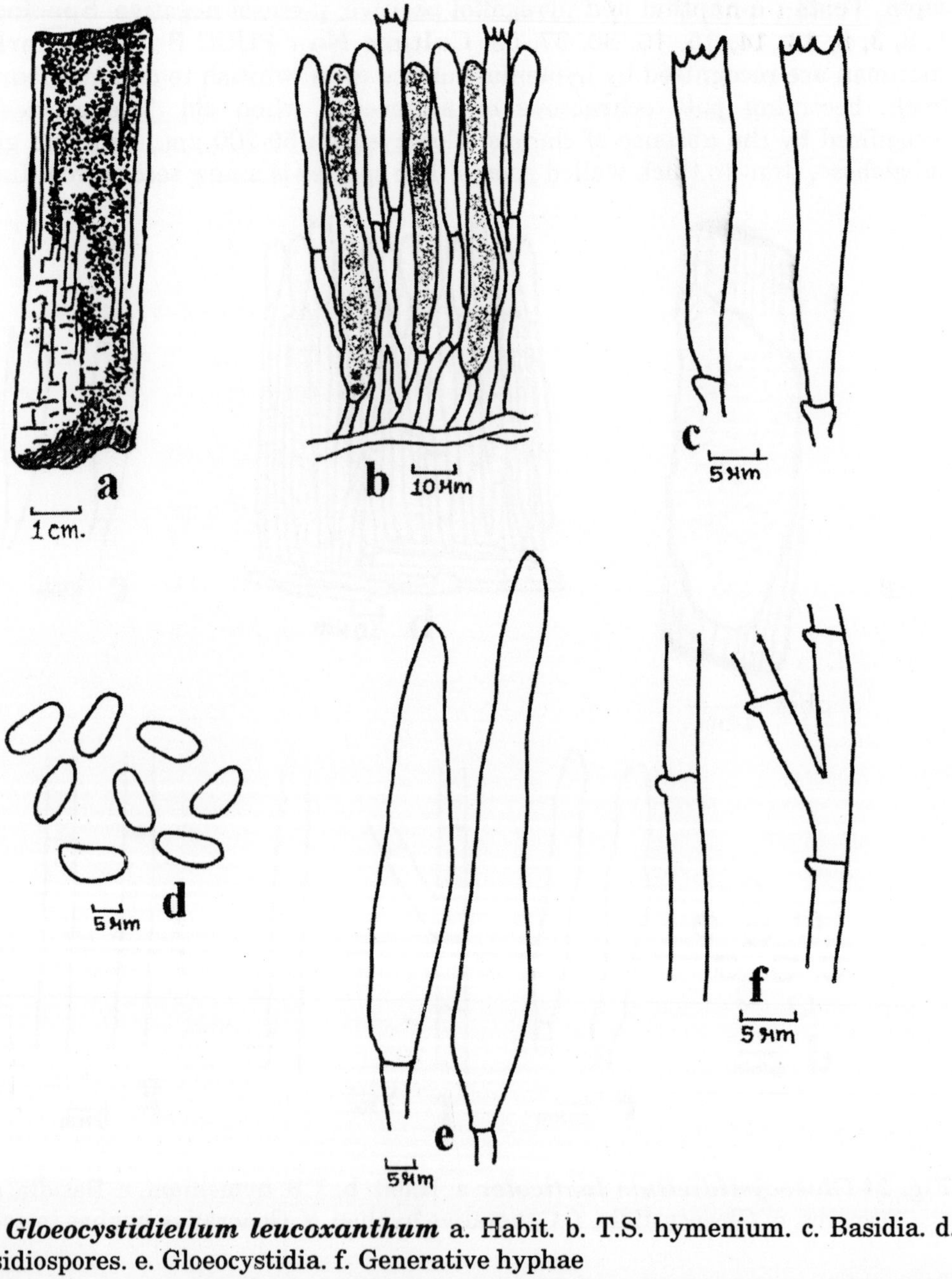

Fig. 25 ***Gloeocystidiellum leucoxanthum*** a. Habit. b. T.S. hymenium. c. Basidia. d. Basidiospores. e. Gloeocystidia. f. Generative hyphae

Morphological Characters : Basidiomata effused, 7-9 × 3-5 cms. and up to 500 μm thick, hymenial surface ceraceous, pinkish buff, cinnamon-buff, to clay color, smooth to strongly tuberculate, not cracked or cracked extensively exposing the nearly white subiculum; margin abrupt to slightly pruinose, whitish or concolorous. Hyphal system monomitic. Hyphae 3-5 μm diam, with clamp, thin-walled. Gloeocystidia 60-100 × 10-15 μm, numerous, cylindric to subfusoid, the contents yellow in KOH, granular, sulfo-negative, but many appear to be empty. Basidia 40-60 × 5 -7.5 μm, clavate, with four sterigmata. Basidiospores 15-20 × 5-7 μm, ellipsoid to suballantoid, smooth, easily collapsed, amyloid, with a distinct, blunt apiculus. **Habitat :** On live standing tree. **Distribution :** Africa, Europe, North America. Japan. **Specimen examined :** Benzidine test positive, white rot on *Azadriachata indica* Juss **Location :** Ambaghat, Dist. Satara (19/02/2007), VH-784. **Remarks :** This species is very close to *G. luridum* but the two species differed from each other in the shape and size of spores. In *G. leucoxanthum* the basal layer is almost thicker and looser.

(9-12) *Gloeocystidiellum luridum* (Bresadola) Boidin. *C.R. Acad. Sci. Paris* **233:** 1668, 1951.

=*Corticium luridum* Bres., *Fung. trident.* **2 (11-13) :** 59, 1898.

=*Gloeocystidium luridum* (Bres.) Höhn. & Litsch., *Sber. Akad. Wiss. Wien*, Math. - naturw. Kl, Abt. *1 116:* 770, 1907.

=*Megalocystidium luridum* (Bres.) Jülich, *Persoonia* **10 (1) :** 140, 1978.

=*Vesiculomyces luridus* (Bres.) Boidin & Lanq., *Mycotaxon* **16 (2) :** 493, 1983. (Plate 2;8, Fig. 26)

Morphological Characters : Basidiomata resupinate, adnate, effused, 3-6 × 2-3 cms. and up to 400 μm thick; hymenial surface buff to ochreous, smooth, sometimes cracked when dry; margin concolorous with the hymenial surface, thinning out, indeterminate. In vertical section, subhyaline, membranous to crustose. Hyphal system monomitic; hyphae 2-4 μmwide, smooth, thin- to slightly thick-walled, with clamps. Gloeocystidia numerous, cylindrical, clavate, obclavate or ventricose, 40-80 ×10-15 μm, with a basal clamp, smooth, thin- to thick-walled, projecting up to 10 μm beyond the hymenial surface, positive to sulphobenzaldehyde. Basidia clavate, 30-40 × 5-6.5 μm, with a basal clamp, producing 4 sterigmata. Basidiospores ellipsoid, 6.5-8 × 3.5-4.5 μm, smooth, thin-walled, amyloid. **Habitat :** On live standing tree. **Distribution :** Africa, Taiwan, Europe, North America, Japan. **Specimen examined :** Benzidine test positive, white rot on *Samanea saman* (Jacq.) Merr. **Location :** Film Institute, Dist. Pune (20/10/2005), VH-NPO-267. **Culture characters :** Growth characters: Growth rate slow, advancing zone raised; Mat cottony white; Reverse bleached. Distanced between hyphal tips distant; Colony white; Outline of colony uneven; Aerial mycelium absent. **Microscopic characters:** Generative hyphae branched, thin-walled, 2-3 μm. wide, clamped, hyaline, gloeocystidia, cylindrical, 20-30 ×3-5 μm, smooth, thin-walled. **Tests :** α-napthal and pyrogallol + H_2O_2 positive; p-cresol negative. **Species Code :** 1, 2, 3, 8, 12, 14, 15, 30, 37, 39, 52, 73. **Culture No :** PUCC. B: 11. **Remarks :** This species is very close to *G. leucoxanthum* but the two species differed from each other in the shape and size of spores. In *G. luridum* the basal layer is almost thiner. The species is a new record from India.

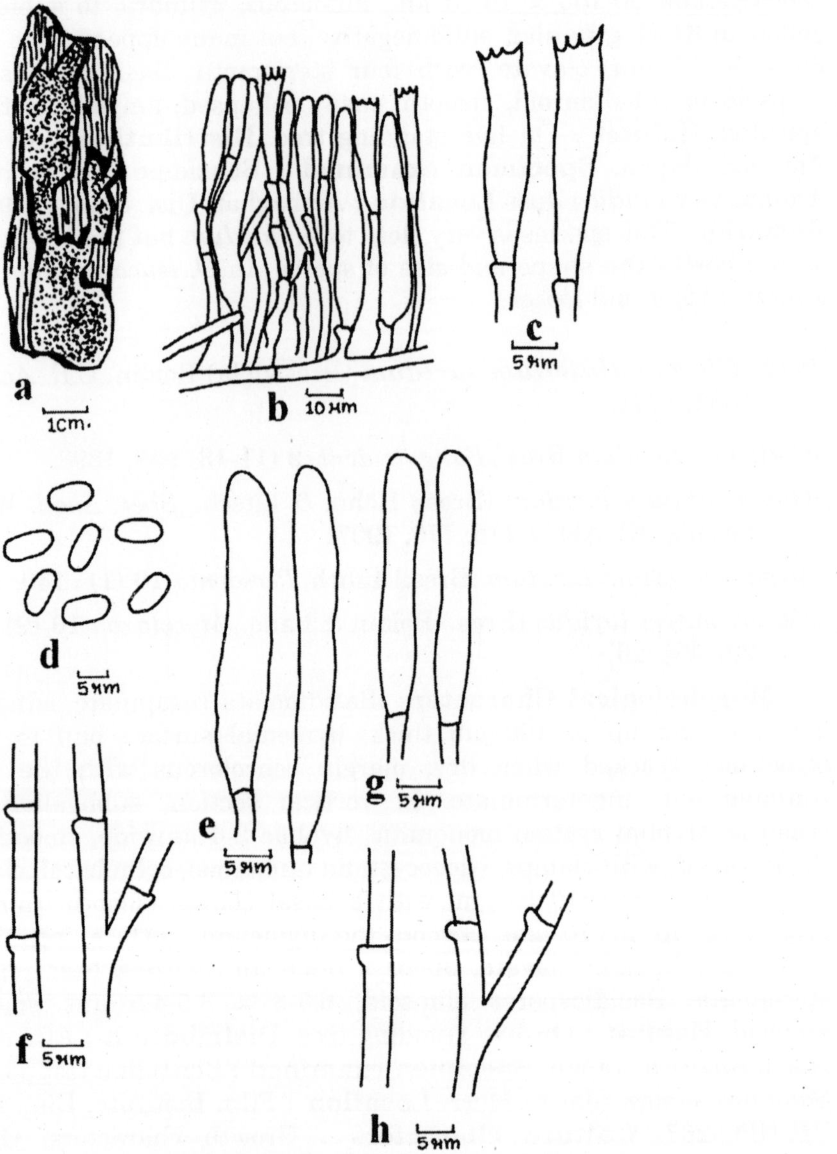

Fig. 26 ***Gloeocystidiellum luridum*** a. Habit. b. T.S. hymenium. c. Basidia. d. Basidiospores. e. Gloeocystidia. f. Generative hyphae. g. Gloeocystidia in culture. h. Generative hyphae in culture

(9-13). *Gloeocystidiellum luteocystidiatum* (Talbot) Boidin. *Cah. Maboké* **4 (1) :** 9, 1966.

=*Corticium luteocystidiatum* Talbot, *Bothalia* 4 (4) : 941, 1948. (Plate 3, 9, Fig. 27)

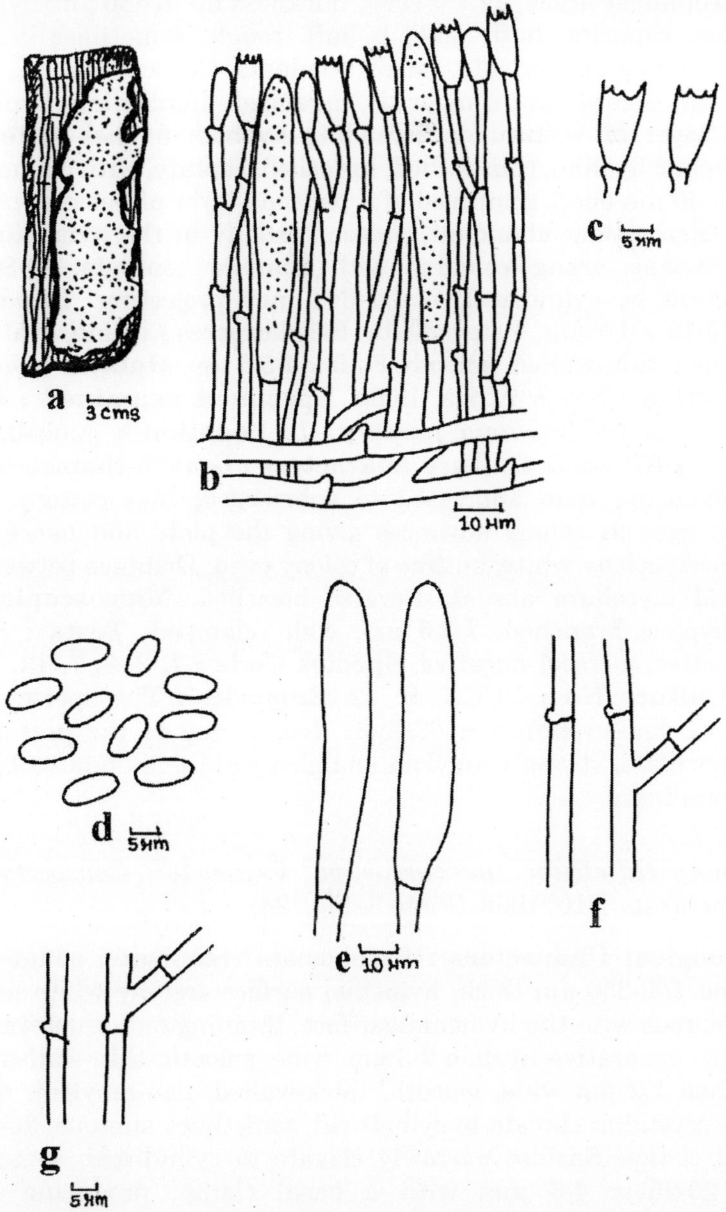

Fig. 27 ***Gloeocystidiellum luteocystidiatum*** a. Habit. b. T.S. hymenium. c. Basidia. d. Basidiospores. e. Gloeocystidia. f. Generative hyphae. g. Generative hyphae in culture

Morphological Characters: Fruit body annual, thin, membraneous, resupinate, effused, small portions separable when moist, sporophores may peel off when dry, arising at first as small circular or irregular colonies which later on coalesce to form linear areas, 22 × 7 cms., thickness up to 450 μm; hymenial surface pinkish yellow, capucine buff, pinkish buff, rough, sometimes cracked; margin concolorous, thinning out, serrate, raised or adnate. Context hyaline, composed of a basal layer of somewhat compactly arranged horizontal hyphae, and an intermediate layer of vertical or ascending hyphae; hyphal system monomitic, generative hyphae hyaline, thin-walled, branched, septate, with clamp, 2-4 μm wide. Hymenium to 45 μm deep, composed of a palisade layer of basidia, paraphyses and gleocystidia. Gleocystidia abundant, present usually in the hymenium, arise from the context hyphae, orange-coloured with granular contents, 50-80 ×10-15 μm, clavate, fusiform or cylindrical, embedded, not projecting. Basidia clavate or subclavate, 10-15 × 4-7 μm, 4 sterigmata. Basidiospores hyaline, oval or cylindrical, strongly amyloid, thin-walled, smooth, 8-10 × 4-5.5 μm. **Habitat :** On live standing tree. **Distribution :** South Africa, India. **Specimen examined :** Benzidine test positive, white rot on *Bauhinia purpurea* L. **Location :** Amboli, Dist. Satara (19/02/2007), VH-NPO-808. **Culture characters :** Growth characters: Growth rate moderate; advancing zone appressed to submerged; Mat cottony white getting submerged in agar as colony advances giving the plate and hence the colony, a transparent look; colony white; outline of colony even; Distance between hyphal tips distant; Aerial mycelium absent; Reverse bleached. **Microscopic characters:** Generative hyphae branched, 2-3.5 μm, wide, clamped. **Tests :** α-napthal and pyrogallol positive; p-cresol negative. **Species Code :** 1, 2, 3, 7, 13, 15, 21, 30, 37, 39, 52, 53. **Culture No :** PUCC. B: 12. **Remarks :** The specimen collected is identified as *G. luteocystidiatum* (Talbot) Boidin, due to the presence of oval or cylindrical, apiculate, strongly amyloid and gleocystidia abundant. The species is a new record from India.

(9-14) ***Gloeocystidiellum percuriosum*** Parmasto. *Conspectus Systematis Corticiacearum:* 210, 1968. (Plate 3, Fig. 28)

Morphological Characters : Basidiomata resupinate, adnate, effused, 7 × 2.5-5 cms. and 100-250 μm thick; hymenial surface creamy white to buff, smooth; margin concolorous with the hymenial surface, thinning out, indeterminate. Hyphal system dimitic; generative hyphae 2-3 μm. wide, smooth, thin-walled, with clamps; skeletal hyphae 1-2 μm wide, smooth, thick-walled, non-amyloid, non-dextrinoid. Cystidia (gloeocystidia) clavate to cylindrical, sometimes sinuous, 25-35 × 7-10 μm, with a basal clamp. Basidia narrowly clavate to cylindrical, sometimes slightly constricted, 20-25 × 4-6 μm, with a basal clamp, producing 4 sterigmata. Basidiospores ellipsoid, 4.5-5 ×3 (-3.5) μm, verrucose, thin-walled, amyloid. **Habitat :** On live standing tree. **Distribution :** Europe, North America, Japan. **Specimen examined :** Benzidine test positive, white rot on *Annona squamosa* L. **Location :** Pune university, Dist. Pune (18/08/2006), VH-NPO-510. **Remarks :** The species is recognized by the hyphal system dimitic; generative hyphae thin-walled, with clamp; skeletal hyphae thick-walled, nonamyloid, non-dextrinoid; basidiospores ellipsoid, verrucose, thin-walled and amyloid. The culture could not be obtained from this speciemen. The species is a new record from India. The cultures could not obtain from this specimen.

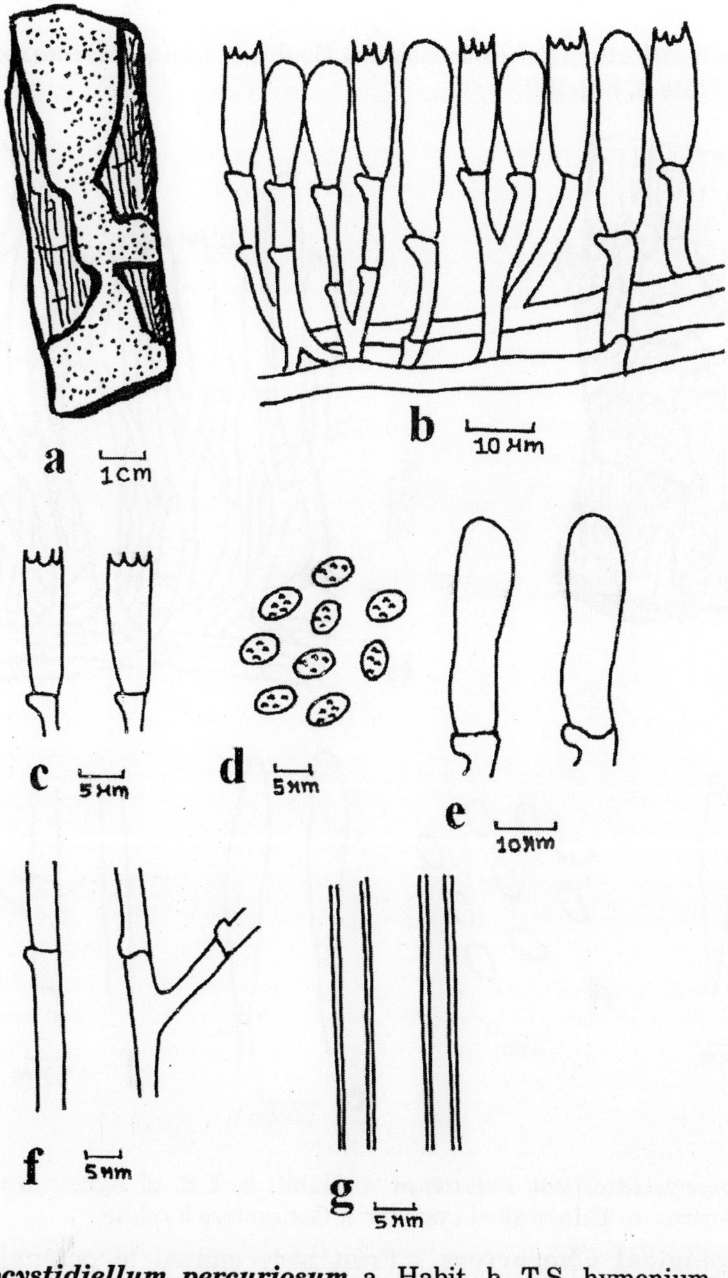

Fig. 28 ***Gloeocystidiellum percuriosum*** a. Habit. b. T.S. hymenium. c. Basidia.d. Basidiospores. e. Gloeocystidia. f. Generative hyphae. g. Skeletal hyphae

(9-15). *Gloeocystidiellum sulcatum* (Rehill & Bakshi) Boidin. *Cah. Maboké* **4 (1)** : 10, 1966.

=*Corticium sulcatum* Rehill & B.K. Bakshi, *Forest Bull.*, Dehra Dun, n.s. **242**:15, 1965.

=*Vesiculomyces sulcatus* (Rehill & Bakshi) Boidin & Lanq., *Mycotaxon* **16 (2)** : 493, 1983. (Plate 3, Fig. 29)

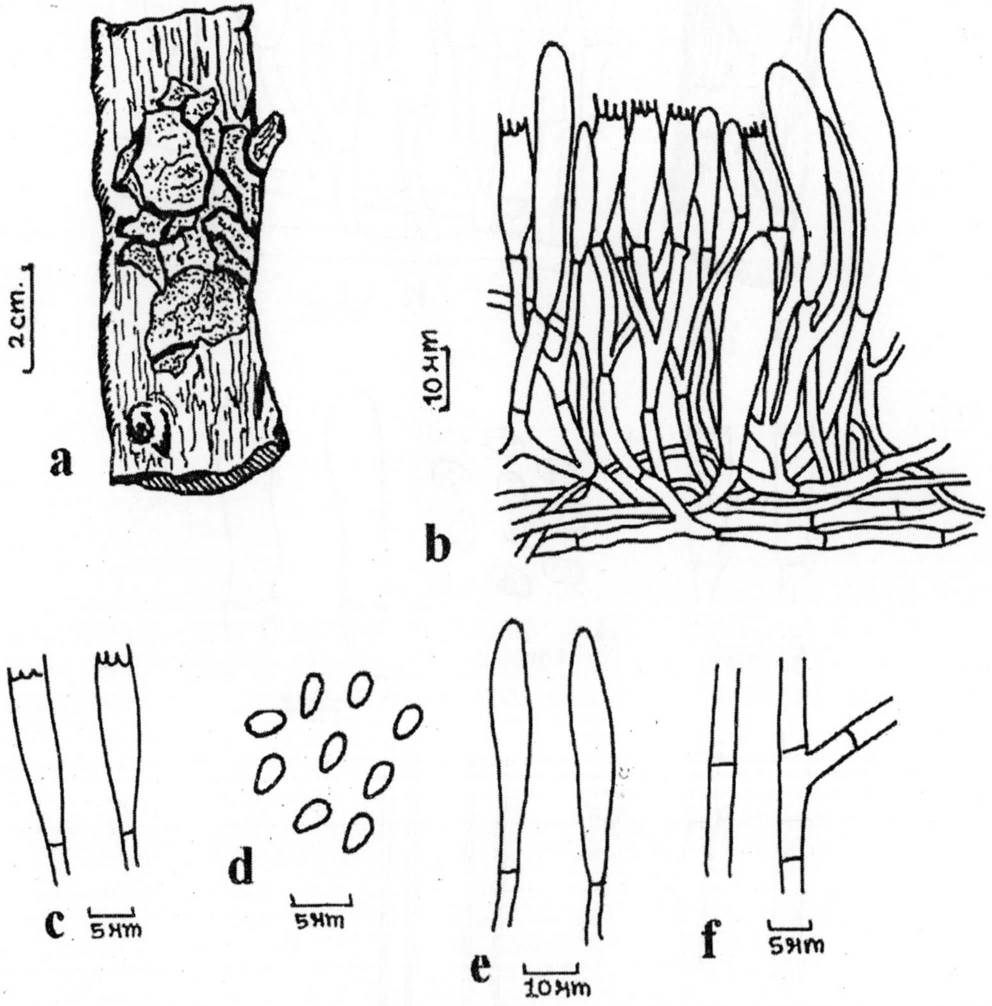

Fig. 29 *Gloeocystidiellum sulcatum* a. Habit. b. T.S. of hymenium. c. Basidia. d. Basidiospores. e. Thin-walled cystidia. f. Generative hyphae

Morphological Characters : Fruit body annual to perennial, resupinate, adnate, membranous-ceraceous, somewhat brittle, stratose, widely effused, 6 × 3.5 cms. and up to 2 mm. thick in section; hymenial surface cream to cream-yellow when young changing to yellowish brown, brown to finally greyish brown or cream grey with age, ail colours may be present in a single patch, smooth to prominentely tuberculate; margin thinning to abrupt, paler concolorous to concolorous, adnate. Context subhyaline to light brown in section. Hyphal system monomitic, hyphae 3-4 μm. wide, branched, septate, clampless, the walls thin, subhyaline, often collapsing and agglutinating and thus indivisual hyphae are difficult to discern in older parts of the context. Cystidia absent. Gloeocystidia 40-80 × 5-9 μm, abundant, clavate-cylindrical to cylindrical with obtuse apices, empty but more often with shinny granular contents staining deeply with phloxine, thin to slightly thick-walled especially in the basal part, immersed or projecting up to 15 μm out of the hymenium. Basidia 15-25 × 3.5-4.5 μm, clavate, with 4 sterigmata. Basidiospores 3. 5-5.5 × 2-3 μm, broadly ellipsoid, thin wall, subhyaline, smooth, amyloid. **Habitat :** On live standing tree. **Distribution :** India. **Specimen examined :** Benzidine test positive, white rot on *Annona squamosa* L. **Location :** Pune university, Dist. Pune (19/08/2006), VH-NPO-513. **Remarks :** This species shows great variation in the morphological features such as colour of basidiocarp, colour varies from cream to cream gray or grayish brown in single patch. The cultures could not be obtained, besides repeated trials

(9-16) ***Gloeocystidiellum turpe*** Freeman. *Syst. Bot.* **6:** 430, 1981. =*Gloiothele turpis* (Freeman) Hjortstam, *Windahlia* **17:** 58, 1987. (Plate 3, Fig. 30)

Morphological Characters : Basidiomes broadly effuse, 6-7 × 2.5-3 cms. and up to 250 μm thick, membranous, adherent, hymenial surface pale yellow-orange to light ochraceous-buff when dry, smooth, not cracking to slightly areolate in places; margin abrupt, pruinose, concolorous or nearly white. Subiculum very thin, indistinct. Hyphal system monomitic. Subiculum a very thin layer of parallel hyphae next to substrate, hyphae compact, agglutinated, indistinct. Hyphae 2-3 μm diam, clampless, frequently branched, thin wall, hyaline. Gloeocystidia arising at various levels in the basidiome, 50-120 × 8-10 μm, numerous, sinuous, cylindric or clavate, thin-walled, the contents globular to granular or crystalline, sulfo-positive. Hyphidia lacking. Basidia 25-35 × 5-9 μm, clavate, thin-walled, hyaline, with four sterigmata. Basidiospores 6-7×4-5 μm, broadly ellipsoid to subglobose, thin wall, hyaline, smooth, weakly amyloid, acyanophilous. **Habitat :** On live standing tree. **Distribution :** United States. **Specimen examined :** Benzidine test positive, white rot on *Gliricidia speium* (Jacquni.) Stend. **Location :** Panchwati; Dist: Pune (23/09/2005), VH-NPO-66. **Remarks :** The specimen is recognized by pale yellow-orange to light ochraceous-buff hymenial surface. Gloeocystidia arising at various levels in the basidiome, numerous, sinuous, cylindric or clavate, the wall thin at first, then thick, the contents globular to granular or crystalline, sulfo-positive. The species is a new record from India. The culture was not obtained besides repeated trials.

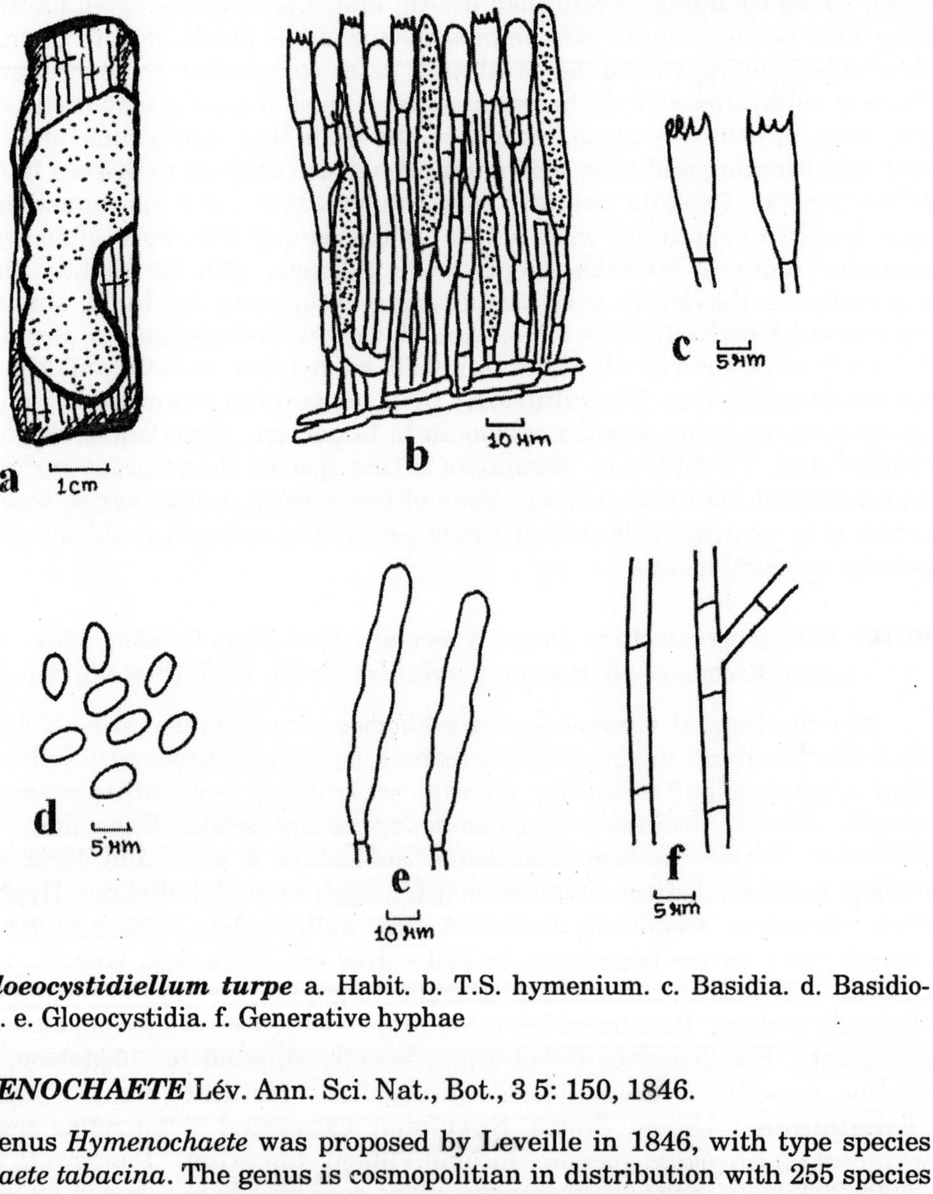

Fig. 30 ***Gloeocystidiellum turpe*** a. Habit. b. T.S. hymenium. c. Basidia. d. Basidiospores. e. Gloeocystidia. f. Generative hyphae

(10) ***HYMENOCHAETE*** **Lév. Ann. Sci. Nat., Bot., 3 5: 150, 1846.**

The genus *Hymenochaete* was proposed by Leveille in 1846, with type species *Hymenochaete tabacina*. The genus is cosmopolitian in distribution with 255 species recorded. Many species are tropical to subtropical in distribution. *Hymenochaete* belongs to a group of Aphyllophoraceous fungi which Donk (1964) has grouped into the family *Hymenochaetaceae.* A nomenclatural history of *Hymenochaete* Lev. and its present status has been reviewed by Donk (1957). If Donk's interpretation of *Hymenochaete* proves acceptable, then to preserve nomenclatorial stability, it will be necessary to conserve *Hymenochaete* Lev. against *Hymenochaete* Beauv. and *Hymenochaete* Ness. Other nomenclatorial discussion revolves around the choice of a type for the genus. Clements and Shear (1931) chose *Stereum tabacinum*

(Sowerby) Fries a suggestion followed by Cunningham (1947). Donk (1957) prefers *Stereum rubiginosum* (Schrader ex Gmelin) Gray lists *Hymenochaete Lev.* twice, once with *S. rubiginosum*, as the selected type, and once with *Thelephora undulate* Fr. [now *Cotylidia undulata* (Fries) Karsten.]. We follow Roger's suggestion that "among species equally eligible, the preference shall be given to the first known, to have been designated as the type". Thus, Clement and Sheer choice of *S. tabacinum* is followed now.

Hymenochaete is characterized by basidiocarp stipitate, dimidiate effuso-reflexed or effused, stratose or not, usually some shade often to brown (but red pigments may also be present), papery to coriaceous to hard; hymenial surface smooth, warted, folded or cracked, rarely granular, margin in effused basidiocarps determinate or indeterminate; cuticle when present, composed of dark, parallel or tightly interwoven hyphae often cemented together; context with or without crystals or mucilage like material, constructed of longitudinally parallel or interwoven hyphae, or of hyphae perpendicular to the substrate, or duplex; hyphae usually thick walled sparsely septate, dark, without clamps connections; setae brown, subulate to ventricose, thick walled, arranged in one to several layers naked or ensheathed with thin-walled hyaline hyphae, sometimes tipped with crystals; hymenium of basidia and basidioles, sometimes perforated by, or producing, accessory structures, basidia cylindrical, cylindrical–clavate, subclavate, or clavate, projecting slightly beyond the hymenial level, four spored; spores allontoid, sub-allontoid, ovoid, or spherical, smooth, non –amyloid, hyaline, thin-walled.

Few macroscopic characters can not be considered as a full-proof character towards classification as they vary widely within the species due to position, age etc. For e.g. form of basidiocarp varies widely from resupinate to effuso-reflexed, as it is probably dependpent upon its position relative to ground. Secondly colour of the basidiocarp also is another variable character. Almost all species of *Hymenochaete* are some shades of brown, but wide colour variations occur with different population because of the basidiocarp, the darker it is. Abundance of setae also plays an important role determining the colour of basidiocarp. The dark brown setae impart a red-brown lusture to the hymenial surface and as they increase in number, the hymenial surface becomes darker. Among effused species, the margin may sometimes be a useful character. It may gradually thin out at margins, or it may end abruptly.

Microscopically, Hymenochaete shows three tissues zones in sections; Cuticle, context and hymenium. The presence of accessory structures in some species can be quite helpful in classification. A cuticle may be present or absent. When present, it can be recognized as a region of tightly interwoven or compactly parallel hyphae at the abhymenial surface of the basidiocarp. Cuticular tissue is darker brown than contextual tissue. A tomentum, or scattered hyphae or setae may arise from the free surface of the cuticle in effuso-reflexed or pileate forms. The context usually forms the bulk of the basidiocarp. Hyphae of the context have thickened; brownish walls and simple septa, and they form branches. They may be arranged in a longitudinally parallel manner, loosely interwoven, or more or less perpendicular to the substrate. The hymenium is composed of a close palisade of basidia and basidioles. Basidia are usually subclavate and project slightly beyond the basidioles. They bear four spores, hyaline, thin-walled, non-amyloid spores on short, needle like sterigmata. Basidioles differ from basidia in being slightly smaller and lacking sterigmata; Spores offer very slight variation and rarely are edifying to the taxonomist in search of characters. Several kinds of accessory structures develop in the basidiocarp of Hymenochaete spp. e.g. dendrophyses; dicophyses; unbranched, slightly thick-walled hyphae projecting beyond the hymenium; crystalline or mucilage like material within their contexts; cystidia etc. There were four specimen collected from the investigation areas.

KEY TO THE SPECIES OF *HYMENOCHAETE*

1. Hyphal system monomitic .... ***H. cinnamomea***

1´. Hyphal system dimitic 2

2. Spores allantoid, spores 5-8×1.5-2 μm .... ***H. tabacina***

2´. Spores ellipsoid to obovate or oval 3

3. Spores ellipsoid to obovate, spores 6-8×1-2.5 μm .... ***H. rubiginosa***

3´. Spores oval, spores 5-6.5 × 4-5 μm .... ***H. attenuate***

(10-1) ***Hymenochaete attenuata*** (Lév.) Lév. *Ann. Sci. Nat., Bot.*, 3 5: 152, 1646. (Plate: 3; Fig. 31)

Morphological Characters : Fruitbody pileate, pilei, effuso-reflexed or applanate, flabelliform connate, deeply, lobed and plicate, upper surface slightly velvety in appearance, concentrically sulcate, easily separable, soft, coriaceous, 4-5 × 2-4 cms., 250-300 μm. thick; hymenophore yellowish brown to light brown; reflexed pilei region brown with loose tomentose hairs and concentric lines; hymenial surface smooth, with abundant dark brown setal apices; margin thin, free reflexed entire to lobed imbricate, if lobed, concolorous; cortex and cuticle absent; abhymenial hairs present; context yellowish brown with compactly and parallel arranged intertwined hyphae, up to 100 μm., thick, homogenous. Hyphal sysrem dimitic, generative hyphae hyaline, thin to thick walled, septate, clampless, branched, 3-4 μm., wide; skeletal hyphae brown, thick-walled, 5-7.5 μm., wide; tissue xanthochroic. Setae abundant, thick-walled, dark brown, with pointed apices, subhymenial in origin, 1-2 layers, overlapping, hymenial in distribution, projecting up to 50 μm., above the hymenium with narrow lumen; setae conical to fusoid, with narrow hyphae like setal bases, 80-110 × 17-20 μm., hymenium, a euhymenium, continuous. Cystidia none. Paraphysis mostly cylindrical, thin-walled, 35-40 × 3-4 μm, distributed along with the basidia and basidioles in the hymenial palisade layer. Basidia cylindrical to clavate, 15-20 × 3-4 μm, with four sterigmata. Spores hyaline, ellipsoid to cylindrical, thin-walled, smooth, non-amyloid, acyanophilous, 5-6.5 × 2.5-4 μm. **Habitat :** On fallen branches. **Distribution:** Argentina, Sweden, Java, Japan, Borneo,

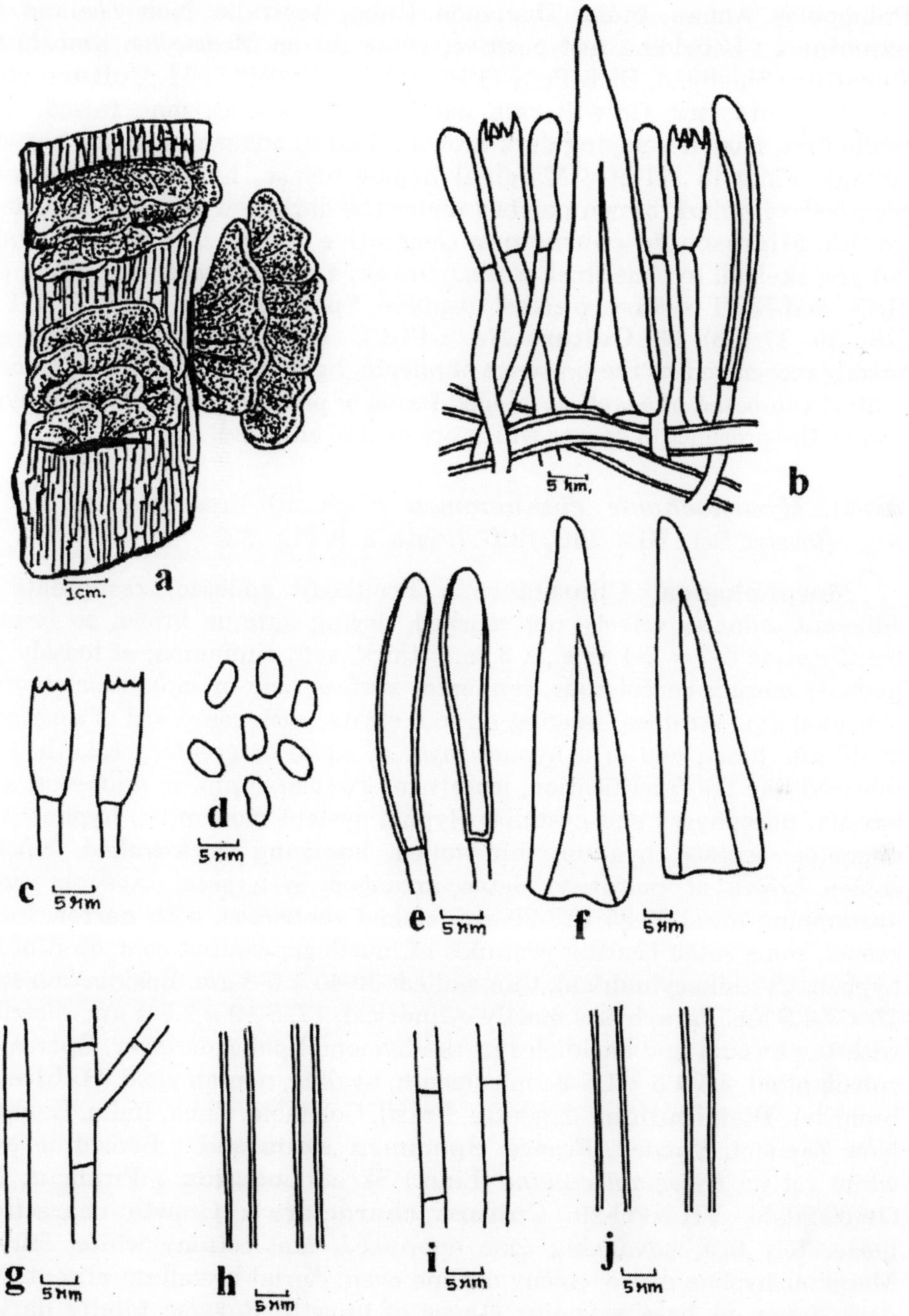

Fig. 31 ***Hymenochaete attenuata*** a. Habit. b. T.S. of hymenium. c. Basidia. d. Basidiospores. e.Paraphysis. f. Setae. g. Generative hyphae. h. Skeletal hyphae. i. Generative hyphae in culture. j. Skeletal hyphae in culture.

Philippines, Annan, India, Thailande, Cuba, Australia, New Zéaland. **Specimen examined :** Benzidine test positive, white rot on *Memecylon umbellatum* Burm. **Location:** Singhgad, Dist. Pune (13/01/2007), VH-NPO-564. **Culture characters :** Growth characters: Growth rate moderate; advancing zone raised; mat cottony white first, later developing dark brown colour in advancing stage; aerial mycelium cottony white to velvety; Marginal hyphae dense, hyphal tips distant; Reverse bleached with dark brown patches under the darkened mat in advancing stages of growth. **Microscopic characters:** Generative hyphae thin-walled, hyaline, septa, 3-4 μm; skeletal hyphae thick-walled, brown, 4-5 μm. **Tests:** α-napthal, pyrogallol + H_2O_2, and KOH positive; p-cresol negative. **Species Code :** 1, 2, 3, 4, 8, 14, 15, 21 (26), 30, 37 (38), 53. **Culture No. :** PUCC. B: 13. **Remarks :** This specimen is readily recognized by the presence of pileate, light coloured fruitbody, cortex absent; context composed of a well developed tissue of parallel or intertwined hyphae; septa and by the presence of spores which are oval to ellipsoid.

(10-2). *Hymenochaete cinnamomea* (Persoon) Bresadola. *Atti I. R. Accad. Roveret.* Sci., III 3: 110, 1897. (Plate: 3, 9; Fig. 32)

Morphological Characters : Fruitbody apileate, resupinate sessile or adherent, adnate, velvety not cracked, drying antique brown to brussels-brown; fructification 5-7 × 2-4 cms., 1-3 mm. thick, soft, cinnamon; of loosely intertwined hyphae; margin concolorous; hymenial surface clay or cinnamon; hymenial layer 500-1000 μm., stratose, ranging up to 6 strata, each composed of a setigerous layer 30-46 μm. broad and of a hyphal layer of equal or greater breadth, with hyphae coloured like the fructification, loosely interwoven, 3 μm. in diameter; a palisade of basidia, paraphyses and cystidia. Hyphal system monomitic, hyphae 3-3.5 μm. in diameter, septate, hyaline, thin-walled, becoming thick-walled, 3.5-4 μm., and golden brown at maturity. Setae arranged in a setal layer of two to three overlapping rows, 60-85 ×17-20 μm., naked ventricose, with narrow lumen, golden brown, some setae bearing granules of mucilage; context composed of intertwined hyphae. Cystidia cylindrical, thin-walled; 30-40 × 5-8 μm. Basidia sub-clavate, 12.5-15 × 3-4.5 μm. Paraphyses mostly cylindrical, 17. 5-19 × 3.5-4 μm; distributed along with the basidia and basidioles in the hymenial palisade layer. Spores elliptical or suballantoid, 3.5-4.5 × 1.5-2 μm., smooth, hyaline, non-amyloid. **Habitat :** On fallen branches. **Distribution :** Argentin, Brazil, Colombie, Cuba, India, Equateur, Africa, New Zéaland, Sweden, France. **Specimen examined :** Benzidine test positive, white rot on *Syzygium cumini* (Linn.) Skeel. **Location :** Pirangut; Dist : Pune (16/09/2005), VH-NPO-29. **Culture characters :** Growth characters: Growth moderately fast; advancing zone appressed, Mat cottony white, felty or woolly; Marginal hyphae dense; colony outline even; Aerial mycelium absent. Mat turning dark brown at base at latter stages of growth. Reverse totally darkened (dark brown). Microscopic characters: Generative hyphae hyaline, thin-walled, 2.5-3 μm; unbranched, few hyphae shows encrustation on them. **Tests :** α-napthal, pyrogallol + H_2O_2 and KOH positive; p-cresol negative. **Species Code:** 1, 3, 4, 7, 13, 15, 16, 30,

38, 52, 53, 57. **Culture No. :** PUCC. B: 14. **Remarks :** This species is characterized by its velvety, cinnamon hymenial surface; two to three segiterous layers 30-45 μm. thick, usually narrower than the adjoining hyphal layers, and spores up to 4. 5-5 μm. long.

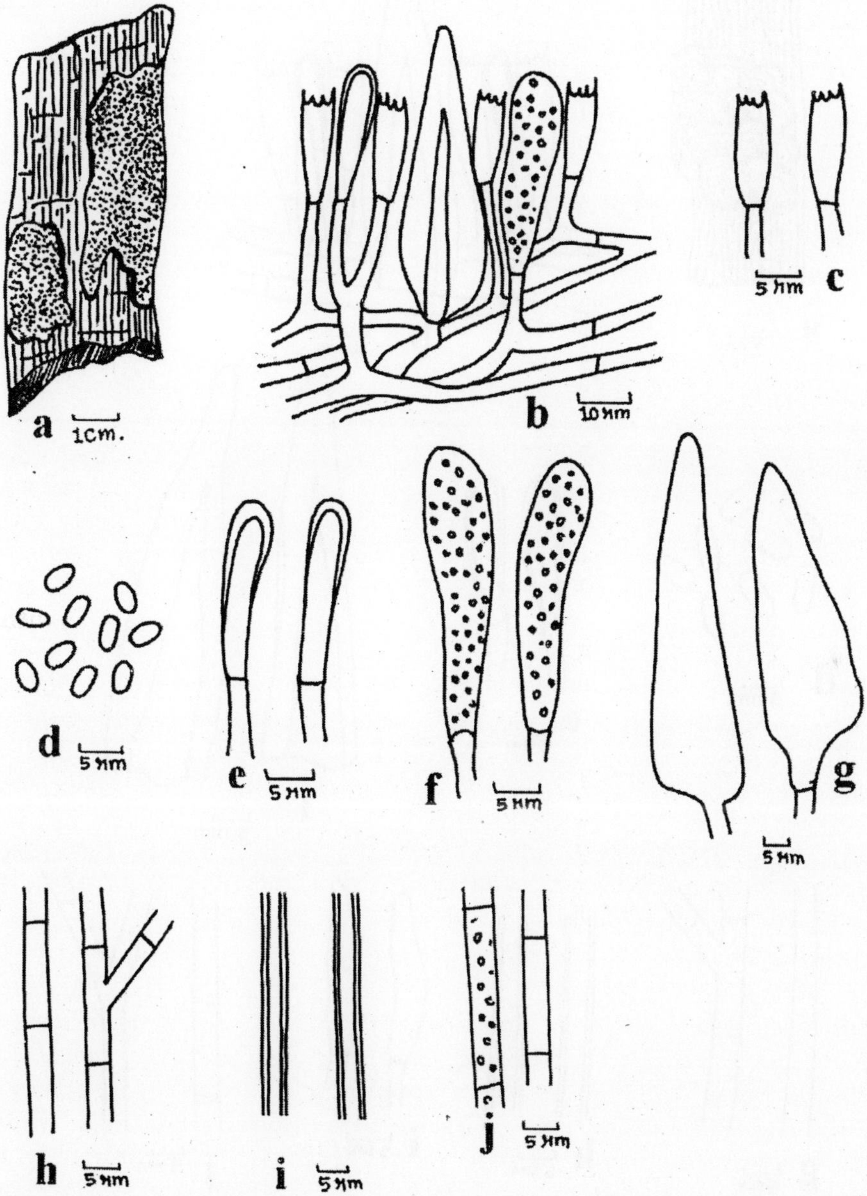

Fig. 32 ***Hymenochaete cinnamomea*** a. Habit. b.T.S. hymenium. c. Basidia. d. Basidiospores. e. Paraphyses. f. Encrusted cystidia. g. Setae, h. Generative hyphae. i. Skeletal hyphae. j. Generative hyphae in culture showing encrustation

(10-3) *Hymenochaete rubiginosa* (Dickson) Léve. *Ann. Sci. Nat., Bot.*, **3 5:** 151, 1846. (Plate 3; Fig. 33)

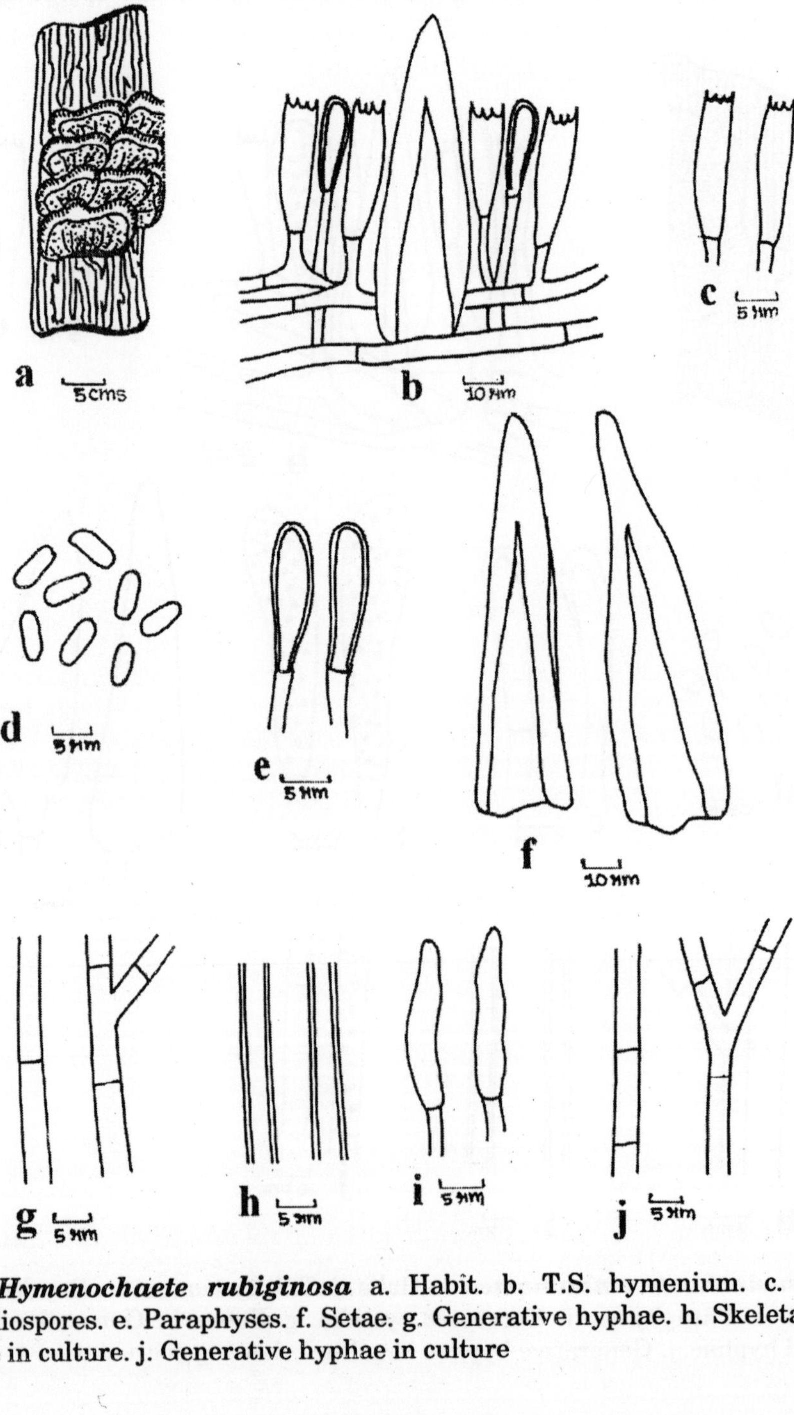

Fig. 33 ***Hymenochaete rubiginosa*** a. Habit. b. T.S. hymenium. c. Basidia. d. Basidiospores. e. Paraphyses. f. Setae. g. Generative hyphae. h. Skeletal hyphae. i. Setae in culture. j. Generative hyphae in culture

Morphological Characters : Fruit body coriaceous, rigid, effuso-reflexed or sometimes wholly resupinate, 15-17 ×7-11 cms., separable, somewhat fasciate above, concentrically sulcate, velvety, brussels's brown, finally glabrous, fuscous black; margin ochraceous-tawny; hymenium conspicuously setulose under a lens, somewhat colliculose, bister; hymenial layer 500-700 μm., thick, with a broad dense, dark, setigerous layer 160 μm. thick and with the intermediate layer composed of longitudinally arranged coloured hyphae 2.5-3 μm. in wide, and bordered above by narrow, dense, dark zone; hymenial surface chestnut, with fawn margins; pileate; Pilei commonly umbonate-sessile. Hyphal system dimitic, generative hyphae 3-3.5 μm. wide, hyaline, septate, branched, thin-walled; skeletal hyphae 3. 5-4.5 μm. wide, thick-walled, golden brown, unbranched, septate. Setae 100-120 ×12-20 μm., naked, ventricose at base, crowded together in all parts of the 160 μm. thick setigerous layer; setae golden brown with a narrow lumen; Hymenial layer a close palisade of basidia and paraphyses. Basidia sub-clavate, 15-20 ×2.5-5 μm. bearing four spores. Paraphyses cylindrical, thick-walled, 17-25 ×3-5 μm. Spores 6-8 ×1.5-2. 5 μm., smooth thin-walled, hyaline. **Habitat :** On dead standing tree. **Distribution :** Mexico, El Salvador, Costa Rica, Cuba, Puerto Rico, India, Colombie, Equateur, Pérou, Argentina, Sweden, Belgique, Sweden, France, Africa, Chili, New Zéaland.

Specimen examined : Benzidine test positive, white rot on *Mallotus philippensis* (Lamk.) Muell. **Location :** Dongerwadi, Dist. Pune (13/01/2007), VH-NPO-583. **Culture characters :** Growth characters: Growth rate moderate; advancing zone appressed to submerged; Mat cottony white initially turning reddish partly and finally brown in advanced stages; Outline of colony even; Aerial mycelium present, cottony woolly, Marginal hyphae distinctively inequivalently branched; Reverse bleached. **Microscopic characters:** Generative hyphae thin-walled, hyaline, branched, 3-4 μm.; culture shows the presence of setae, 15-18 × 2-3 μm., light brown cylindrical to ventricose. **Tests :** α-napthal, pyrogallol + H_2O_2 positive; p-cresol negative. **Species Code :**1, 3, 4, 7, 13, 21, 30, 34, 37, 53, 70. **Culture No. :** PUCC. B: 15. **Remarks :** This species is easily recognized by its rigid pileus, velvety concentrically sulcate and brussels brown colour, even resupinate specimens may be recognized at sight by being separable form the substratum, and by the form and colour of the hymenium and the contrasting bright margin. The species produces a pocket heart rot in angiospermic wood.

(10-4). ***Hymenochaete tabacina*** (Sowerby) Lév. *Ann. Sci. Nat., Bot.*, **35:** 152, 1846.

= *Helvella nicotiana* Boltan, *Hist. Funguses*, 174, pl. 174. 1789.

= *Auricularia tabacina* Sowerby, *British Fungi*. pl. 25, 1797.

= *Thelephora tabacina* (Sow.) Fries., *Syste. Myc.* **1 :** 437, 1821.

= *Hymenochaete imbricata* Sacc. *Syll. Fungo.* **6 :** 593, 1888. (Plate 3;9, Fig. 34)

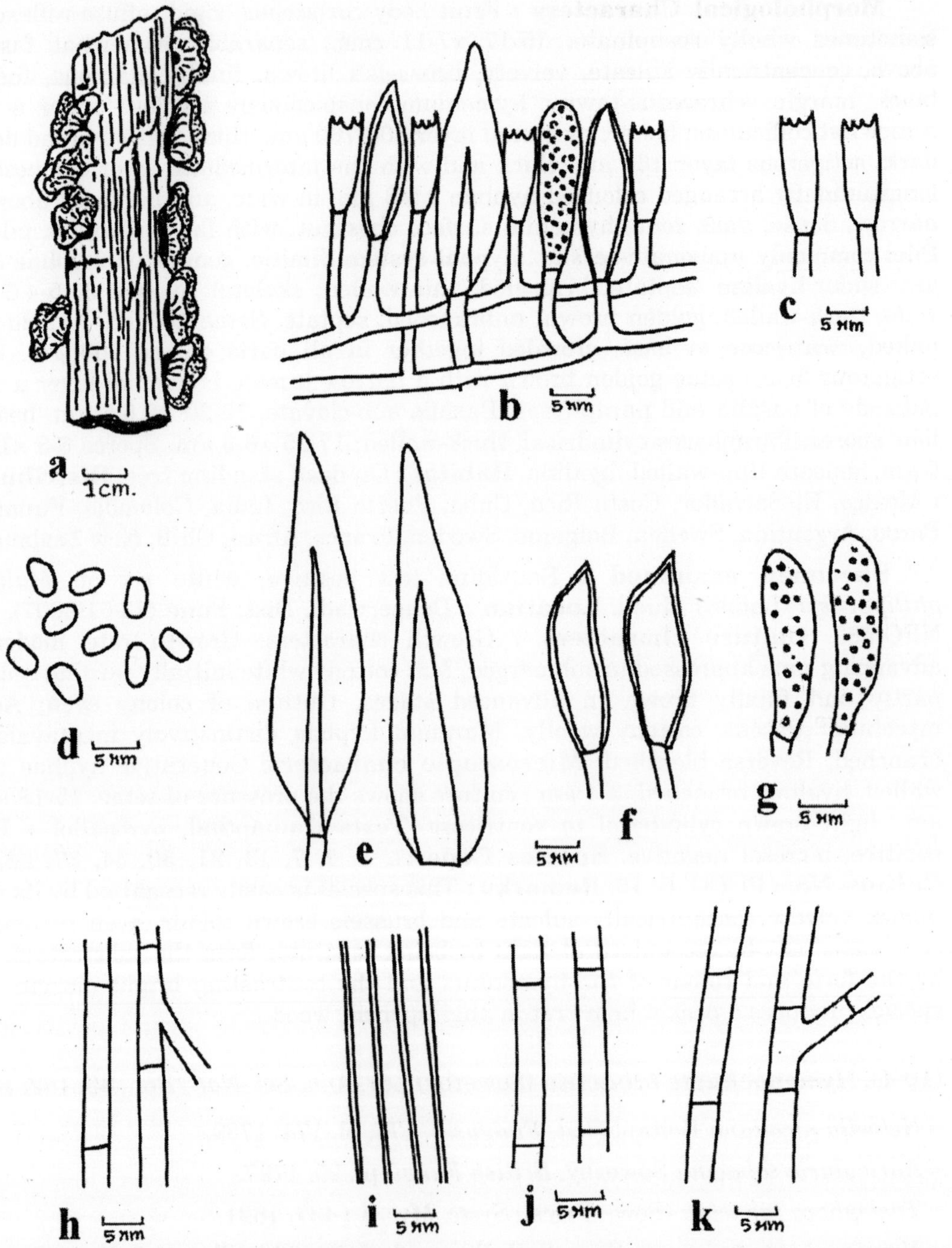

Fig. 34 ***Hymenochaete tabacina*** a. Habit. b. T.S. hymenium. c. Basidia. d. Basidiospores. e. Setae. f. Paraphyses. g. Encrusted cystidia h. Generative hyphae. i. Skeletal hyphae. j. Aerial hyphae in culture. k. Marginal hyphae in culture

Morphological Characters : Fruit body coriaceous, effuso-reflexed, often imbricated, sometimes wholly resupinate, thin, ceraceous, antique brown, at length becoming glabrous and deep brownish drab; 1-3 ×1-1.5 cms.; hymenial surface snuff brown to tobacco brown or ferruginous, often deeply cracked where resupinate, with a series of radial anastomosing cracks for each centimeter of area; margin fulvous; pileus tomentose; hymenium 340 to 600 µm. thick with the setigerous layer 100 µm. thick, and with the intermediate layer bordered on each side by a narrow, dark, dense zone. Hyphal system dimitic, generative hyphae, 2-3 µm, thin-walled, branched, hyaline, septate, skeletal hyphae, 3-4 µm., thick-walled, golden-brown, unbranched. Setae 45-90 ×10-12., emerging up to 50 µm, originating from all parts of the setigerous layer although chiefly from the dark, subhymenial zone; hymenium layer a close palisade of basidia, cystidia and paraphyses. Cystidia cylindrical, thin-walled, with encrustation, 22-25 ×5-7 µm. Basidia sub-clavate, 13-15 ×4-5.5 µm. Paraphyses cylindrical with acuminate apices, 20-22 ×3.5-5 µm.. Spores allantoid, hyaline, smooth, 5-8 ×1.5-2 µm. **Habitat :** On fallen branches. **Distribution :** Canada, Mexico, Costa Rica, Panama, Argentina, India, Norway, Danemark, Sweden, France, Africa, New-Zéaland. **Specimen examined :** Benzidine test positive, white rot on: *Azadriachata indica* A. Juss. **Location :** NCL park, Dist. Pune (19/09/2005), VH-NPO-51. **Specimen examined :** Benzidine test positive, white rot on: *Allophylus cobbe* (L.) Raeusch. **Location :** Pune university, Dist. Pune (08/10/2005), VH-NPO-109. **Culture characters :** Growth characters: Growth moderately rapid; advancing zone appressed; mat cottony white initially turning brown; Mat even; colony out line smooth; reverse bleached. Microscopic characters: Aerial hyphae hyaline thin-wall, simple septa, 1.5-2.5 µm; marginal hyphae hyaline, branched, simple septa, 3-5 µm wide. **Tests :** α-napthal, pyrogallol + H_2O_2 and KOH positive, p-cresol negative. **Species Code :** 1, 3, 4, 7, 13, 15, 21, 22, 30, 34, 37, 52, 53. **Culture No. :** PUCC. B: 16. **Remarks :** This species is characterized by its rust brown, grayish or dark brown hymenial surface. Hymenium smooth but uneven, usually concentrically zonate and sulcate. Pilei confluent, imbricate, dimidiate or short and broad, pileal surface radiately frbrillose, silky, glabrous.

(11). ***HYPHODERMA*** Wallroth. *Flora Cryptogamica Germaniae* **2:** 576, 1833.

The genus was created with its type species *H. setigerum* (Fr.) Donk. The genus is not clearly delimited, thus constitutes a heterogeneous assemblage of species which are likely to be subdivided in smaller genera in future (Eriksson and Ryvarden, 1975). The species of *Hyphoderma* are separated in to twelve different groups by Parmasto in 1968. The genus is widespread with 157 species. The genus is characterized by hymenial surface smooth, tuberculate, granor odontoid.; basidia narrowly clavate to subcylindrical, usually longer than 20 µm; basidiospores subglobose, ellipsoid, cylindrical or allantoid and non-amyloid. There was one specimen collected from Pune. The specimen with characteristic cylindrical to suballantoid, non-amyloid spores was identified as *H. radula* Fries.

(11-1) *Hyphoderma radula* (Fries) Donk. *Fungus* 27: 15, 1957.

=*Radulum orbiculare* Fr., *Syst. orb. veg.* (Lundae) : 81, 1825. (Plate 3, 9, Fig. 35)

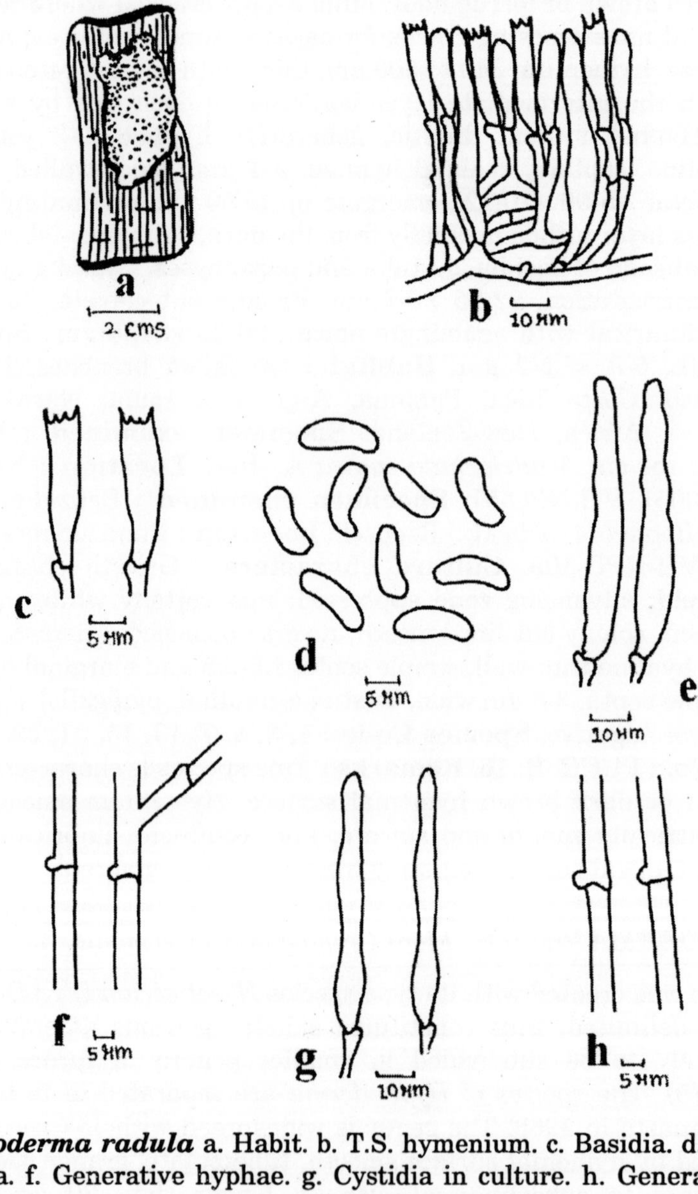

Fig. 35 *Hyphoderma radula* a. Habit. b. T.S. hymenium. c. Basidia. d. Basidiospores. e. Cystidia. f. Generative hyphae. g. Cystidia in culture. h. Generative hyphae in culture

Morphological Characters: Basidiomata resupinate, adnate, at first orbicular and then becoming confluent, 4 × 3 cms. and 100-800 μm thick; hymenial surface white to pale yellowish white, smooth, tuberculate or raduloid, ceraceous, sometimes cracked when dry; margin white, thinning out. In vertical section, subhyaline, membranous, with a thin basal layer composed of horizontal hyphae. Hyphal system monomitic, hyphae 2-4 µm. wide, smooth, thin-walled, with clamp. Cystidia (leptocystidia) cylindrical, sometimes with several constrictions, 40-70 × 5-7 µm, smooth, thin-walled, with a basal clamp. Basidia suburniform, 25-35 × 4-6 µm, with a basal clamp, producing 4 sterigmata. Basidiospores cylindrical to suballantoid, 7.5-12 × 3-3.5 µm, smooth, thin-walled, non-amyloid. **Habitat :** On fallen branch. **Distribution :** India, Japan, Australasia, Europe, North, South America. **Specimen examined :** Benzidine test positive, white rot on unkown host. **Location :** Ambghat; Dist: Pune (19/02/2007), VH-NPO-781. **Specimen examined:** Benzidine test positive, white rot on *Grevilla robusta* Cunn. **Location :** Pirangut, Dist. Pune (19/08/2006) VH-NPO-524. **Culture characters :** Growth characters: Growth rate fast; advancing zone appressed to slightly raised; Mats white, thin to moderately thin, subfelty or silky, then becoming slightly raised, silky to downy toward margins, margins even to slightly uneven, no odor; reverse bleached. **Microscopic characters:** Generative hyphae branched, 2.5-5 µm, wide, thin walled, nodose septate, sparsely branched. Cystidia cylindrical to obclavate, 35-60 × 2.5-10 µm, thin-walled, terminal, moniliform, scattered to numerous throughout mats. **Tests :** α-napthal and pyrogallol + H_2O_2 positive; p-cresol negative. **Species Code :** 1, 2, 3, 6, 12, 13, 15, 30, 37, 39, 52, 53, 72. **Culture No :** PUCC. B: 17. **Remark**: *Hyphodrema radula* is referred to a genus of its own by Nobles (*loc. cit.*). The hyphae and cystidia agree with many other species of the genus. The homogenus spores seem to be an important character; as such spores are not also orbicular on bark. One morphological character may be restriocted to *H. radula* it seem as if initial basidia in young hymenium do not develop to maturity as soon as they are formed but first when a small area of hymenium is developed. Then many basidia become mature at about the same time. This way of development makes the hymenium continuous from the beginning whole in other species of the genus, as in most other Corticiaceae, the hymenium is first open and then eventually becomes continous. The main reason for us to keep the species in *Hyphodrema* is the problem of separation, i.e. to decide which species should follow *H. radula* to the seprate genus.

(12) *HYPHODONTIA* John Erikss. *Symb. Bot. Ups.* **16:** 1, 1958.

The genus was erected by Schroter in 1888 by studying the type species *Hypochnella violacea* (Auersw.) Schroet. The genus is widespread with 131 species studied under the genus. The genus is characterized by resupinate to effused fruitbodies, adnate, fibrous but soft and easily squeezed when fresh and wet, more friable or tough when dried, whitish or pale ochraceous; hymenophore varying from smooth or finely tuberculate to odontioid with aculei; cystidia or cystidiole organs of some kind always present but varying in nature, conspicuously thick-walled, originating deeply in the hyphal texture; hyphae richly branched, with small distinct clamps; Basidia small to medium in size, sub-clavate to sub-cylindrical, more less constricted in a sub-uniform way; spores thin-walled, smooth, non-amyloid, varying in shape, either allantoid or sub-globose-ellipsoid. The specimen with is characteristic pale-yellowish white, buff to pale ocherous fruitbodies, ellipsoid, non-amyloid spores was identified as *H. alienata* (Lundell) Eriksson.

(12-1). *Hyphodontia alienata* (Lundell) Eriksson. *Symb. Bot. Upsal.* **16 (1) :** 104, 1958.

=*Grandinia alienata* (Lundell) Jülich, *Int. J. Mycol. Lichenol.* **1 (1) :** 35, 1982.

=*Kneiffiella alienata* (Lundell) Jülich & Stalpers, *Verh. K. ned. Akad. Wet., 2 Sectie* 74: 130, 1980.

=*Peniophora alienata* Lundell, *Fungi Exsiccati Suecici* **21-22:** 28, 1941. (Plate 3, 9; Fig. 36).

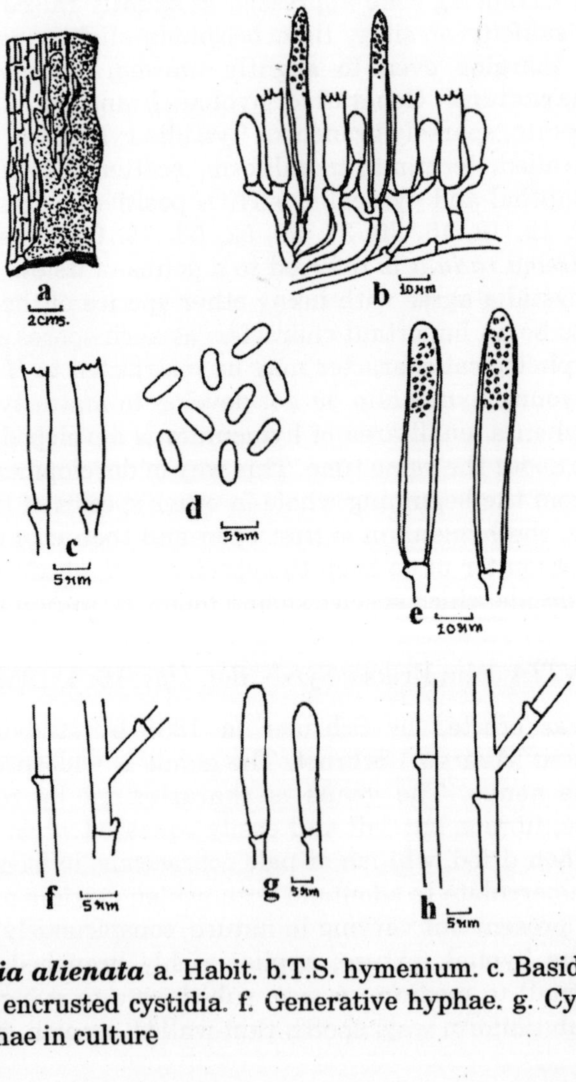

Fig. 36 *Hyphodontia alienata* a. Habit. b.T.S. hymenium. c. Basidia. d.Basidiospores. e. Thick-walled encrusted cystidia. f. Generative hyphae. g. Cystidia in culture. h. Generative hyphae in culture

Morphological Characters : Basidiomata resupinate, loosely adnate, effused, 10-12 × 2-5 cms., thin; hymenial surface pale yellowish white, buff to pale ocherous, smooth, margin concolorous with the hymenial surface, thinning out, indeterminate. In vertical section, subhyaline, submembranous, the subicular tomentose; margin concolorous with the hymenial surface, thinning out, indeterminate. Hyphal system monomitic, hyphae 3-5.5 μm wide, smooth, thin- to thick-walled, with clamps. Cystidia (hyphocystidia) numerous, cylindrical to narrowly fusiform, 50-80 × 6-9 μm, with a basal clamp, smooth, thick-walled, projecting up to 50 μm beyond the hymenial surface, Basidia subcylindrical, 15-20 × 5-6 μm, with a basal clamp, producing 4 sterigmata. Basidiospores ellipsoid, 5-6 × 3-4 μm, smooth, thin-walled, non-amyloid. **Habitat :** On live standing tree. **Distribution** : Africa, Europe, North and South America, Japan. **Specimen examined :** Benzidine test negative, Brown rot on *Syzygium cumini* (Linn.) Skeel. **Location :** Baneshwar; Dist:Pune (13/01/2007), VH-NPO-724. **Culture characters :** Growth characters: Growth rate slow; advancing zone appressed. Mat cottony white getting submerged in agar as colony advances giving the plate and hence the colony, a transparent look; colony white; outline of colony even; Distance between hyphal tips distant; Aerial mycelium absent; Reverse bleached. Microscopic characters: Generative hyphae branched, 3-5 μm, wide, with clamps, cystidia cylindrical to narrowly fusiform, 30-40 × 4-7 μm, with a basal clamp, smooth, thin-walled. **Tests :** α-napthal and pyrogallol + H_2O_2 positive; p-cresol negative. **Species Code :** 1, 2, 3, 6, 13, 14, 15, 16, 30, 37, 39, 53, 72. **Culture No :** PUCC. B: 18. **Remarks :** *Hyphodontia alienata* has beenn confused with *H. abieticola* but differs from this species in several respects. The fruitbody is almost smooth while odontioid in *H. abieticola*, the Cystidia sre scattered, not grouped together. the cystidia are morever encrusted and they differ in shape and size; shorter and slightly fusiform in *H. alienate*, long and cylindrical in *H. abieticola*. The species is a new record from India.

(13). ***HYPOCHNICIUM*** Eriksson. *Symb. Bot. Upsal.* 16 (1) : 100, 1958.

The genus was erected by John Eriksson in 1958 by studying the type species *Hypochnicium bambycinum* (Sommerf. ex. Fr.) John Erikss. The genus is widespread in distribution with 35 species so far known under this genus. The genus is characterized by resupinate, adnate, effused basidiomata; smooth, tuberculate or irregularly odontoid hymenial surface; monomitic hyphal system; hyphae thin- to thick-walled, with clamps; ellipsoid to globose, thick-walled, smooth or warted, cyanophilous, non-amyloid basidiospores. There were three specimen collected from the investigation areas. The specimen with characteristic subulate, thickwalled; somewhat sinuous, clamped at the base cystidia; subglobose, thickwalled, cyanophilous, non-amyloid spores was identified as *H. cymosum* (Rogers & Jackson) Larsson & Hjortstam. The specimen collected from Baneshwar, Dist. Pune is identified as *H. eichlerie* (Bresadola ex Saccardo) Eriksson & Ryvarden, due to the presence of verruocose, ellipsoid to sub-globose spores, and the presence of mostly enclosed cystidia. The specimen collected from the Baneshwar,

Dist. Pune is recognized by grayish white to pale yellowish white hymenial surface, cystidia lacking, obovate to globose and thick-walled basidiospores the specimen was identified as *H. globosum* Wu.

KEY TO THE SPECIES OF *HYPOCHNICIUM*

1. Cystidia lacking spores obovate to globose .... ***H. globosum***

1´. Cystidia present spores broadly ellipsoid to subglobose 2

2. Cystidia subulate, spores subglobose, 5 × 4 μm .... ***H. cymosum***

2´. Cystidia cylindrical to subfusiform, spores broadly ellipsoid to subglobose, 7-8. 5 ×5. 5-6. 5 μm .... ***H. eichleri***

(13-1). *Hypochnicium cymosum* (Rogers & Jackson) Larsson & Hjortstam. *Mycotaxon* 5 (2) : 477, 1977.

=*Peniophora cymosa* Rogers & Jacks. 1948 (Plate 3, Fig. 37)

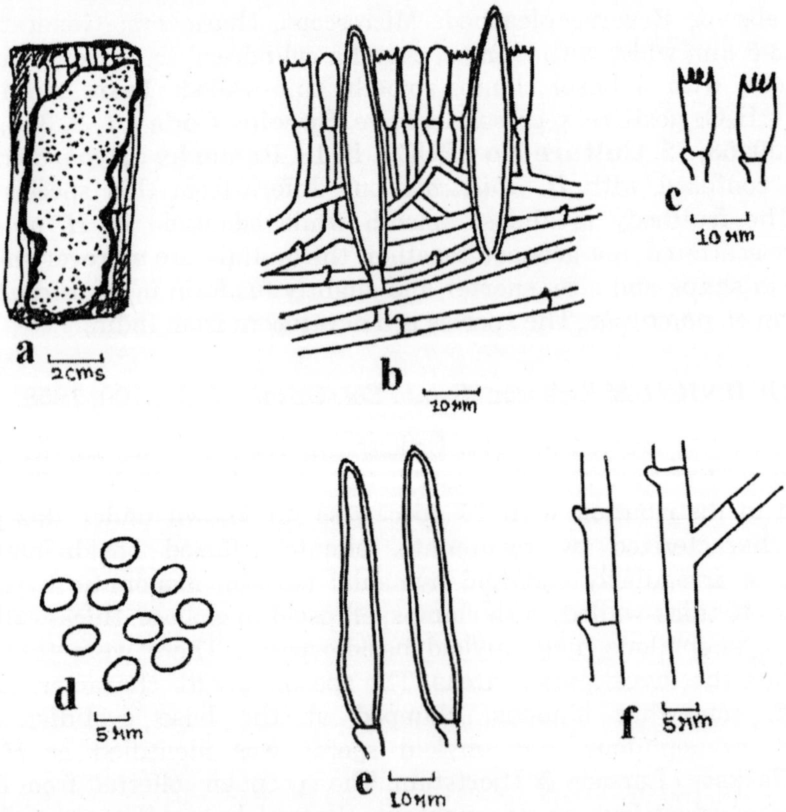

Fig. 37 *Hypochnicicum cymosum* a. Habit. b. T.S. hymenium. c. Basidia. d. Basidiospores. e. Cystidia. f. Generative hyphae.

Morphological Characters : Fruit body resupinate, effused, 6-8×2-3 cms. often very thin and rather inconspicuous, whitish, smooth. Hyphal system monomitic, basal hyphae more or less distinctly thickwalled, 4-7 μm wide, subhymenial hyphae thin-walled or slightly thickening, all hyphae with clamps. Cystidia subulate, thickwalled, somewhat sinuous, clamped at the base, 40-60 × 7-10 μm. Basidia subclavate, slightly sinuose, with clamped base, 10-15 ×4-6 μm, with 4 sterigmate. Spores subglobose, thick-walled, cyanophilous, 5 ×4 μm, non-amyloid. **Habitat :** On fallen branch. **Distribution :** North America, Europe. **Specimen examined :** Benzidine test positive, white rot on *Terminalia crenulata* W. & Prodr. **Location :** Baneshwar, Dist. Pune (13/01/2007), VH-NPO-661. **Remark:** Juilich (1975) incorporated the species in *Lagarobasidium* with type species *Odontia pruinosa* Bres. Eriksson and Ryvarden (1976) in the Corticiaceae of North Europe make it clear that this species is the same as *Peniophora detritica* Bourd. & Galz. and they refer the species to *Hypochnicium* owing to its thickwalled and cyanophilous spores. *Peniophora cymosa* is closely related to other species in *Hypochnicium*, preferably in having thickwalled and cyano-philous spores and in the shape of the basidia. It therefore ought to be referred to this genus. The species is a new record from India. The culture was not obtained besides repeated trials.

(13-2) *Hypochnicium eichleri* (Bres.) Erikss. & Ryv. *Syll. Fung.* **16:** 194, 1902.

= *Peniophora eichleri* Bres. in Sacc. and Syd., *Syll. fung.* **16**: 194, 1902.

= *Kneiffia eichleri* Bres., *Ann. mycol.* **1**: 100, 1903.

= *Hypochnus albo – stramineus* Bres., *Ann. mycol.* **1**: 109, 1903. (Plate: 3, 9; Fig. 38).

Morphological Characters : Basidiocarps resupinate, effused, 1-2×0.5-1 cms, adnate, cracking into small blocks on drying; hymenial surface cream to ochraceous, smooth; margin thinning out. Hyphal system monomitic; generative hyphae with clamps, thin-walled, with branching, 3-5 μm wide Gloeocystidia abundant, cylindric to ventricose, apex rounded to fusoid, highly refractive in Melzer's reagent, imbedded or projecting slightly, 35-50 ×7.5-12 μm, with a basal clamp. Basidia clavate, 4-sterigmate, 30-40 ×7.5-10 μm. Basidiospores ellipsoid, smooth, hyaline, non amyloid, 8-10.5 ×5-7 μm. **Habitat :** On fallen branch. **Distribution :** Africa, Europe, North America, Japan. **Specimen examined :** Benzidine test positive, white rot on *Mangifera indica* Linn. **Location :** Panshet; Dist: Pune (13/01/2007), VH-NPO-595. **Culture characters :** Growth characters: Growth rate very fast; advancing zone raised; Mat cottony white turning ash coloured; Mat gives cushion appearance, filling the petriplate with a cushion of mat; Aerial mycelium absent; colony whitish to ash coloured; Reverse slightly bleached, with deep dark patches of brown colour in it; outline of colony even; Distance between hyphal tips dense. Microscopic characters: Generative hyphae thin-walled, hyaline, with clamps at septa, branched, 2-3.5 μm. wide, gloeocystidia cylindric, 20-30 × 5-7 μm, with a basal clamp. **Tests :** α-napthal, pyrogallol + H_2O_2 positive, p-cresol negative. **Species**

Code : 1, 2, 3, 6, 12, 14, 15, 16, 30, 37, 39, 52, 53, 73. **Culture No. :** PUCC. B: 19. **Remarks :** The specimen is recognized by the verruocose, ellipsoid-subglobose spores, and the presence of mostly enclosed cystidia. it is near to *H. punctulatum,* but then has smaller spores, in length.

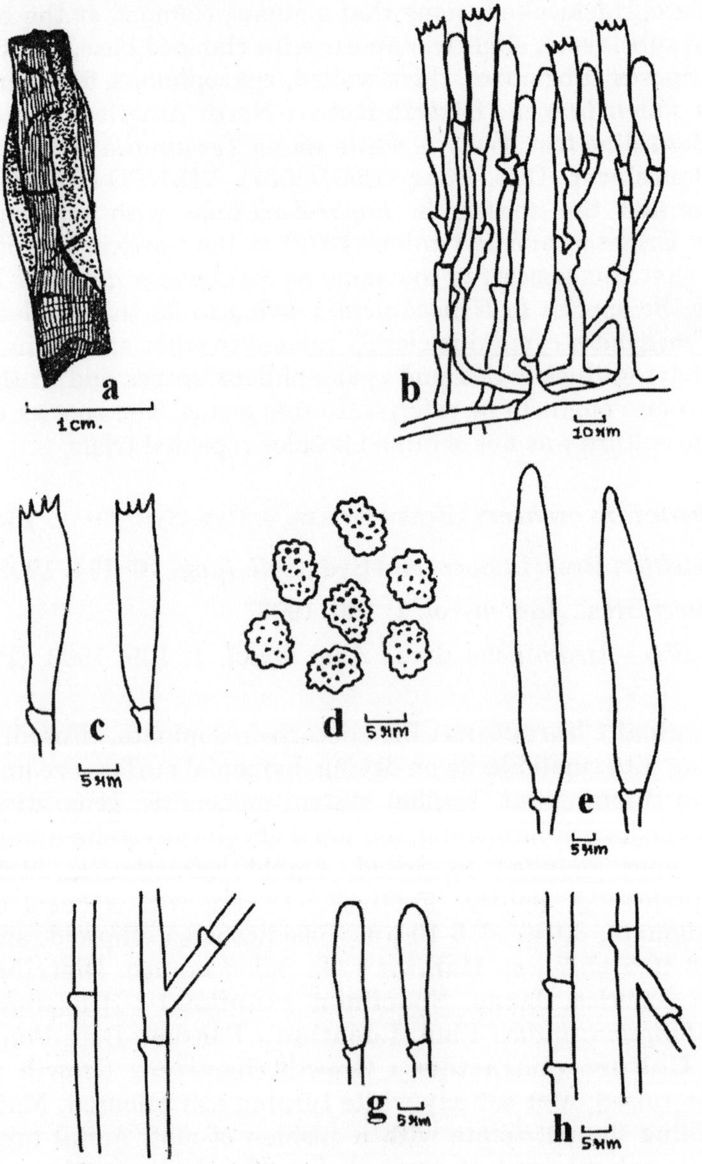

Fig. 38 ***Hypochnicicum eichlerie*** a. Habit. b. T.S. hymenium. c. Basidia. d. Basidiospores. e. Gloeocystidia. f. Generative hyphae. g. Gloeocystidia in culture. h. Generative hyphae in culture

(13-3). *Hypochnicium globosum* Wu. *Acta Bot. Fenn.* 142: 110, 1990. (Plate 3, 9; Fig. 39)

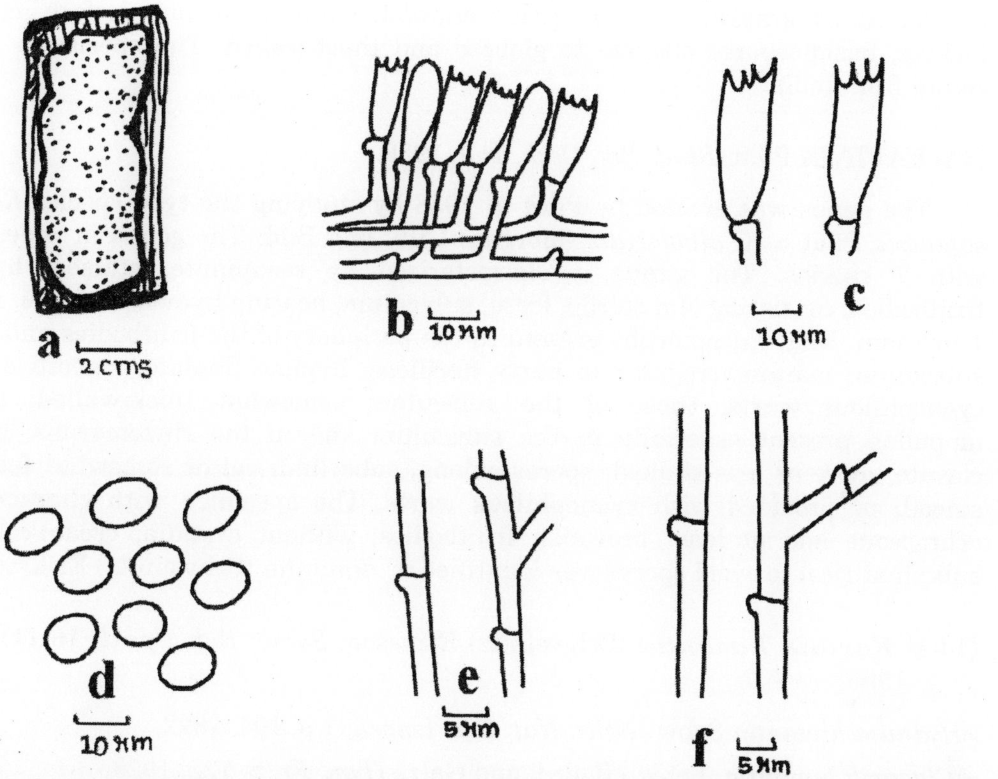

Fig. 39 *Hypochnicicum globosum* a. Habit. b. T.S. hymenium. c. Basidia. d. Basidiospores. e. Generative hyphae. f. Generative hyphae in culture.

Morphological Characters : Basidiomata resupinate, adnate, effused, 7-9×3-5 cms., hymenial surface grayish white to pale yellowish white, smooth; margin concolorous with the hymenial surface, thinning out, indeterminate. In vertical section, subhyaline, membranous, with a basal layer composed of horizontally parallel hyphae. Hyphal system monomitic; hyphae 4-5 μm wide, smooth, thin- to slightly thick-walled, with clamps Cystidia lacking. Basidia subclavate to suburniform, sometimes sinuous, 30-40 × 9-10.5 μm, with a basal clamp, producing 4 sterigmata. Basidiospores obovate to globose, 9-12 ×8-9.5 μm wide, smooth, thick-walled, non-amyloid, non-cyanophilic. **Habitat :** On fallen branch. **Distribution :** Africa, Europe, North America, Japan. **Specimen examined :** Benzidine test positive, white rot on *Pavetta indica* Linn. **Location :** Singhgad, Dist. Pune (13/01/2007), VH-NPO-495. **Culture characters :** Growth characters: Growth rate moderate; advancing zone appeased to submerged; Mat cottony white, cottony to wooly, mycelium white; Marginal hyphae distant to dense; outline of colony uneven to fringed slightly; Reverse bleached. Microscopic characters: Generative hyphae, thin-walled hyaline, branched, clamped, 3-5 μm., wide. **Tests :** α-napthal, pyrogallol+ H_2O_2 and KOH positive; p-cresol negative. **Species Code :** 1, 2, 3, 7, 13, 14, 15, 21, 22, 30, 37, 39, 53. **Culture No :** PUCC. B: 20. **Remarks:** The specimen is recognized by grayish white to pale yellowish white hymenial surface, cystidia lacking, basidiospores obovate to globose and thick-walled. The species is a new record from India.

(14) *KAVINIA* Pilat. *Stud. Bot. Cech.* **1:** 3, 1938.

The genus was erected by Pilat in 1938 by studying the type species *Kavinia sajanesis* Pilat = *K. alboviridis* (Morgan) Gilb. and Bud. The genus is widespread with 7 species. The genus is characterized by resupinate, effused, hydnoid fruitbodies, consisting of a sterile, loose, subiculum, bearing hymenial aculei, mostly 1 to 5 mm. long; rhizomorphs present in the periphery of the fruitbodies and in the subiculum; margin irregular to party fibrillose; hyphae fibulate, smooth or with cyanophilous warts, those of the subiculum somewhat thick-walled; hyphal ampullae present especially in the subiculum and in the rhizomorphs; basidia clavate, more or less stalked; spores oblong, subcylindrical or somewhat fusiform, smooth or provided with cyanophilous warts. The specimen with characteristic ochraceous and at least brownish fruitbodies; without cystidia; cream-coloured, subcylindirical to ovoid spores was identified *K. himantia* (Schweinitz) Eriksson.

(14-1) *Kavinia himantia* (Schweinitz) Eriksson. *Symb. Bot. Upsal.* **16 (1) :** 160, 1958.

=*Hydnum himantia* Schw., *Schr. Nat. Ges. Leipzig* 1 p. 104, 1822.

= *Clavaria himantia* (Schw.) Bourd. and Galz., *Hym. Fr.* p. 122, 1928.

= *Hydnocristella himantia* (Schw.) Petersen, *Ceska Mykol.* 25 p. 130, 1971. (Plate 4, 9, Fig. 40)

Morphological Characters : Fruitbody resupinate, effused, loosely adnate, 20-23 × 4-6 cms., consisting of a loose, ochraceous and at last brownish;hymenial aculei., cylindrical, slightly tapering towards the apex which is rather obtuse in the mature fungus, margin white, variable, often fibrillose;rhizomorphs present in the periphery and in the subiculum. Hyphal system monomitic;hyphae 3-4 μm, fibulate, distinct, thin walled in young parts, somewhat thick walled in mature hyphae, as a rule with dense cyanophilous warts. Cystidia absent. Basidia clavate, 25-35 ×6-8 μm, with 4 sterigmata and basal clamp; oildrops in the protoplasm. Spoers hyalin or cream-coloured, subcylindrical to narrowly ovoid, smooth, thin-walled, 8-10 (-12) ×4-5 μm. **Habitat :** On live standing tree. **Distribution :** Ontario, British Columbia, southwestern U. S. A. **Specimen examined :** Benzidine test positive, brown rot on *Artocarpus heterophyllus* Lam. **Location :** Karad, Dist. Satara (12/10/2005), VH-NPO-197. **Culture characters :** Growth characters: Growth rate moderate; advancing zone; raised; Mat cottony white; Reverse slightly bleached; Aerial mycelium absent; Colony white; Outline of colony even; Distance between the hyphal tips dense. Microscopic characters: Generative hyphae, hyaline, smooth, clamped 3-3.5 μm, wide, thick-walled hyphae wide, 3.5-4 μm., smooth, branched. **Tests :** α-napthal and pyrogallol + H_2O_2 positive; p-cresol negative. **Species Code** :1, 2, 3, 7, 12, 14, 15, 16, 30, 37, 39, 53. **Culture No :** PUCC. B:21. **Remarks :** Distinguished from *K. alboviridis* preferably in the smooth spores and asperulate hyphae. The hymenium aculei are smaller and more subulate in *K. alboviridis*. The species is recorded for the first time from India.

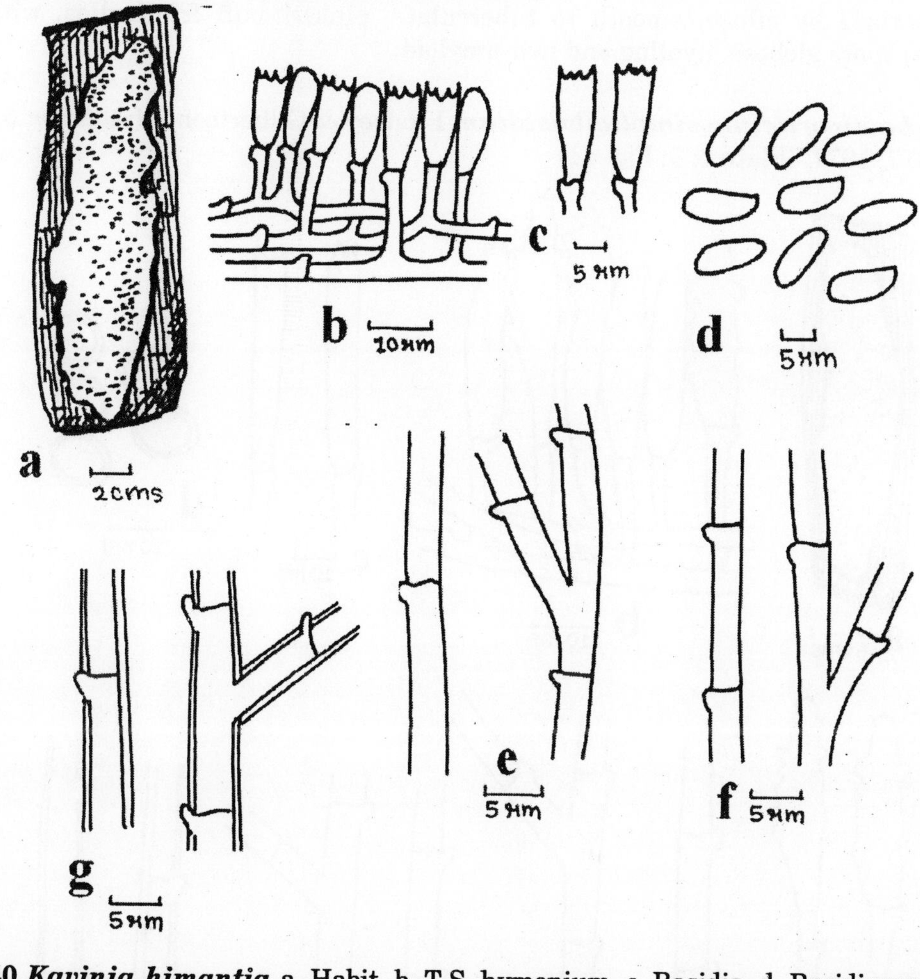

Fig. 40 ***Kavinia himantia*** a. Habit. b. T.S. hymenium. c. Basidia. d. Basidiospores. e. Generative hyphae. f. Generative hyphae in culture. g. Thick-walled generative hyphae in culture

(15). *LAETICORTICIUM* Donk. *Fungus* **26:** 16, 1956.

The genus *Laeticorticium* was erected by Donk in 1956 with the type species *Laeticorticium roseum* (Fr.) Donk. The genus was erected to encompass those taxa of corticiaceous fungi that possess a catahymenium consisting of dendrohyphidia, basidia that at an early stage in their development produce a thick to thin-walled probasidium, basidia originating deep in subiculum as probasidia, becoming greatly elongated while growing through a previously developed layer of dendrohyphidia with more or less intricate terminal branching. It is also characterized by clamped hyphae, smooth hyaline, non-amyloid, ellipsoid to broadly ellipsoid spores. There are 31 species recorded under the genus. The specimen collected from Pune, is characterized by effuse, smooth to tuberculate, pinkish-buff fruitbodies, without cystidia; spore globose, hyaline and non-amyloid.

(15-1). *Laeticorticium simplicibasidium* Lindsey & Gilbertson. *Mycotaxon* **5 (1) :** 317, 1977. (Plate 4, 9; Fig. 41)

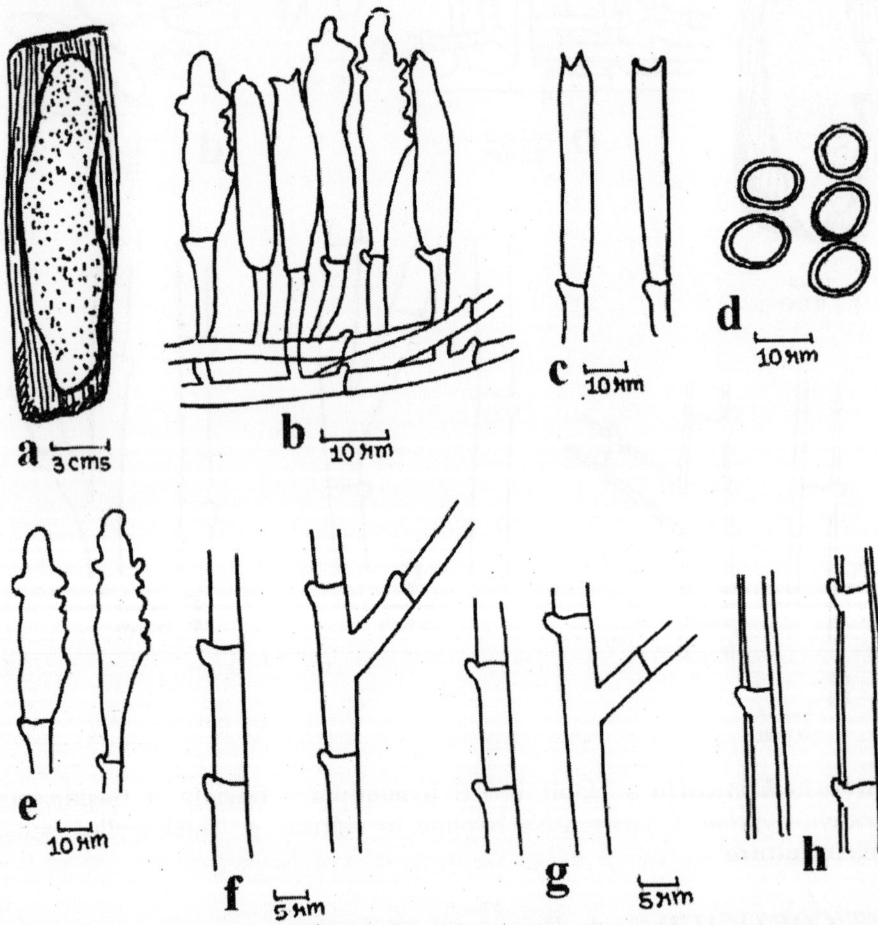

Fig. 41 *Laeticorticium simplicibasidium* a. Habit. b. T.S. hymenium. c. Basidia. d. Basidiospores. e. Generative hyphae. f. Marginal and aerial hyphae in culture. g. Submerge hyphae in culture. h. Hyphae with clamp

Morphological Characters : Basidiocarp effused, 12-15 × 3-5 cms., adnate; hymenial surface pinkish-buff at first, becoming reddish-purple, smooth to tuberculate; margin thinning out, pink ish-buff, entire to minutely fimbriate; subiculum dark purplish with a pale pinkish-buff layer next to the subtratum, less than 1 mm thick. Hyphal system monomitic; thin-walled, with clamp, with occasional branching, 3-5 μm diam. Cystidia none. Dendrohyphidia abundant in catahymenium, thin-walled, branched, with occasional clamps, 60-80 × 10-15 μm. Basidia elongated, narrow, unbranched or occasionally bifurcate, thin-walled, 70-90 × 3-5 μm, tapering at the apex to a single sterigma. Basidiospores subglobose, thin-to thick-walled, smooth, 9-11 μm diam, non-amyloid. **Habitat :** On fallen branches. **Distribution :** U. S. A. **Specimen examined :** Benzidine test positive, white rot on *Pongamia pinnata* (Linn.) Pierre. **Location :** Pune university, Dist. Pune (02/09/2006), VH-NPO-534. **Culture characters :** Growth characters: Growth rate moderate; advancing zone raised; Mat cottony white, Aerial mycelium absent; Distance between hyphal tips distant; Colony white. Outline of colony uneven; Reverse unchanged. Microscopic characters: Marginal and aerial hyphae same, branched, with clamps, 3.5-5.5 μm. wide, submerg hyphae thin to thick-walled, hyaline, clamped, 3-6 μm. wide. **Tests :** α-napthal and pyrogallol + H_2O_2 positive; p-cresol negative. **Species Code :** 1, 2, 3, 8, 12, 14, 15, 16, 30, 39, 53, 54. **Culture No :** PUCC. B: 22. **Remarks :** The species is placed in *Laeticorticium* because the thick-walled, hyphae with clamp, catahymenial development, dendrohyphidia, thick walled probasidia, and general macroscopic appearance are all typical of that genus. The species is a new record from India.

(16). *LAXITEXTUM* Lentz. *U S. Dep. Agric. Monograph* **24:** 18, 1955.

The genus *Laxitextum* was described by Lentz in 1955 with its type species *Laxitextum bicolor* (Persoon) Lentz. The genus was later emended and restricted by Boidin in 1958. The type species *L. bicolor* is widespread and illustrared by Eriksson and Ryvarden in 1976. Talbot reported *L. bicolor* from S. Africa, Nigeria, Kenya and Uganda. The genus is widespread in distribution with seven species known under the genus. The genus is characterized by effuse to effuse reflexed basidiomata, typically several cm long and broad; smooth, pale yellow to grayish white hymenial surface; pallid to brown, thin to nearly 1 mm thick, densely cottony, duplex subiculum; typically arranged in a layer parallel to and adjacent to the substrate generative hyphae, then becoming perpendicular to the substrate and slightly woven between the gloeocystidia and developing hymenium, with clamp; cylindrical, sinuous, the contents oily to granular and pale yellow in KOH gloeocystidia; ellipsoid to broadly ellipsoid, amyloid, acyanophilous, verruculose to warty spores. The specimen collected from NCL Park, Dist. Pune is identified as *L. lutescens* Hjortstam & Ryvarden, due to the presence of deep yellow to pale brown velutinate, azonate pileus; numerous, arising from the tramal layer, fusiform gloeocystidia; sub-globose to ellipsoid, echinulate, strongly amyloid spores.

(16-1) *Laxitextum lutescens* Hjortstam & Ryvarden. *Mycotaxon* **13 (1)** : 40, 1981. (Plate 4, 9 Fig. 42)

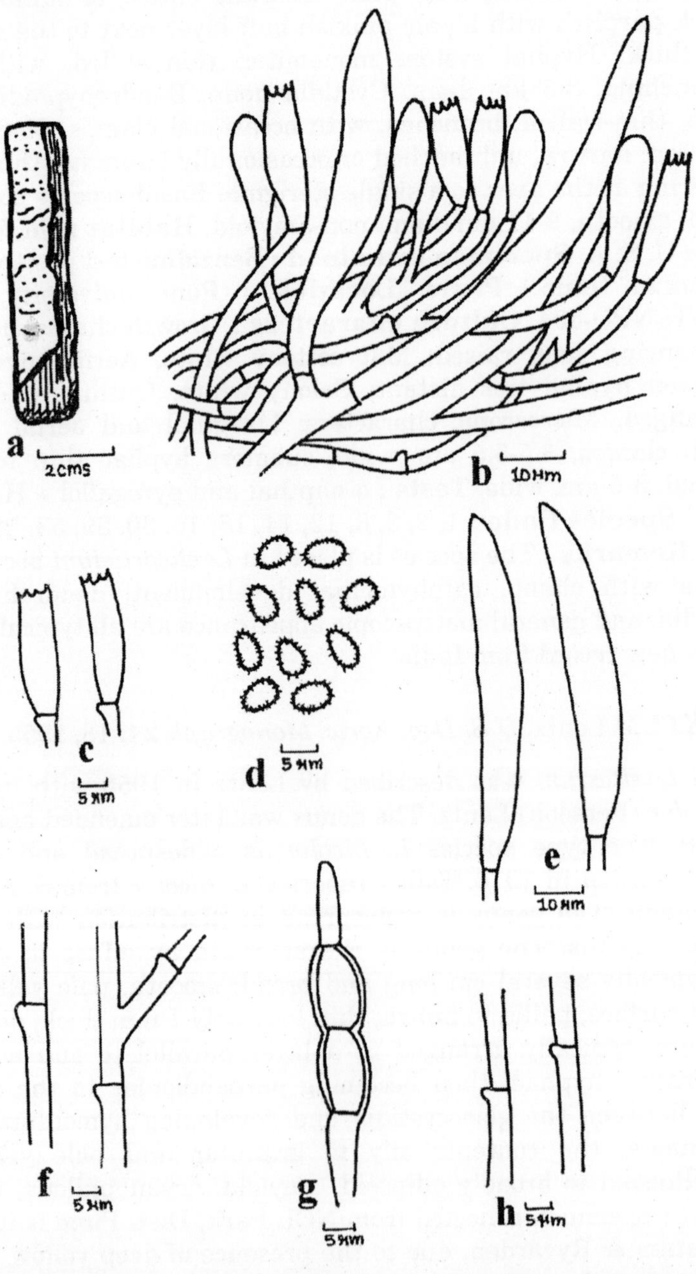

Fig. 42. *Laxitextum lutescens* a. Habit. b. T.S. hymenium. c. Basidia. d. Basidiospores. e. Gloeocystidia. f. Generative hyphae. g. Chlamydospores in culture. h. Generative hyphae in culture

Morphological Characters : Fruitbody resupinate to distinctly reflexed, 10-12 × 2-3 cms., and 0. 4-0. 6 mm. thick, pileus deep yellow to pale brown velutinate, azonate, trama well developed, pale yellowish brown, hymenium cream-yellowish to strawcoloured with a light purple-brown tint, slightly cracked. Hyphal system monomitic, pale yellowish brown, 3-5 μm wide, smooth and with clamps. Gloeocystidia numerous, arising from the tramal layer, fusiform, with the apex obtuse, now and then with schizopapilles, 80-100 × 7-9 μm, distinctly amyloid, but negative in benzaldehyde. Basidia narrowly clavate, 20-25 × 4 μm, with four sterigmata. Spores subglobose to ellipsoid, echinulate, 4 × 3 μm, strongly amyloid. **Habitat :** On live standing tree. **Distribution :** Africa, Ghana **Specimen examined :** Benzidine test positive, white rot on *Poamia pinnata*. **Location :** NCL park, Dist. Pune (19/09/2005), VH-NPO-46. **Culture characters:** Growth characters: Growth rate moderate; advancing zone appressed to submerged; Mat downy to low cottony white getting submerged in agar as colony advances giving the plate and hence the colony, a transparent look; colony white; outline of colony even; Distance between hyphal tips distant; Aerial mycelium absent; Reverse bleached. **Microscopic characters:** Generative hyphae branched, 4-5 μm, wide, clamped, aerial hyphae narrow, 2-3 μm wide, Chlamydospores present intercalary, 10-15 × 5.5-8 μm. **Tests :** α-napthal and pyrogallol + H_2O_2 positive; p-cresol negative. **Species Code :** 1, 8, 9, 13, 17, 19, 21, 26, 30, 31, 39, 52, 85. **Culture No :** PUCC. B: 23. **Remarks:** *Laxitextum lutescens* seems to be wel located in the genus and is similar to L. bicolor, but delimited by its amyloid gloeocystidia and echinulate spores. The species is a new record from India.

(17). ***LEPIDOMYCES*** Jülich. *Persoonia* 10 (3) : 329, 1979.

The genus was erected by Jülich in 1979. The genus is with only one species. The genus is characterized by resupinate, effused, thin, white fruitbodies; monomitic, thin-walled, clampless, densely branched hyphal system; absent cystidiole; abundant, mostly thick-walled, oblong, strongly encrusted cystidia. The specimen with characteristic thin-walled, clampless, densely branched hyphae and abundant, mostly thick-walled, oblong, strongly encrusted cystidia was identified as *L. subcalceus* (Litschauer) Jülich.

(17-1) ***Lepidomyces subcalceus*** (Litschauer) Jülich. *Persoonia* **10 (3) :** 330, 1979.

=*Peniophora subcalcea* Litsch., *Öst. bot. Z.* **88:** 119 (1939)

=*Phlebia subcalcea* (Litsch.) M. P. Christ., *Dansk bot. Ark.* **19 (2) :** 171 (1960)

=*Xenasma subcalceum* (Litsch.) Oberw., *Sydowia* **19:** 42 (1966) [1965] (Plate 4; Fig. 43)

Morphological Characters : Fruitbidy resupinate, effused, 7-15×2-3 cms., thin, white. Hyphal system monomitic, hyphae 2-3 μm. wide, thin-walled, clampless, densely branched, hyaline. Cystidiol absent. Cystidia abundant, mostly thick-walled, strongly encrusted in the upper part, 30-60×4-5 μm. Basidia clavate, 20-25×4-5 μm. Spores cylindrical, 7-8 × 2-3 μm., hyaline, thin-walled, smooth and non-amyloid. **Habitat :** On fallen branches. **Distribution :** Denmark, Austria. **Specimen examined :** Benzidine test positive, white rot on *Eugenia jambolana* Lam. **Location :** Sawantwadi, Dist. Satara (19/02/2007), VH-NPO-785. **Remarks :** The species is recognized by thin-walled, clampless, densely branched hyphae and abundant, mostly thick-walled, oblong, strongly encrusted cystidia. The species is a new record from IndiaThe culture was not obtained besides repeated trials.

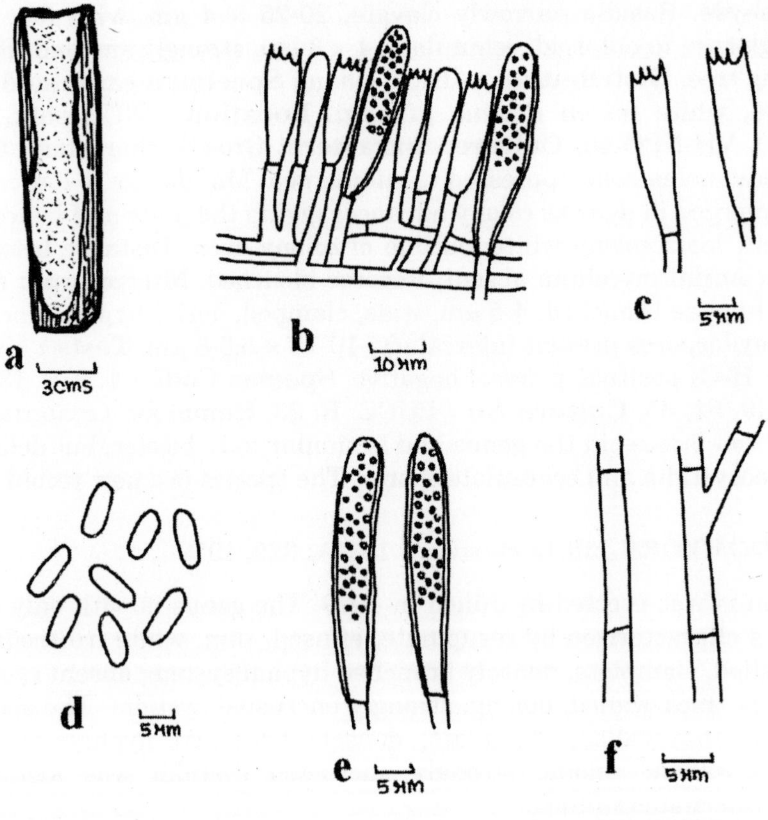

Fig. 43 ***Lepidomyces subcalceus*** a. Habit. b. T.S. hymenium. c. Basidia. d. Basidiospores. e. Thick-walled encrusted cystidia. f. Generative hyphae

(18) *LEUCOGYROPHANA* Pouz. *Ceska Myk.* 12: 32, 1958.

The genus was erected by Pouzer in 1958 by studying the type species *Merulius molluscus* Fr. The genus is widespread with 19 species known. The delimitation of *Leucoggyrophana* in the opinion of Paramasto is problematic. The only characters in common with corticiaceae seem to be an allantoid fruitbodies with fibulate hyphae, ellipsoid, thick-walled, mostly cyanophilous spores about the size of 5-7 μm. which react with Melzer's reagent. The genus is characterized by fruitbodies resupinate, effused, more or less athelioid i.e., hymenial part forming a thin, continuous, easily detached layer, smooth or merulioid, white-yellow to orange red; all hyphae fibulate, subicular ones distinct, straight, sparsely branched, subhymenial hyphae dense, richly branched; cystidia lacking or sometimes thin-walled, cylindrical cystidia present, basidia clavate, normally with four sterigmata; spores broadly or narrowly ellipsoid, smooth, thick-walled, cyanophilous in all species, in Melzer's unchanged, grey or yellow brown, not blue or violet as in strictly amyloid species; spores of the generic type dextrinoid. The specimen collected from Pune, is characterized by resupinate, effused, thin, loosely attached to the substrate, when fresh and alive often soft and meruloid, when dried smooth and brittle, herbarium material mostly pale ochreaceous, spores narrowly elliptical, smooth, adaxial side mostly straight, rarely somewhat concave, and grey in Melzer, obvious especially when several spores lay together, mature spores yellowish in KOH.

(18-1) *Leucogyrophana mollis* (Fries) Parmasto. *Eesti NSV Tead. Akad. Toim. Biol.* **16:** 385, 1967.

=*Thelephora mollis* Fr., *Syst. mycol.* (Lundae) **1:** 443, 1821. (Plate 4, 9, Fig. 44)

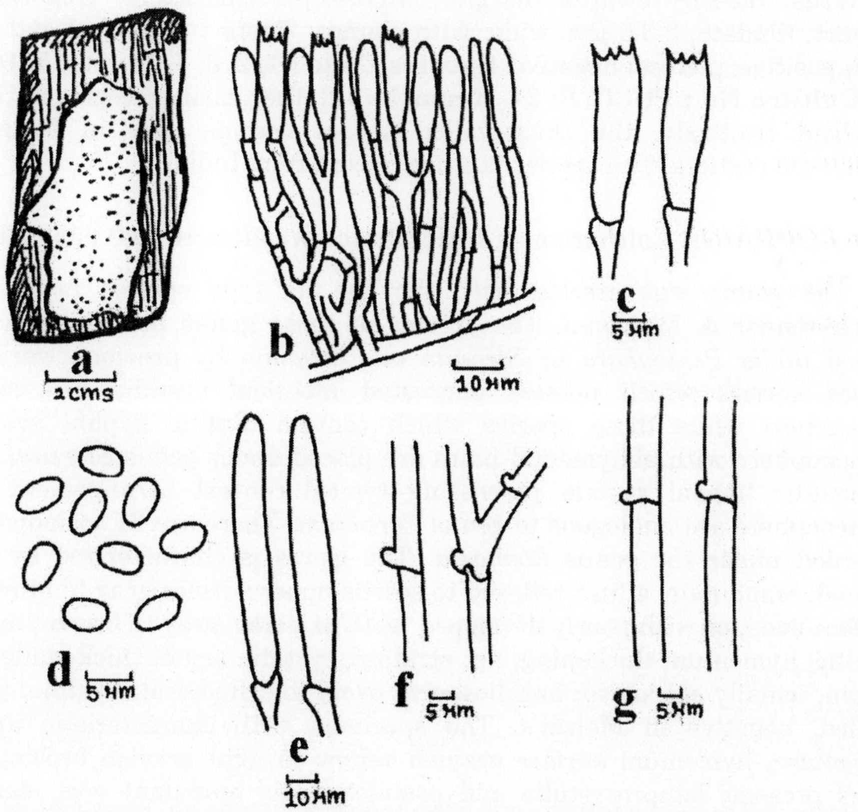

Fig. 44 *Leucogyrophana mollis* a. Habit. b. T.S. hymenium. c. Basidia. d. Basidiospores. e. Cystidia. f. Generative hyphae. g. Generative hyphae in culture

Morphological Characters : Fruitbodies resupinate, effused, 5-6 × 2-3 cms., thin, loosely attached to the substrate, when fresh and alive often soft and meruloid, when dried smooth and brittle, in the living state as young be whitish, later greyish or pale ochreaceous, sometimes with a reddish tint;hymenium in the mature fungus continuous, when dried often cracking, thinning out into the arachnoid margin. Hyphal system monomitic, hyphae 3-6 μm wide. Cystidia few, sometimes difficult to find, maybe even lacking, cylindrical, obtus, with thin or somewhat thickened walls, 75-100 × 6-9 μm. Basidia clavate, 20-25 × 7-10 μm. Spores 5. 5-7 × 2.5 - 3.5 μm, narrowly elliptical, smooth, with thickened walls, grey in Melzer. **Habitat :** On live standing tree. **Distribution :** Sweden. **Specimen examined :** Benzidine test negative, brown rot on *Ficus bengamina* L. **Location :** Film Institute; Dist: Pune (20/10/2005), VH-NPO-263. **Culture characters :** Growth characters: Growth rate moderately fast; advancing zone appressed; Mat cottony white to grayish lilac, plumose, floccose slightly at margin tips, prostrate over surface; Distance between the hyphal tips dense. Outline of colony bayed; Aerial mycelium absent; Reverse bleached with dark brown patches concentrically at centre and margins radiating outwards, darker towards margin. Microscopic characters: Generative hyphae distinct, fibulate, 3-3.5 μm. wide, with clamps. **Tests :** α-napthal and pyrogallol + H_2O_2 positive; p-cresol negative. **Species Code :** 1, 2, 3, 7, 13, 14, 15, 16, 30, 37, 39, 53. **Culture No :** PUCC. B: 24. **Remarks :** Rather easily recognized thanks to the athelioid fruitbody, the thick-walled spores become grey in Melzer and the cylindrical cystidia. The species is a new record from India.

(19). *LOPHARIA* Kalchbrenner & MacOwan. *Grevillea* **10 (54) :** 58, 1881.

The genus was erected with the the its type species *Lopharia lirellosa* Kalchbrenner & Macowan. The species of those genus have been several times placed under *Peniophora* or *Stereum* or *Corticium* by previous workers because thoses species which possess encrusted metuloid cystidia are considered as *Peniophora* while those species which contain dimitic hyphal system, pileate hymenophore with abhymenial hairs are placed under genus *Stereum* species with monomitic hyphal system possessing typical context arrangement of corticoid hymenophore are analogous to genus *Corticium*. There are 31 cosmopolitan species recorded under the genus *Lopharia.* The genus is characterized by basidiomata annual; resupinate, effuso-reflexed to sessile;upper surface gray to brown; hymenial surface even, or with poorly developed teeth or pores; gray to brown; hyphal system dimitic; hymenium thickening, not stratose; cystidia acute, thick-walled, hyaline to brown, usually encrusted; basidiospores ovoid to cylindrical, hyaline, smooth; thin-walled, negative in Melzer's. The specimen with characteristic upper surface tomentose, hymenium surface grayish yellow to light grayish brown; abhymenial hairs present; lamprocystidia and pseudocystidia abundant was identified as *L. fulva* (Lév.) Boidin.

(19-1). *Lopharia fulva* (Lév.) Boidin. *Bull. Mens. Soc. Linn. Lyon* **28:** 213, 1959.
=*Thelephora fulva* Lév., *Annls Sci. Nat.*, Bot., sér. **3 5:** 149, 1846. (Plate 4, 9, Fig. 45)

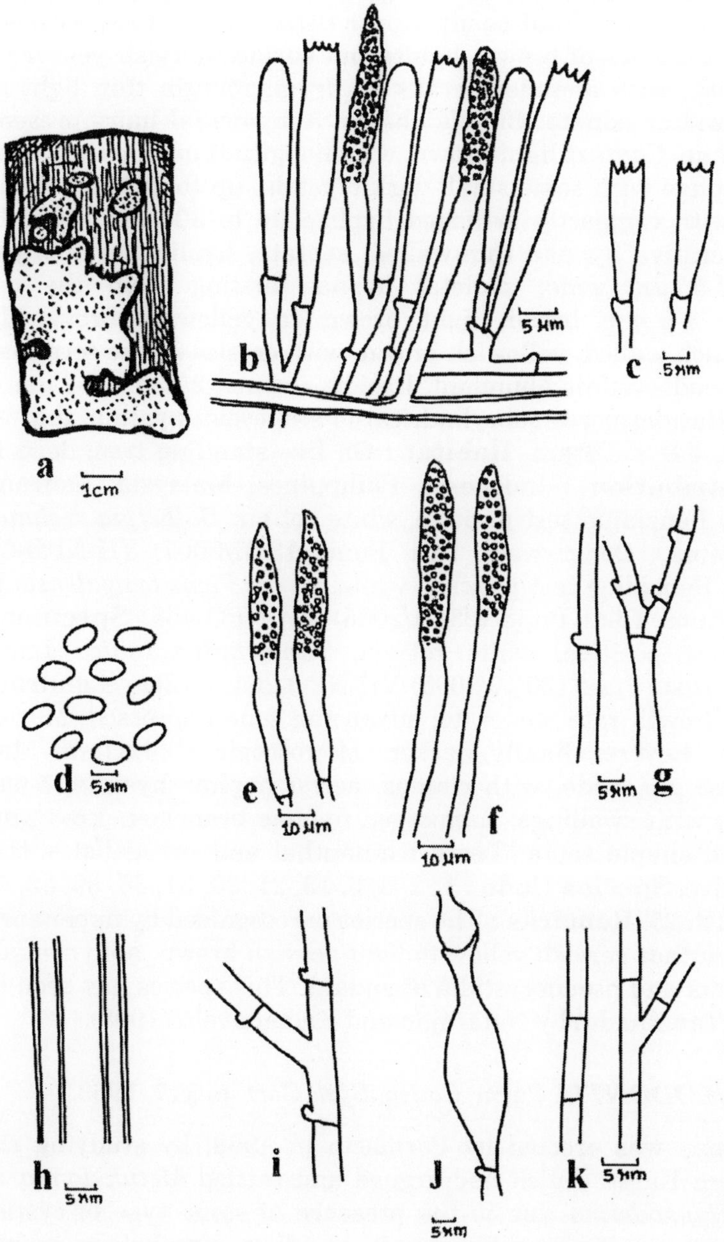

Fig. 45 *Lopharia fulva* a. Habit. b. T.S. hymenium. c. Basidia. d. Basidiospores. e. Cystidia. f. Pseudocystidium. g. Generative hyphae. h. Skeletal hyphae. i. Marginal hyphae in culture. j. Aerial hyphae in culture. k. Submerge hyphae in culture

Morphological Characters : Basidiomata annual, effused to effuso-reflexed, 25-30 × 4-7 cms., and less than 1 mm. thick, firmly attached, memberaneous to coriaceous, soft to leathery, appearing in the form of small patches initially and later becoming a large frout boady, upper surface tomentose, with concentric lines, with different shades of brown; hymenium surface geryish yellow to light geryish brown, smooth, with abundant craks on drying;margin thin light geryish yellow, free or reflexed or adnate. Cuticle absent. Abhymenial hairs present, mostly with skeletal hyphae. Context light brown, with liongitudinally, parallely ande compactly arranged hyphae with some small dark crystals, up to 250 μm; subhymenium and hymenium with compactly arranged hyphae up to 300 μm thick. Hyphal system dimitic; generative hyphae thin-walled, septate, hyaline to subhyaline, clamped, branched, 3-5 μm wide, skeletal hyphae arising from the context, mostly unbranched, 3-6 μm broad, light brown to yellowish brown. Lamprocystidia abundant, thich walled, yellowish brown, with crystal in the terminal region, 40-60 × 6-8 μm. Pseudocystidia abundant. Basidia clavate, 25-35 × 6-7 μm, sterigmata 2-4 in number. Basidiospores subcylindrical to elliposoid, hyaline, thin-walled, smooth, non-amyloid, 6-8 × 3-4 μm. **Habitat :** On live standing tree, dead tree and fallen branch. **Distribution :** Indonesia, Philippines, Malaysia, Tanzania. **Specimen examined :** Benzidine test positive, white rot on: *Dalbergia melanoxylon* Guill. & Perr. **Location :** Dongerwadi; Dist: Pune (13/01/2007) VH-NPO-624, **Specimen examined :** Benzidine test positive, white rot on: *Ficus bengalensis* Linn **Location :** Film Institute; Dist: Pune (23/01/2006) VH-NPO-455. **Specimen examined :** Benzidine test positive, white rot on: *Tamarindus indica* Linn. **Location :** Baneshwar; Dist: Pune (20/10/2005) VH-NPO-264. **Culture characters :** Growth characters: Growth rate moderate; advancing zone appressed; Mat cottony whitish to creamish. Reverse finally darker. Microscopic characters: Marginal hyphae branched, 3-5 μm, wide, with clamps, aerial hyphae hyaline, 3-6μm, wide, thin-walled often with swellings; submerged hyphae branched, 1.5-3 μm, wide, clamps lacking, with simple septa. **Tests :** α-napthal and pyrogallol + H_2O_2 positive; p-cresol negative. **Species Code :** 1, 2, 3, 6, 13, 21, 30, 31, 38, 39, 52, 53, 54. **Culture No :** PUCC. B: 25. **Remarks :** The species is recognized by upper surface tomentose, hymenium surface geryish yellow to light geryish brown, abhymenial hairs present, lamprocystidia and pseudocystidia abundant. This species has been earlier reported from India (Tamilnadu) by Natarajan and Kolandavelu (1998).

(20). *METULODONTIA* Parm. *Consp. Syst. Cort.* p. 117, 1968.

The genus was erected by Parmasto in 1968, by studying the type species *Kneiffia nivea* Karst. Julich had argued and settled *Metulodontia* as synonyms of the genus *Hyphoderma* due to the presence of same type of cystidia in both the genera and no significant difference in basidiole morphology, as pre him. But on critical observation, basidia do differ from those of *Hyphoderma*, in being smaller, narrowly clavate and not very slightly constricted. The spores too have none of the species of *Metulodontia* has different cystidia. There are 11 species recorded under the genus. The genus characterized by resupinate, adnate, effused fruitbodies; cystidia varying in size and number, strongly encrusted; thin walled gloeocystidia, turning blue in sulfovanilline (sulfocystidia); basidia small; spores ellipsoid, non-amyloid, non-cyanophilous. The specimen with characteristic "Buff", grayish yellow to "Ochreous" hymenial surface; generative hyphae clampless, ellipsoid to broadly ellipsoid, adaxially flattened spores was identified *M. flavidoalba* (Cooke) Malençon & Bertault.

(20-1). *Metulodontia flavidoalba* (Cooke) Malençon & Bertault. *Acta phytotax. barcin.* **19:** 34, 1977.

=*Peniophora flavidoalba* Cooke, *Grevillea* **8** (**45**) : 21, 1879. (Plate 4, 9 Fig. 46)

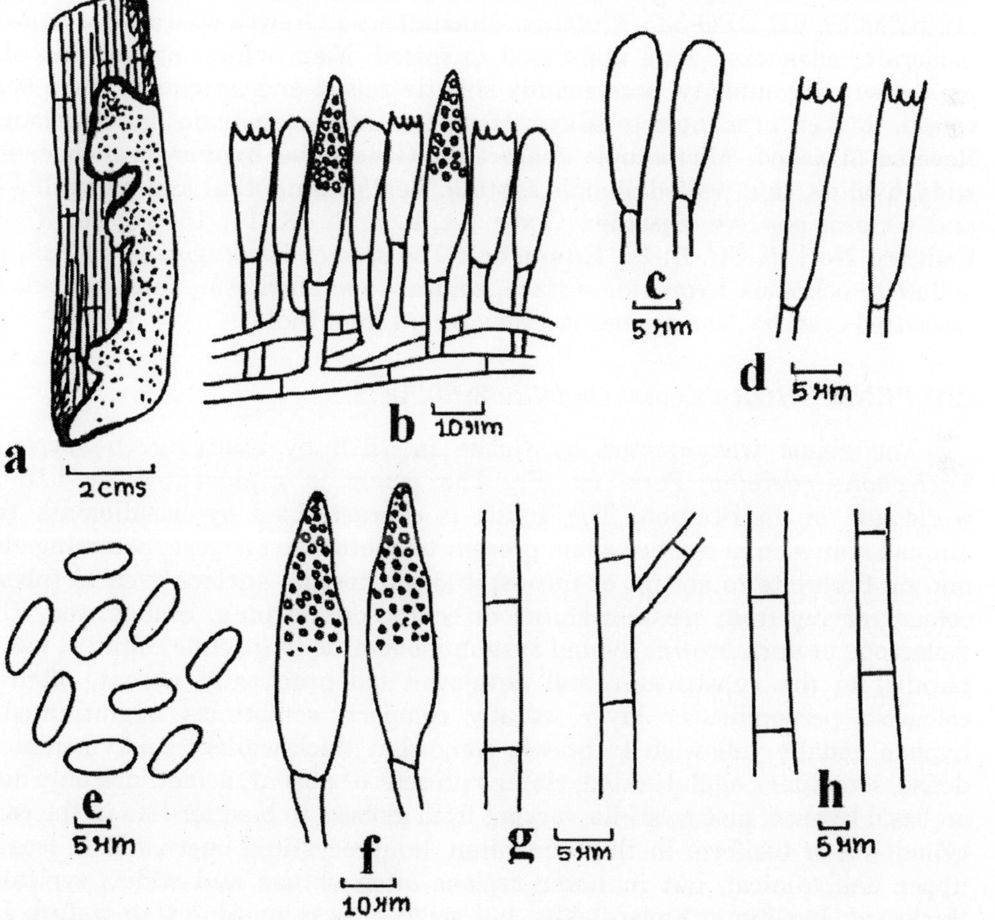

Fig. 46 *Metulodontia flavidoalba* a. Habit. b. T.S. hymenium. c. Basidiole. d. Basidia. e. Basidiospores. f. Encrusted cystidia. g. Generative hyphae. h. Generative hyphae in culture

Morphological Characters : Basidiomata resupinate, adnate, effused, 7-9 × 1-2 cms. and 100-250 μm thick: hymenial surface buff, grayish yellow to ochreous, smooth, margin concolorous with the hymenial surface, thinning out, indeterminate, without hyphal strands. In section subhyaline, membranous to subceraceous, with a compact subhymenium composed of agglutinated and thin-walled hyphae. Hyphal system monomitic, hyphae 2-6 μm. wide, smooth, clampless, thin- to thick-walled. Cystidia (lamprocystidia) conical to subfusiform, 55-80 × 9-14 μm, smooth, thick-walled, heavily encrusted with subhyaline crystalloid. Basidia clavate, 25-35 × 5-6 μm, producing 4 sterigmata. Basidiospores ellipsoid to broadly ellipsoid, 5.5-7.5 ×3. 5-4 μm, smooth, thin-walled, non-amyloid;basidioles clavate, up to 25 × 5 μm. **Habitat :** On live standing tree. **Distribution :** Southeastern United States, Texas, North Carolina, South America. **Specimen examined :** Benzidine test positive, white rot on *Memecylon umbellatum* Burm. **Location :** Dapoli, Dist. Ratnagiri (12/10/2005), VH-NPO-335. **Culture characters :** Growth characters: Growth rate moderate; advancing zone appressed to raised; Mats white, around inocula thin, appressed, and subfelty, occasionally slightly raised and arachnoid, then becoming raised, thicker, arachnoid to silky-cottony toward margins.; outline of colony even;; Reverse bleached. Microscopic characters: Generative hyphae branched, 6-8 μm, wide, hyaline, thin walled, simple septate. **Tests :** α-napthal and pyrogallol + H_2O_2 and p-cresol positive. **Species Code :** 1, 2, 3, 8, 13, 14, 15, 16, 30, 37, 54, 55. **Culture No :** PUCC. B: 26. **Remarks :** The species is recognized by buff, grayish yellow to ochreous hymenial surface, conical to subfusiform, thick-walled, heavily encrusted cystidia. The species is a new record from India.

(21) *PENIPOHORA* Cooke. *Grevillea* **8:** 20, 1879.

The genus was erected by Cooke in 1879 by studying the type species *Thelephora quercina* Pers. ex. Fr. The genus is represented by 510 species worldwide in distribution. The genus is characterized by basidiomata typically annual; abhymenial surface when present tomentose to strigose, becoming glabrous; margin fimbriate to abrupt or inconspicuous;hymenial surface even to tuberculate; colour varying from cream-coloured or ochraceous to pink, orange, red, lilaceous, violaceous or dark brown; hyphal system monomitic to (pseudo) dimitic; basal layer parallel to the substratum well developed too practically absent, often darker coloured; perpendicular layer usually compact, sometimes agglutinated; basal hyphae usually yellowish to brown, somewhat thick-walled; subhymenial hyphae dense, sometimes agglutinated; clamps present or absent, sometimes only dispersed on basal hyphae; gloeocystidia varying from globose to bladder-like in the context to cylindrical or fusiform in the hymenium; lamprocystidia encrusted at least in the upper half, conical, but in lower regions often obtuse and wider, typically with thickened, hyaline to brown walls, but sometimes remaining thin-walled. It is not always possible to distinguish between young not yet encrusted lamprocystidia and gloeocystidia; dendrohyphidia sometimes present, hyaline to brownish, thin- to thick-walled, smooth or covered with crystalline material; basidia subclavate to subcylindrical, sometimes sinuous or with an indistinct constriction; spores hyaline, thin-walled to rarely slightly thick-walled, smooth, ellipsoid to cylindical, reniform or allantoid, not amyloid;spores in mass usually with pinkish or orange tinge. There were seven specimen collected from the investigation areas. The specimen with characteristic hyphae without clamps was identified as *P. laurentii* Lundell. The specimen with characteristic hyphae with clamps; globose to subglobose spores was identified as *P. gladiola* Cunningham. The specimen is identified as *P. farinosa* (Bresadola) Höhnel & Litschauer, due to the presence of grayish white hymenial surface, conical to sub-cylindrical tapering toward the apex and heavily encrusted cystidia. The specimen with characteristic presence of clamps;ellipsoid spores was identified as *P. aurantiaca* (Bresadola) Höhnel & Litschauer. The specimen with characteristic hymenium of the growing fungus reddish grey but variable with age and water content; cylindrical-suballantoid spores, spore-print light red was identified as *P. nuda* (Fries) Bresadola. The specimen with characteristic reddish or violaceous grey, when wet darker, sometimes bluish violaceous hymenial surface; allantoid, hyaline spores but reddish in spore print. was identified as *P. pithya* (Persoon) Eriksson. The specimen with characteristic very variable in surface colour, allantoid spores, spore-print pale red was identified as *P. violaceolivida* (Sommerfelt) Massee.

KEY TO THE SPECIES OF *PENIPHORA*

1. Generative htphae without clamps .... ***P. laurentii***

1′. Generative htphae with clamps 2

2. Spores globose to subglobose.............................. ***P. gladiola***

2′. Spores others shape 3

3. Spores ellipsoid 4

3′. Spores suballantoid to allantoid 5

4. Spores 15-20 × 10 µm. .... ***P. aurantiaca***

4′. Spores 6.5-7.5 × 4.5-5.5 µm. .... ***P. farinosa***

5. Sulfocystidia numerous, 20 µm width .... ***P. nuda***

5′. Sulfocystidia fewer, 10 µm width 6

6. Spore-print pale red, spores 7.5-9 × 3 µm .... ***P. violaceolivida***

6. Spore-print reddish, spores 6.5-7.5 × 2.5-3 µm .... ***P. pithya***

(21-1). *Peniophora aurantiaca* (Bresadola) Höhnel & Litschauer. *Sitzungsber. Akad. Wiss. Wien, Math. -Naturwiss. Kl.* **115:** 1583, 1586, 1906.

=*Byssomerulius* aurantiacus (Bres.) Gilb., *Fungi dec. Pond. Pine*: 44 (1974)

=Corticium aurantiacum Bres., Fung. trident. **2 (8-10) :** 37 (1892)

=*Kneiffia aurantiaca* (Bres.) Bres., *Annls mycol.* **1(1/2):**104(1903) (Plate 4,9, Fig. 47)

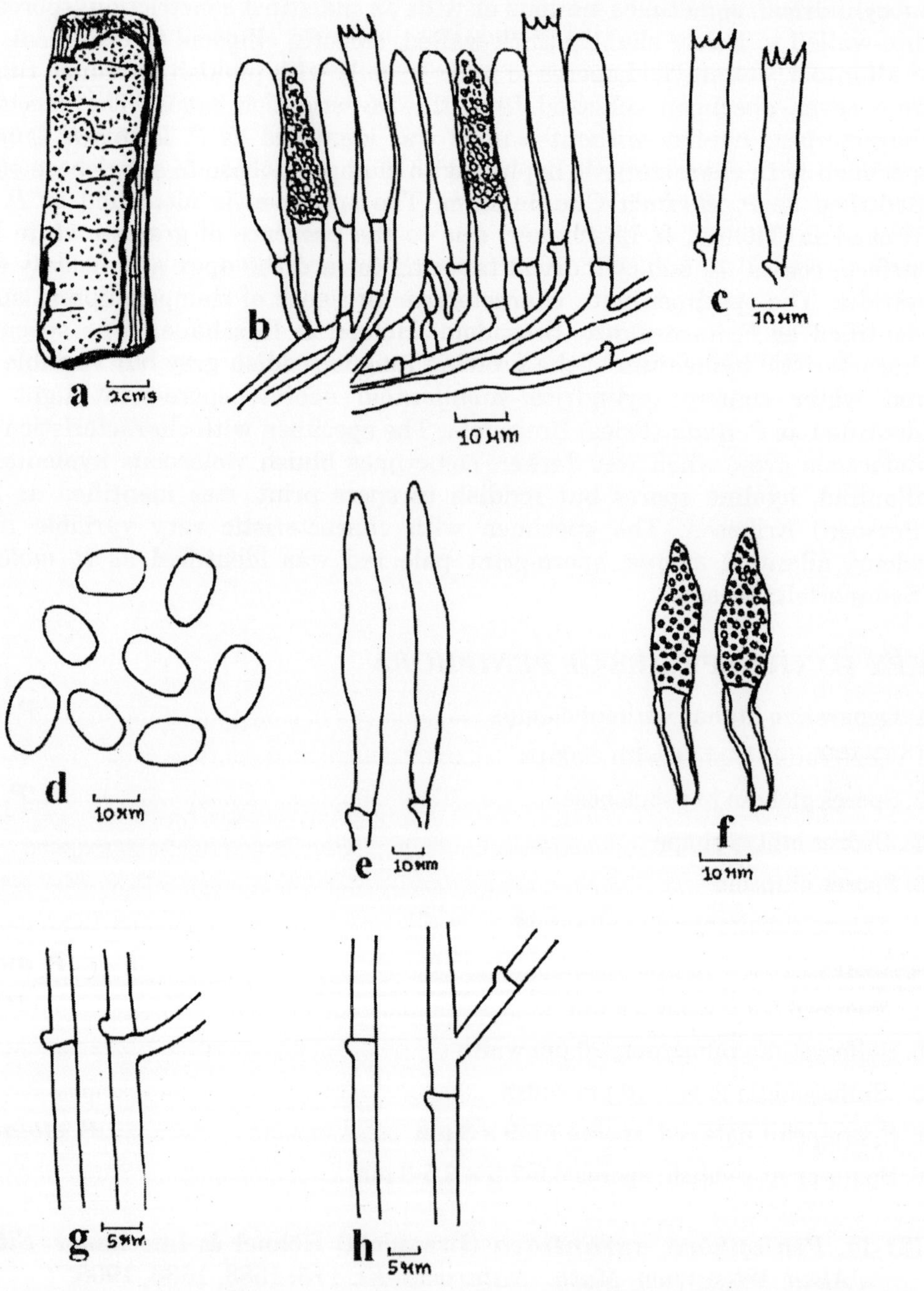

Fig. 47 ***Peniophora aurantiaca*** a. Habit. b. T.S. hymenium. c. Basidia. d. Basidiospores. e. Sulfocystidia. f. Encrusted cystidia. g. Generative hyphae. h. Generative hyphae in culture

Morphological Characters : Fruitbody resupinate, mostly totally adnate, 16 × 2.5-3 cms., only when old somewhat loosening from the substrate, at first orbicular round lenticels and other holes in the outer bark, then confluent and covering undersides of branches, more or less tuberculate especially in the centre, more in the tubercles, lighter or darker orange-red depending on age; consistency ceraceous in young fruitbodies, membranaceous in older ones; margin lighter, pale yellow or white, fibrillose in young specimens. Hyphal system monomitic; hyphae with thin or slightly thickened walls, with clamps, 3-4 μm wide, branched. Cystidia of two kinds a) encrusted thick-walled metuloids, encrusted part 30-50 × 8-12 μm, basal, not encrusted part of varying length, of ten with adventitious septation; b) thin-walled sulfocystidia, reaching a length of 100-150 ×10-15 μm, with granular protoplasm with reaction to sulfovanillin, elongate, obtuse, or tapering to an apex, numerous. Basidia varying but mostly cylindrical, more or less sinuous and constricted, reaching a size of 60-90 × 10-15 μm, with 4 sterigmata. Spores ellipsoid, smooth, thin-walled, mostly 14-16 ×8-10 μm, hyaline but in spore-print yellow or light orange. **Habitat :** On fallen branches. **Distribution :** N. Europe, N. America **Specimen examined :** Benzidine test positive, with rot on *Cassia pinnata* Hort. ex. Voigt. **Location :** Baneshwar; Dist: Pune (13/01/2007), VH-NPO-627; **Specimen examined :** Benzidine test positive, with rot on Unkown host. **Location :** Koyna, Dist. Satara (11/02/2007), VH-NPO-718. **Culture characters:** Growth characters: Mats white to Light Buff, thin, appressed and subfelty throughout or limited around inocula, then becoming raised, arachnoid to sparsely woolly toward margins, margins even, slightly raised; odor absent or sweet, reverse unchanged. Microscopic characters: Generative hyphae 3-5 μm wide, thin walled, with clamps, sparsely branched. **Tests :** α-napthal and pyrogallol + H_2O_2 positive; p-cresol negative. **Species Code :** 1, 2, 3, 6, 12, 15, 26, 30, 37, 39, 53. **Culture No :** PUCC. B: 27. **Remarks :** Very close to *P. erikssonii* from which is distimguished by the presence of clamps. A slide of the holotype of *P. lepida* Bres. studied and its identity with *P. aurantiaca* is confirmed. The species is a new record from India.

(21-2) *Peniophora farinosa* (Bresadola) Höhnel & Litschauer. *Sitzungsber. Akad. Wiss. Wien, Math. -Naturwiss.* Kl. **117:** 1095, 1908.

=*Kneiffia farinosa* Bres., *Annls mycol.* **1 (1/2) :** 105, 1903. (Plate 4, Fig. 48)

Morphological Characters : Basidiomata resupinate, adnate, effused, ceraceous, 5-7 × 1-1.5 cms. and up to 200 μm thick; hymenial surface grayish white, smooth; margin concolorous with the hymenial surface, thinning out indeterminately. Context in vertical section subhyaline, membranous. Hyphal system monomitic, hyphae 3-4.5 μm wide, smooth, thin-walled, with clamp. Cystidia (lamprocystidia) conical to sub-cylindrical tapering toward the apex, 50-80 × 8-10 μm. smooth, at first thin-walled and then thick-walled with a basal clamp, heavily encrusted. Basidia suburniform, 25-30 × 6-7 μm. with a basal clamp, producing 4 sterigmata, containing oily drops. Basidiospores ellpsoid, 6.5-7.5 × 4.5-5.5 μm, smooth, slightly thick-walled, non-amyloid. **Habitat :** On live standing tree.

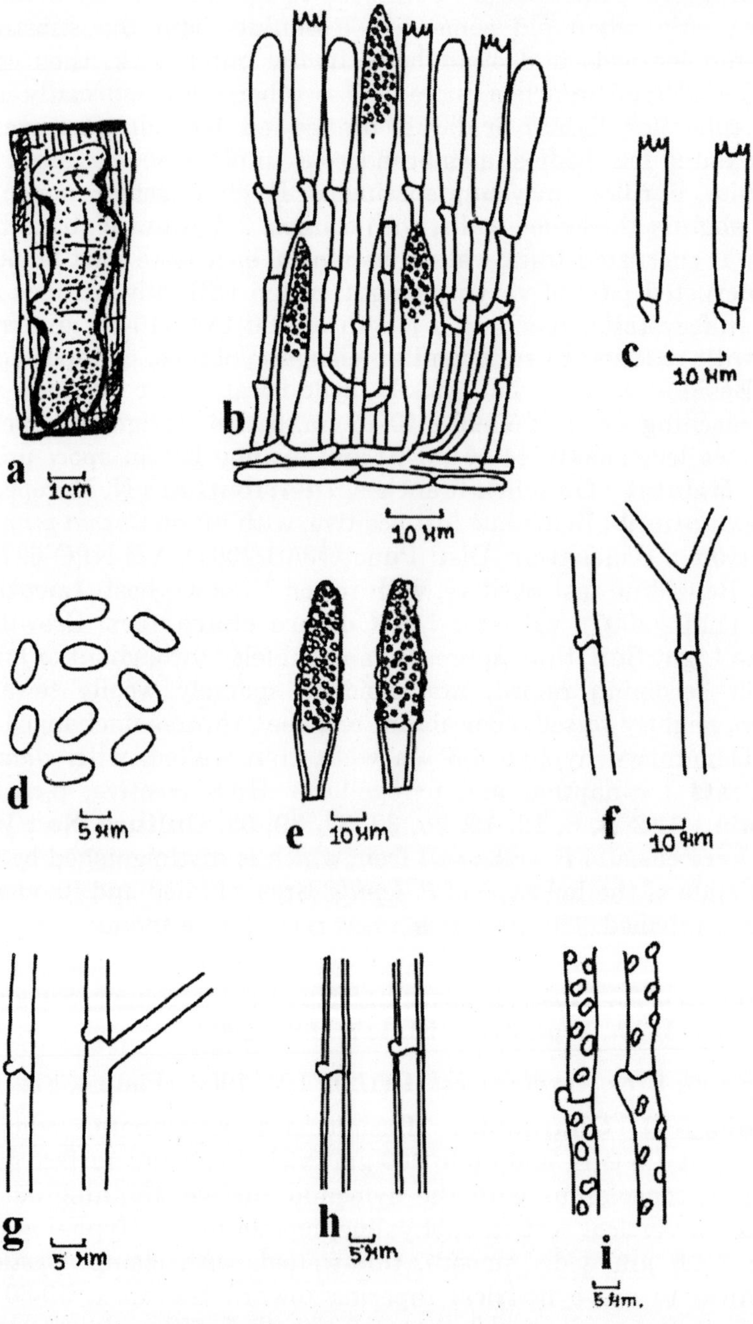

Fig. 48 ***Peniophora farinosa*** a. Habit. b. T.S. hymenium. c. Basidia. d. Basidiospores. e. Encrusted cystidia. f. Generative hyphae. g. Marginal hyphae in culture. h. Submerged hyphae in culture. i. Aerial hyphae in culture

Distribution : U.S.A. Specimen examined : Benzidine test positive, white rot on *Cassia pinnata* Hort. ex. Voigt. **Location :** Kondhona, Dist. Pune (13/01/2007), VH-NPO-563. **Culture characters :** Growth characters: Growth rate moderate; advancing zone appressed; Mats white, around inoculum thin, appressed, subfelty, then becoming thicker, slightly raised to raised, downy to cottony toward margins; outline of colony even; appressed; odor sour; Reverse bleached. Microscopic characters: Marginal hyphae 4-6 μm wide, thin-walled, with clamps, sparsely branched, branches usually arising below clamp connections. Submerged hyphae 2-6 μm wide, thin walled to slightly thick walled, nodose septate, moderately to frequently branched. Aerial hyphae similar to submerged hyphae except segments heavily encrusted with small hyaline crystals. **Tests :** α-napthal and pyrogallol + H_2O_2 positive; p-cresol negative. **Species Code :** 1, 2, 3, 6, 13, 14, 15, 16, 30, 37, 39, 52, 53, 54. **Culture No :** PUCC. B: 28. **Remarks :** This species is recognized by grayish white hymenial surface, conical to sub-cylindrical tapering toward the apex and heavily encrusted cystidia. The species is a new record from India.

(21-3). *Peniophora gladiola* Cunningham. *Trans. Roy. Soc. New Zealand* **83 (2) :** 283, 1955. (Plate 4, 10, Fig. 49)

Morphological Characters : Hymenophore annual, membranous, adnate,, effused, 12-15 ×3-5 cms.; surface dingy white or pallid cream, velutinate; margin thinning out, membranous, adnate, sometimes tending to lift. Context white, 50-130 μm thick, generative hyphae 3-5 μm diameter, hyaline, branched, septate, with inconspicuous small clamp connexions. Hymenial layer to 35 μm deep, an irregular palisade of basidia, paraphyses, hyphae and cystidia. Paraphyses scanty, subclavate, 18 ×7 μm. Cystidia arising from the base of the intermediate layer, emerging for the greater part of their length, subulate, 90-150 ×12-16 μm, bases rounded and inflated, apices long-acuminate, thickwalled, rugulose-roughened or etched. Basidia subclavate or subcylindrical, 20-25 × 7-9 μm, with four sterigmata. Spores globose, subglobose, oval sometimes broadly elliptical, walls smooth, hyaline, 7-11 ×6-9 μm. **Habitat:** On live standing tree. **Distribution:** New Zealand. **Specimen examined:** Benzidine test positive, white rot on *Syzygium cumini* (Linn.) Skeel. **Location:** Guhaghar Dist: Ratnagiri; (20/10/2005), VH-NPO-282. **Culture characters:** Growth characters: Growth rate moderate; Advancing zone raised; Mat cottony white; Aerial mycelium absent; Outline of colony bayed; Reverse bleached. Microscopic characters: Generative hyphae, thin –walled, smooth, branched, 4-5 μm. wide, clamped. Paraphysate hyphae 15 ×3 μm. **Tests :** α-napthal and pyrogallol + H_2O_2 positive; p-cresol negative. **Species Code :** 1, 2, 3, 6, 12, 14, 15, 16, 30, 37, 39, 53. **Culture No :** PUCC. B: 29. **Remarks :** The species is differentiated by the large basidia, subglobose or oval smooth spores with granular contents, and presence of paraphysate hyphae. The species is a new record from India.

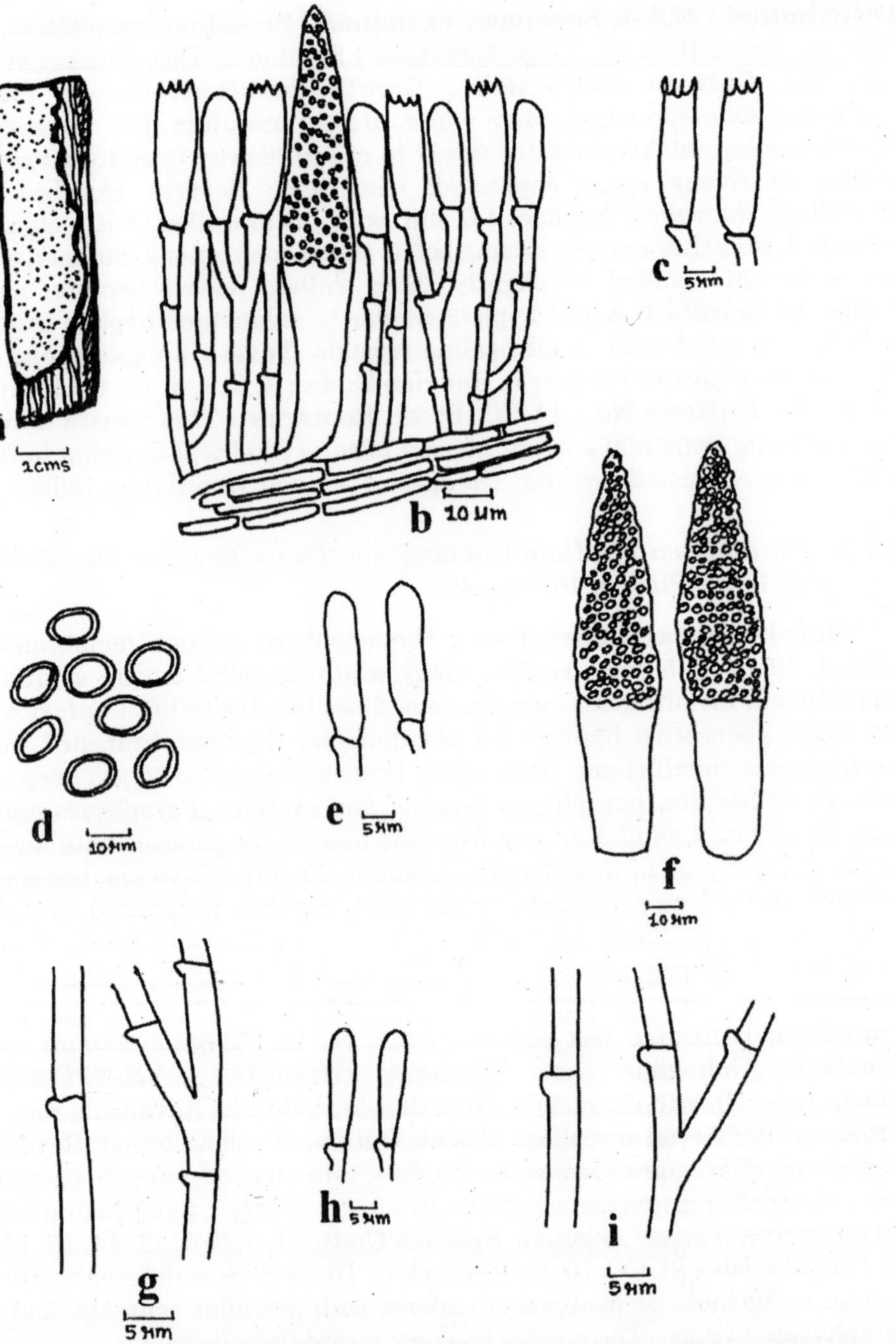

Fig. 49 ***Peniophora gladiola*** a. Habit. b. T.S. hymenium. c. Basidia. d. Basidiospores. e. Paraphyses. f. Encrusted cystidia. g. Generative hyphae. h. Paraphyses in culture. i. Generative hyphae in culture

(21-4) *Peniophora laurentii* Lundell. *Fungi exsicc. suec.* 1342, 1946.

=*Corticium lepidum* (Romell) Bourdot & Galzin, *Hyménomyc. de France* (Sceaux) : 187, 1928.

=*Merulius lepidus* Romell, *Ark. Bot.* **11 (3)** : 29, 1911. (Plate 4, Fig. 50)

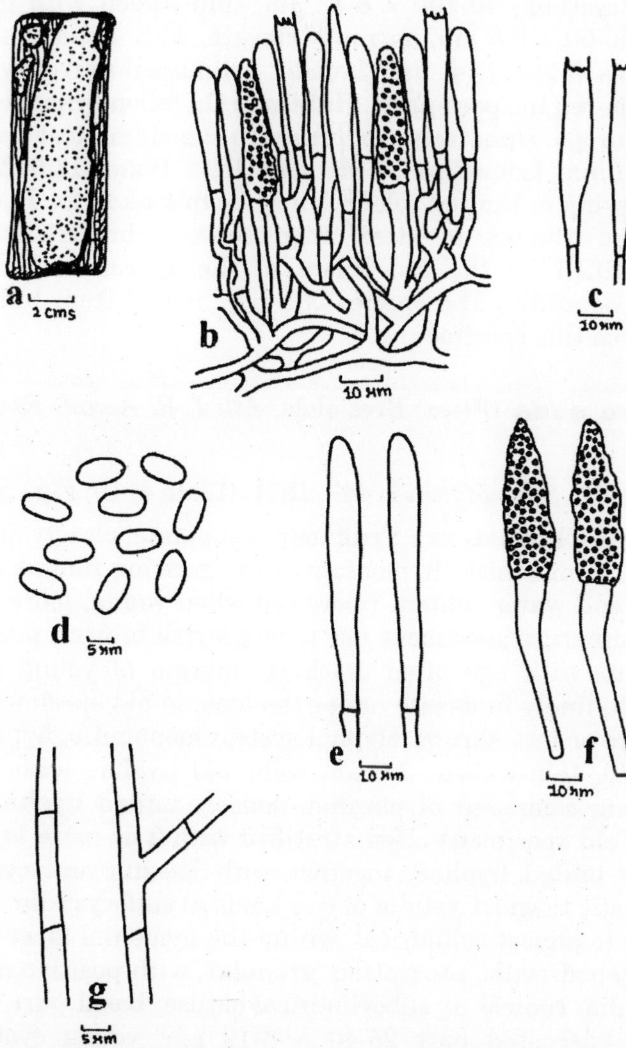

Fig. 50 *Peniophora laurentii* a. Habit. b. T.S. hymenium. c. Basidia. d. Basidiospores. e. Cystidia. f. Encrusted cystidia. g. Generative hyphae

Morphological Characters : Fruitbody resupinate, on smooth bark orbicular and adnate but usually confluent-effused, more or less detachable, 12 × 2.5-3 cms.; hymenium at first smooth, then tuberculate, plicate or merulioid, bright orange-red, fading to ochraceous red; subiculum white and often forming a white marginal zone;

no rhizomorphs. Hyphal system monomitic; hyphae 3-5 μm wide, without clamps, subicular ones thick-walled, sparsely branched in all directions, forming it White, open texture; subhymenial hyphae thinwalled, richly branched and united into a dense texture, giving the layer a ceraceous membranaceous structure. Cystidia of two kinds (a) encrusted pseudocystidia, thick-walled, 80-100 × 10-15 μm strongly encrusted (b) sulfocystidia, 70-100 × 8-12 μm, thin-walled with grainy plasmatic contents. Basidia 50-60 × 6-8 μm, narrowly clavate, with 4 sterigmata and without basal clamp. Spores 9-13 × 4-5 μm, broadly cylindrical-narrowly ellipsoid, thin-walled, smooth, pale-red in spore-print. **Habitat :** On fallen branches. **Distribution :** Sweden, North Europe. **Specimen examined :** Benzidine test positive, white rot on *Caryota urens Linn.* **Location :** Singhgad, Dist. Pune (13/01/2007), VH-NPO-615, **Remarks :** Owing to the red colour it may be mistaken for *P. incarnate*, but is easily distinguished thanks to clear microscopical chractristics. Its place in *Peniophora* is justified by the red colout of the spores and the sulfoaldehde reactionin the gloeocystidia. The species is a new record from India. The cultures could not obtain from this specimen.

(21-5). *Peniophora nuda* (Fries) Bresadola. *Atti I. R. Accad. Roveret. Sci.*, **III 3:** 114, 1897

= *Thelephora nuda* Fr., *Syst. Mycol.* **I:** 447, 1821. (Plate 4, 10; Fig. 51).

Morphological Characters : Fruitbody resupinate, closely adnate, 10-12 × 2-3 cms. and up to 0.2 mm thick; hymenium of the growing fungus reddish grey but variable with age and water content (more red when young, more violaceous when wet, while old, postmature specimens use to be greyish brown), mostly smooth with a low central wart, with age often cracking; margin of young fruitbodies rose-coloured or whitish, finely fimbriate under the lens, in old specimens not especially differentiated, more or less abrupt. Hyphal system monomitic, hyphae with clamps, young ones hyaline, thin-walled, 3-5 μm wide, old hyphae with thickened walls, more or less brown; composed of parallel, densely united hyphae; subhymenium thickening and in old specimens often stratified with 2 or more layers of vertically arranged, densely united hyphae, together with basidia and cystidia forming a pseudoparenchymatic tissue. Cystidia of two kinds a) sulfocystidia, 30-60 × 8-20 μm, rounded or oblong to almost cylindrical, within the hymenial layer thin, then at the least basally thickened walls; protoplasm granular, with positive aldehyde reaction b) encrusted cystidia, conical or subcylindrical-obtuse, basal part thick-walled and becoming brown, encrusted part 25-40 × 8-12 μm; young cystidia thinwalled, pointed, projecting from the hymenium. Basidia 35-45 × 6-8 μm, thinwalled, subcylindrical, with 4 sterigmata and basal clamp. Spores 7.5-11 × 2.5-3.5 μm, cylindrical -suballantoid, thinwalled, smooth, hyaline, spore-print light red. **Habitat:** On live standing tree. **Distribution** : Denmark, Norway, Sweden, Finland, Scandinavia, Europe, India. **Specimen Examined**: Benzidine test positive, white rot on *Bauhinia purpurea* L. **Location** : Karad, Dist. Satara (12/10/2005), VH-NPO-199. **Culture characters :** Growth characters: Growth rate moderately fast; Advancing zone raised; Mat cottony white turning creamish yellow; Aerial mycelium absent; Outline of colony bayed; Reverse bleached. Microscopic characters: Generative hyphae, thin-walled, smooth, branched, 3-4 μm. wide, a few hyphae pigmented brown. **Tests :** α-napthal and pyrogallol + H_2O_2 positive; p-cresol negative. **Species Code :** 1, 2, 3, 6, 12, 14, 15, 16, 30 (31), 37, 39, 53, 67. **Culture No :** PUCC. B: 30. **Remarks**: *P. nuda* is easily recognized by numerous, rounded sulfocystidia in the basal part of the fruitbody. It has a reddish colour and a marked marginal zone.

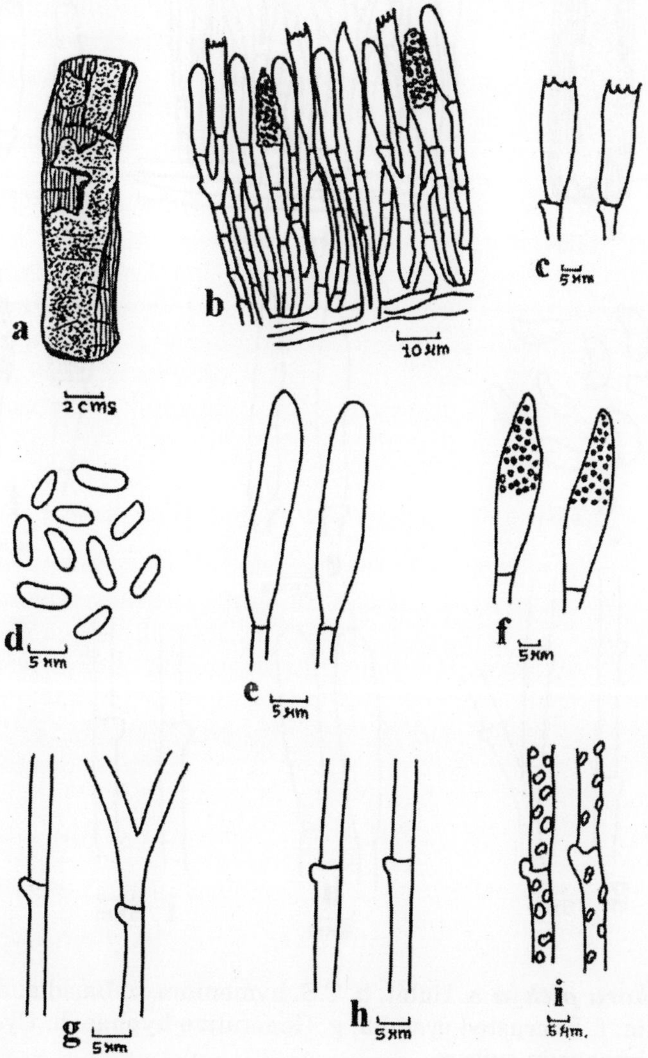

Fig. 51 *Peniophora nuda* a. Habit. b. T.S. hymenium. c. Basidia. d. Basidiospores. e. Sulfocystidia. f. Encrusted cystidia. g. Generative hyphae. h. Generative hyphae in culture. i. Generative hyphae with pigment in culture

(21-6) *Peniophora pithya* (Persoon) J. Eriksson. *Fungi exsicc. suec.* **37-38**:37, 1950.

=*Thelephora pithya* Pers., *Mycol. eur.* (Erlanga) **1**: 146 (1822) (Plate 5, 10, Fig. 52)

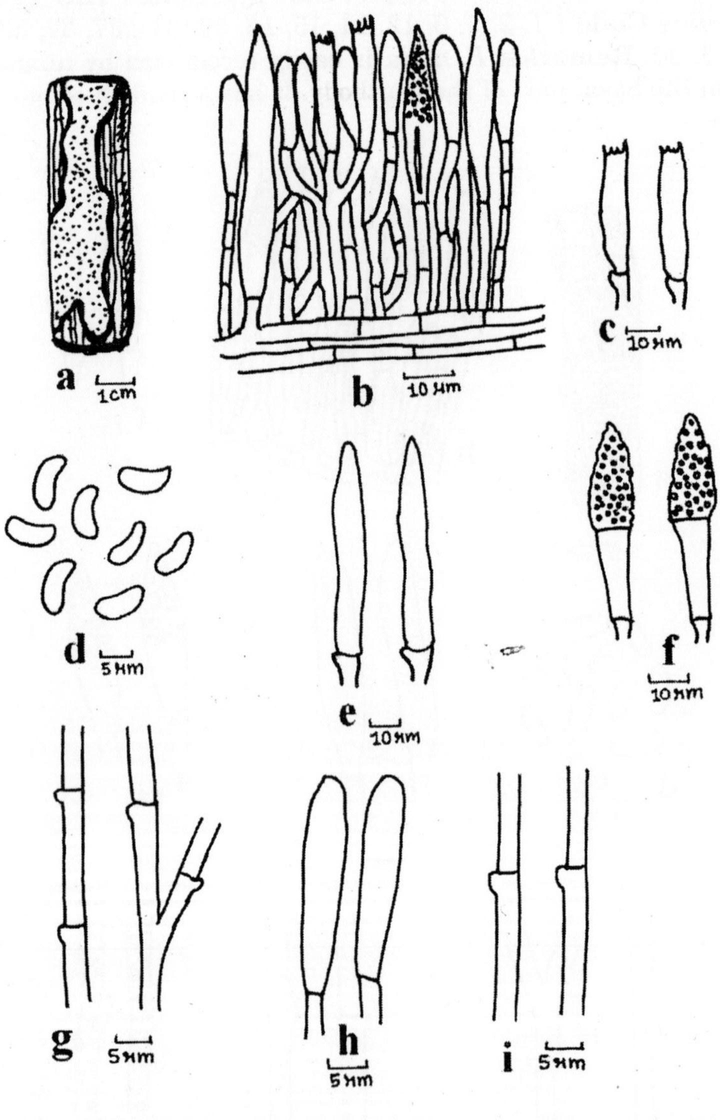

Fig. 52 *Peniophora pithya* a. Habit. b. T.S. hymenium. c. Basidia. d. Basidiospores. e. Sullfocystidia. f. Encrusted cystidia. g. Generative hyphae. h. Cystidia in culture. i. Generative hyphae in culture

Morphological Characters : Fruitbody resupinate, 5-7 × 2-3 cms. and up to 0.2 mm thick, closely adnate, smooth and continuous on wood, more uneven on bark, when dried more or less cracked, reddish or violaceous grey, when wet darker, sometimes bluish violaceous; margin abrupt, narrow sterile zone. Hyphal system monomitic, with clamps, thin-walled, hyaline, subiculum varying in thickness from an inconspicuous layer of horizontal hyphae to a considerable thickness, composed of parallel, dense, dark brown hyphae; subhymenial layer of dense, vertically arranged hyphae, hyaline in young specimens, more or less pigmented in older ones but as a rule sharply contrasting against the darker subiculum, subhymenium generally not stratified but in some specimens two or three layers visible. Cystidia numerous, of two kinds a) encrusted cystidia, generally conical, at first thin-walled and hyaline, then apically encrusted and at least in the basal part brown, 35-65 × 12-15 μm, encrusted part 25-35 μm long, those of the first hymenial layer biggest and basally brown, later cystidia smaller and more or less hyaline b) sulfocystidia oblong to subcylindrical, 50-70 × 8-10 μm, as a rule obtuse, with thin or basally thickened walls; contents continuous, granular. Basidia subcylindrical or subclavate, somewhat constricted or sinuous, 30-40 × 5-7 μm, with 4 sterigmata and with basal clamp. Spores 6. 5-7.5 × 2. 5-3 μm, allantoid, smooth, thinwalled, hyaline but reddish in spore print. **Habitat:** On live standing tree. **Distribution** : Denmark, Norway, Sweden, Finland, Scandinavia, Europe, India. **Specimen Examined**: Benzidine test positive, white rot on *Memecylon umbellatum* Burm. **Location :** Panshet, Dist. Pune (09/02/2006), VH-NPO-488.

Culture characters : Growth characters: Growth rate moderate; advancing zone raised; Mat rosy vinaceous to buff, on cottony mycelium against side wall, at first slightly raised, fairly abundant, woolly to woolly-floccose, somewhat reticulate in appearance, outline of colony uneven, Odor strong, reverse unchanged. Microscopic characters: Generative hyphae hyphae hyaline, with clamps, frequently branched, usually between septa, 2-4 μm. wide. Cystidia hyaline, thin-walled, 35-50 × 5-7μm. **Tests :** α-napthal and pyrogallol + H_2O_2 positive; p-cresol negative.

Species Code : 1, 2, 3, 6, 12, 15, 16, 30, 37, 39, 52, (53), 72. **Culture No :** PUCC. B: 31. **Remarks :** *P. pithya*, like all other *Peniophora*, varies greatly, e.g. in colour, size and thickness of the fruitbody and under the microscopically in the thickness of subiculum, but is in most cases recognized without difficulty. The smaller spores usually afford good help for determination but in some speciemren the spores are found to be larger and intergrade with *P. cinera*. The spore size 5.5-6.5 μm given by Eriksson 1950 is true mainly for N. Sweden, from where mostof this material originated. In 1950 Eriksson did not know the use of the sulfoaldehyde reaction and did not observe the presence of gloeocystidia, which now are found to occur regularly, at least in young specimens. Without this reaction the gloeocystidia are easily mistaken for young, not yet enrusted Cystidia. In old fruitbodies they can not always be found. The very dark subiculum contrasting against the lighter subhymenium and the large basal Cystidia form, together with the substrate and the outer appearance, sufficient chracteristics for the determination also for old, sterile fruitbodies. *Peniophora exima* Jacks. and Deard. (Mycologia 43:1 p. 60, 1951) from Pacific N. America is close to *P. pithya* but differs in having much shorter, almost ellipsoid spores. The species is a new record from India.

(21-7). ***Peniophora violaceolivida*** (Sommerfelt) Massee. *J. Linn. Soc., Bot.* **25:** 152, 1890.

=*Corticium violaceolividum* (Sommerf.) Fr., *Epicr. syst. mycol. (Upsaliae)*: 564, 1838.

=*Gloeopeniophora violaceolivida* (Sommerf.) Y. Hayashi, *Bull. Govt Forest Exp. Stn Meguro* **260:** 19, 1974.

=*Peniophora cinctula* (Quél.) Bourdot & Galzin, *Bull. Soc. mycol. Fr.* **28:** 405, 1913.

=*Terana violaceolivida* (Sommerf.) Kuntze, *Revis. gen. pl.* (Leipzig) **2:** 873, 1891.

=*Thelephora violaceolivida* Sommerf., *Suppl. Fl. lapp.* (Oslo): 283, 1826.

=*Xerocarpus violaceolividus* (Sommerf.) P. Karst., *Revue mycol.,* Toulouse **3(9):** 22, 1881. (Plate 5, 10, Fig. 53)

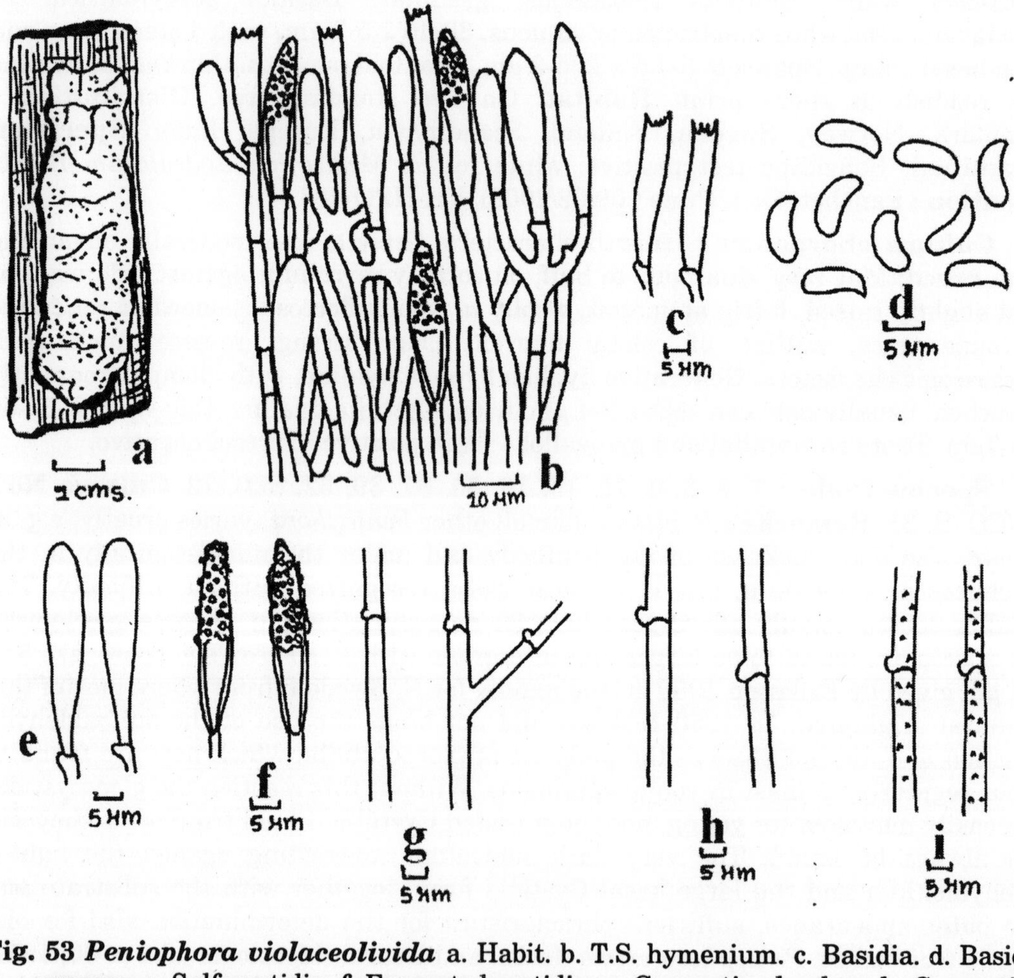

Fig. 53 ***Peniophora violaceolivida*** a. Habit. b. T.S. hymenium. c. Basidia. d. Basidiospores. e. Sulfocystidia. f. Encrusted cystidia. g. Generative hyphae. h. Generative hyphae in culture. i. Encrusted generative hyphae in culture

Morphological Characters : Fruitbody resupinate, totally adnate, 7-9 × 2-3cms. and up to 0.2 mm thick; hymenium sometimes quite smooth but commonly more or less tuberculate, in dry weather cracking in small polygons, greyish with a varying tint of reddish or violaceous, in wet weather darker, more bluish, young fruitbodies fading to pale ochraceous grey; margin indeterminate. Hyphal system monomitic, with clamps, young ones thin-walled, hyaline, older hyphae with brown thickened walls. Cystidia two kinds a) thin-walled, subulate hymenial cystidia, developing to encrusted metuloids, 40-60 × 6-9 μm; b) thin-walled sulfocystidia, at first subclavate, then oblong, fusiform to subcylindrical, filled with granular protoplasm, and varying in length with the thiekness of the fruitbody, 50-90 × 7-10 μm. Basidia subcylindrical-subclavate, 30-40 × 5-6 μm, with 4 sterigmata and basal clamp. Spores allantoid, thin-walled, hyaline, 7.5-9 × 3 μm. Spore-print pale red. **Habitat :** On live standing tree. **Distribution :** India, Japan, Australasia, Africa, Asia, Europe, NorthAmerica, Norway, U. S. A., Canada, Italy, Sweden. **Specimen examined :** Benzidine test positive, white rot on *Tamarindus indica* Linn. **Location :** Koyna; Dist:Satara (11/02/2007), VH-NPO-711. **Culture Characters**

Growth characters: Growth rate moderate; advancing zone raise; Mat downy to low cottony white getting submerged in agar as colony advances giving the plate and hence the colony, a transparent look; colony white; outline of colony uneven; Distance between hyphal tips distant; Aerial mycelium absent; Reverse bleached. Microscopic characters: Generative hyphae branched, 3-4 μm, wide, pigmentation hyphae present, 3-5 μm, wide. **Tests :** α-napthal and pyrogallol + H_2O_2 positive; p-cresol negative. **Species Code :** 1, 2, 3, 6, 12, 15, 16, 30, 37, 39, 52, (53), 67. **Culture No :** PUCC. B: 32. **Remarks :** The species is rather very variable in surface colour, its greyish with a varying tint of reddish or violaceous, in wet weather darker, more bluish, young fruitbodies fading in the herbarium to pale ochraceous grey, old sterile fruitbodies brownish grey. The culture was not obtained besides repeated trials.

(22) *PHANEROCHAETE* P. Karsten. *Bidr. Känn. Finl. Nat. Folk* **48:** 426, 1889.

The genus was erected by Karsten in 1889 by studying the type species *Phanerochaete odorata*. But later on in 1893 he himself suggested *P. sordida* and therefore Erikson and Ryvarden selected *P. sordida* (Karsten) J. Eriksson & Ryvarden as a type species. The genus is represented by 130 species worldwide in distribution.

The genus characterized by usually resupinate, effuso-reflexed fruitbodies; hymenial surface even to odontioid; white, yellow, orange, red to brown; spores subglobose to allantoid; smooth; thin-walled; negative in Melzer's. From the investigation area seven specimens were collected. The specimen with pale ochraceous fruitbodies with narrow light coloured margin;allantoid spores which were usually large for the genus was identified as *P. cacaina*. The specimen with creamish paleochraceous fruitbodies; with subicular hyphae more or less thickwalled, clamps occurring rarely on the basal hyphae;with variable number and shape of cystidia, thin-walled, little encrusted and fewer was identified as *P. sordida*. (Karsten) J. Eriksson & Ryvarden The specimen with sometimes encrusted hyphae; cystidia only apically encrusted was identified as *Phanerochaete sanguinea* (Fries) Pouzar. The specimen with light to dark-ochraceous fruitbodies with grayish hue; cystidia with numerous, encrusted was identified as *P. laevis* (Fries) Eriksson & Ryvarden. The specimen with cinnamon buff to clay colour; without cystidia was identified as *P. tuberculata* (Karsten) Parmasto. The specimen with brownish orange fruitbodies;without cystidia was identified as *P. joseferroirae* (Reid) Reid. The specimen with straw, buff, salmon, saffron to orange fruitbodies;slightly encrusted cystidia was identified as *P. calotricha* (Karsten) Eriksson & Ryvarden.

KEY TO THE SPECIES OF *PHANEROCHAETE*

1. No cystidia 2

1´. With cystidia 3

2. Spores 5-6. 5 ×3-4µm .... ***P. tuberculate***

2´. Spores 7. 5-10 ×2. 5-3 µm .... ***P. jose-ferreirae***

3. Spores longer than 8µm .... ***P. cacaina***

3´. Spores shorter than 8µm 4

4. Fruitbodies cream to reddish, substrate reddish .... ***P. sanguinea***

4´. Fruitbodies white, cream, ochraceous, yellow to pink, substrate not reddish 5

5. Subicular hyphae thick walled, randomly branched and oriented, no rhizomorphs .... ***P. sordida***

5´. Subicular hyphae thin walled, if thick walledmore or less parallel along the substrate, rhizomorphs usually present 6

6. Mature cystidia apically encrusted to at least one third of their Length. .... ***P. laevis***

6´. Mature cystidia naked or only slightly encrusted .... ***P. calothricha***

(22-1). *Phanerochaete cacaina* (Bourdot & Galzin) Burdsall & Gilbertson. *Mycologia* **66:** 781, 1974.

=*Grandiniella cacaina* (Bourdot & Galzin) Burds., *Taxon* **26:** 329, 1977.

=*Peniophora cacaina* Bourdot & Galzin, *Bull. Soc. mycol. Fr.* **28(4)**: 397, 1913. (Plate 5, 10; Fig. 54)

Morphological Characters : Basidiocarp broadly effused, 5-7 × 2.5-5 and up to 0.5 mm thick; fertile area greyish orange to light brown, not changing color in KOH, smooth or cracking slightly on drying and revealing a light tan subiculum, membranous to crustaceous; margin thin, narrow, pubescent to byssoid, concolorous with subiculum. Hyphal system monomitic. Subiculum a textura intricata, hyphae 3.5-6 μm diam, hyaline, smooth, simple septate, frequently branched; subhymenium not well defined, hyphae 3.5-5 μm diam, firm walled, simple septate, hyaline, with

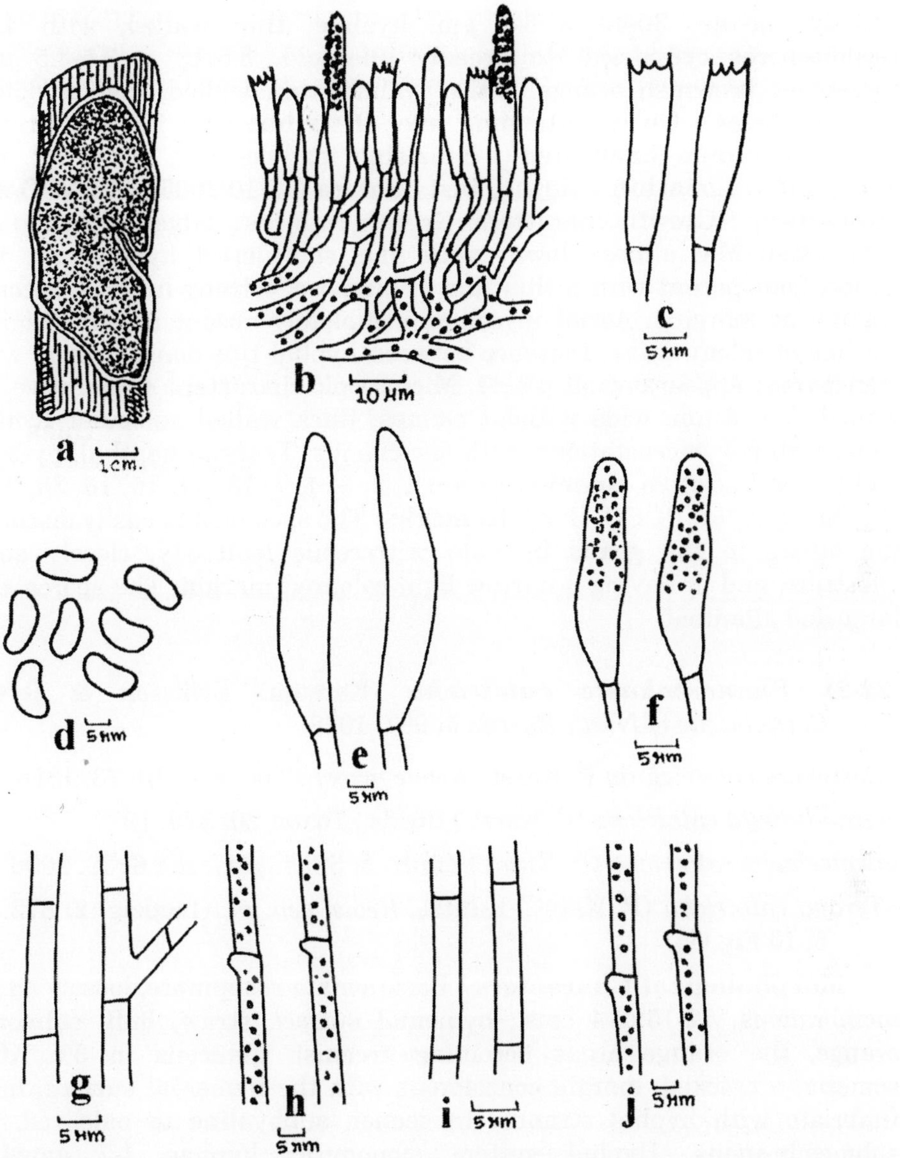

Fig. 54 ***Phanerochaete cacaina*** a. Habit. b. T.S. hymenium. c. Basidia. d. Basidiospores. e. Cystidia. f. Encrusted cystidia. g. Generative hyphae. h. Encrusted generative hyphae. i. Generative hyphae in culture. j. Encrusted generative hyphae in culture.

reddish brown to orange granular material encrusting many of the hyphae, frequently branched; hymenium with orange crystalline material throughout. Cystidia very numerous, thin-walled with protoplasmic content, 50-70 × 9-11 µm.; encrusted, 25-50 × 5-7 µm.; subcylindrical to subfusiform, tapering at apex. Basidia broadly clavate, 30-40 × 8-9 μm, hyaline, thin walled, with 4 sterigmate. Basidiospores cylindrical to broadly allantoid, 8.5-12 × 3.5-4.5 μm, contents sometimes brownish orange, walls hyaline, thin walled, non-amyloid, acyanophilous. **Habitat :** On live standing tree. **Distribution :** France, Arizona (U.S.A.), India. **Specimen examined :** Benzidine test positive, white rot on *Mangifera indica* Linn. **Location :** Koyna, Dist. Satara (12/10/2005), VH-NPO-247. **Culture characters :** Growth characters: Growth rate fast; advancing zone appressed to submerged; Mat almost downy with fine, short erect hyphae; the whole colony almost transparent with a thin film of plumose, cottony hyphae, becoming floccose slightly at margins; Aerial mycelium completely absent Reverse highly bleached; Outline of colony wavy; Distance between hyphal tips dense; colony white giving a transparent appearance at places. Microscopic characters: Generative hyphae thin-walled, 3-3. 5 μm. wide without clamps; thick walled subicular hyphae, 5-6 μm. wide, with few encrustations, with few clamps. **Tests:** α-napthal, pyrogallol + H_2O_2 and p-cresol positive. **Species Code:** 1, 2, 3, 4, 7, 13, 14, 15, 16, 30, 37, 39, 53, 54, 57. **Culture No:** PUCC. B: 33. **Remarks:** The specimen is easily distinguished from the others in the genus by pale ochraceous fruitbody, closely adnate to the substrate, and in having a narrow light-coloured margin. The spores are unusually large and allantoid.

(22-2). *Phanerochaete calotricha* (Karsten) Eriksson & Ryvarden. *The Corticiaceae of North Europe* **5:** 997, 1978.

=*Corticium calotrichum* P. Karst., *Revue mycol.,* Toulouse **10:** 73, 1888.

=*Grandiniella calotricha* (P. Karst.) Burds., *Taxon* **26:** 329, 1977.

=*Leptochaete calotricha* (P. Karst.) Zmitr. & Spirin, *Mycena* **6:** 39, 2006.

=*Terana calotricha* (P. Karst.) Kuntze, *Revis. gen. pl.* (Leipzig) **2:** 872, 1891. (Plate 5, 10 Fig. 55)

Morphological Characters : Basidiomata resupinate, loosely adnate, effused, membranous, 12-15×3-4 cms.; hymenial surface straw, buff, salmon, saffron to orange, the orange areas becoming greenish glaucous in 3% KOH, smooth, sometimes cracked; margin concolorous with the hymenial surface, indeterminate, fimbriate with hyphal strands. In section subhyaline to pale red, pellicular to submembranous. Hyphal system monomitic; hyphae 4-7 μmwide, smooth, clampless, thin- to thick-walled. Cystidia (leptocystidia) cylindrical to subfusiform, 30-50× 4.5-6 μm, smooth, thin- to slightly thick-walled, projecting up to 30 μm beyond the hymenial surface. Basidia narrowly clavate, 20-33 × 4. 5-5.5 μm, without a basal clamp, producing four sterigmata. Basidiospores ellipsoid to subcylindrical, 5-6 × 2.5-3 μm, smooth, thin-walled, non-amyloid. **Habitat :** On Dead standing tree. **Distribution :** China, Japan, Taiwan, Europe, North America. **Specimen examined :** Benzidine test positive, white rot on *Bauhinia purpurea* L. **Location:** Baneshwar, Dist. Pune (13/01/2007), VH-NPO-578. **Culture Characters:** Growth characters: Growth rate moderate; advancing zone raised; Mats white, moderately thin, raised, arachnoid or silky to cottony; Outline of colony even, appressed to raised, fimbriate, odor grasslike. Reverse bleached. **Microscopic characters:** Generative hyphae thin–walled, 3-5 μm., wide, simple septate, sparsely branched, often branches arising immediately below septa. **Tests :** α-napthal, pyrogallol+ H_2O_2 and p-cresol positive. **Species Code :** 1, 2, 3, 8, 12, 14, 15, 30, 36, 37, 39, 54. **Culture No.:** PUCC. B: 34. **Remarks:** Similar to *P. galactites* and *P. sanguniea*, distinguished from the former species by the presence of numerous, thin walled, not encrusted cystidia, from the latter by lack of red coloure and in the shape of the cystidia, which are more or less subulate, while in *P. sanguinea* almost cylindrical. The species has been treated as a taxon of its own by Parmasto (1968) but often considered as only a white form of *P. sanguniea.* The species is a new record from India.

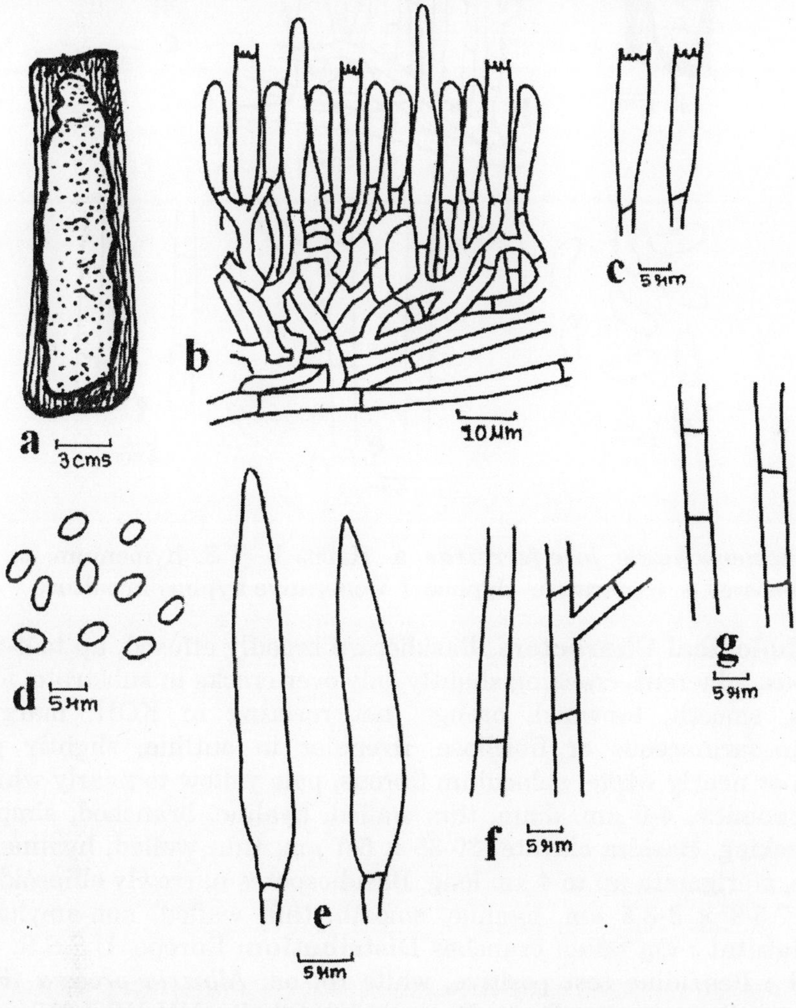

Fig. 55 ***Phanerochaete calotricha*** a. Habit. b. T.S. hymenium. c. Basidia. d. Basidiospores. e. Cystidia. f. Generative hyphae. g. Generative hyphae in culture

(22-3). *Phanerochaete jose-ferreirae* (Reid) Reid. Acta bot. croat. **34:** 135, 1975.
=*Byssomerulius jose-ferreirae* (Reid) Zmitr. & Malysheva, *Mycena* **6:** 31, 2006.
=*Corticium jose-ferreirae* Reid *Revta Biol.*, Lisb. **5(1-2)**:140,1965. (Plate 5,10 Fig. 56)

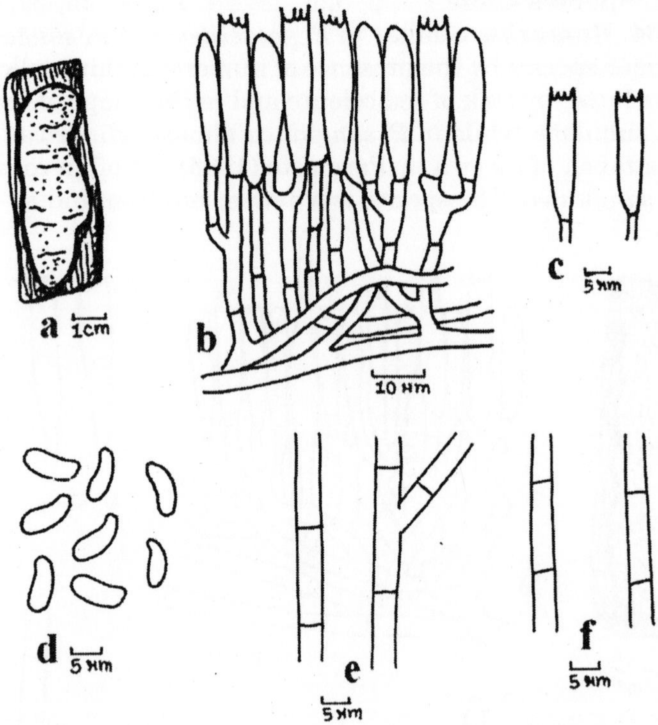

Fig. 56 *Phanerochaete jose-ferreirae* a. Habit. b. T.S. hymenium. c. Basidia. d. Basidiospores. e. Generative Hyphae. f. Generative hyphae in culture

Morphological Characters : Basidiocarp broadly effused, up to 4-6 × 1-2 cm, membranous, adherent, cracking slightly only over cracks in substrate; fertile areas continuous, smooth, brownish orange, not reacting in KOH; margin abrupt, pruinose to farinaceous or fibrillose, irregular in outline, slightly paler than hymenium or nearly white; subiculum fibrous, pale yellow to nearly white. Hyphal system monomitic, 4-6 μm. diam, thin walled, hyaline, branched, simple septate. Cystidia lacking. Basidia clavate, 30-35 × 5-7 μm, thin walled, hyaline, with four sterigmate, sterigmata up to 4 μm long. Basidiospores narrowly ellipsoid to broadly allantoid, 7.5-9 × 3-3.5 μm, hyaline, smooth, thin walled, non-amyloid, acyanophilous. **Habitat :** On fallen branches **Distribution:** Europe, U.S.S.R. **Specimen examined :** Benzidine test positive, white rot on: *Albizzia procera* (Roxb.) Bah. **Location :** Baneshwar, Dist. Pune (13/01/2007), VH-NPO-617. **Specimen examined :** Benzidine test positive, white rot on: *Cassia fistula* L. **Location :** Kondhona, Dist. Pune (13/01/2007), VH-NPO-582. **Culture characters:** Growth characters:Growth rate moderate; advancing zone appressed to submerged; mat white to grayish; subfelty to felty, then becoming thin, slightly raised, colony white to grayish; outline of colony slightly uneven to bayed, odor sharp and pungent, reverse unchanged. Microscopic characters: Generative hyphae thin-walled, 3-5 μm., wide, without clamps with rare single clamps, moderately branched. **Tests :** α-napthal, pyrogallol + H_2O_2 and p-cresol positive. **Species Code :** 1, 2, 3, 8, 13, 15, 30, 36, 37, 39, 53. **Culture No :** PUCC. B: 35. **Remarks :** The distinctive characters of the species are the relatively large allantoid spores, the lack of cystidia, and the vinaceous color of the hymenial surface The species is a new record from India..

(22-4) *Phanerochaete laevis* (Fries) Eriksson & Ryvarden. *The Corticiaceae of North Europe* **5:** 1007, 1978.

= *Thelephora laevis* Fr. *Syst. Mycol.* **I:** 451. 1821.

= *Thelephora laevis* Pers., *Myc. Eur.* **I:** 130, 1822

= *Peniophora affinis* Burt. Ann. Miss. *Bot. Gard.* **12:** 266, 1926. (Plate 5, Fig. 57)

Morphological Characters : Basidiomata resupinate, loosely adnate, effused, 10-12 × 6-8 cms. and up to 300 μm thick; hymenial surface buff, ochreous to cinnamon, smooth, margin buff, thinning out. In section subhyaline, membranous. Hyphal system monomitic; hyphae 3-6 μm wide, smooth, thin- to thick-walled, clampless. Cystidia numerous, narrowly cylindrical and tapering to the acute apex, first thin-walled and not encrusted 75-80 × 9-11 μm., walls later thickening and finally the cystidium gets encrusted in the apical half or else, completely, 55-60 × 9-11 μm. The encrustation normally coarse and it easily falls off, as seen in few sections;very young cystidia show the presence of many oil-drops in the protoplas. Basidia narrowly clavate, 30-42 × 5-6 μm. without a basal clamp, producing 4 sterigmata. Basidiospores narrowly ellipsoid to subcylindrical. 5-6.5 × 2.5-3 μm, smooth, thin-walled, non-amyloid. **Habitat :** On live standing tree. **Distribution :** The temperate regions of North America and Europe. The United States, South America, North western Himalayas, India. **Specimen examined :** Benzidine test positive, white rot on: *Ficus bengiamina* L. **Location :** Dabhol; Dist:Ratnagiri (12/10/2005), VH-NPO-231. **Culture characters :** Growth characters: Growth rate moderate; advancing zone raised; Mat cottony white; Marginal hyphae ending in thick tufts, flat on the substratum; Outline of colony bayed; hyphal tips dense; Distance between the hyphal tips dense; Aerial mycelium absent; Reverse bleached. Microscopic characters: Generative hyphae 3-4 μm. wide, hyaline, smooth thin-walled to thick walled with rare clamps; Cystidia coarsely encrusted, cylindrical, 25-40 × 5-9 μm. **Tests :** α-napthal, pyrogallol + H_2O_2 and p-cresol positive. **Species Code :** 1, 2, 3, 4, 8, 12, 13, 14, 15, 16, 30, 37, 39, 53, 72. **Culture No :** PUCC. B: 36. **Remarks :** The specimen is easily recognized by numerous cystidia which are narrow, more tapering in simple thin-walled cystidia, clamps on subicular hyphae, and basidia narrowly clavate.

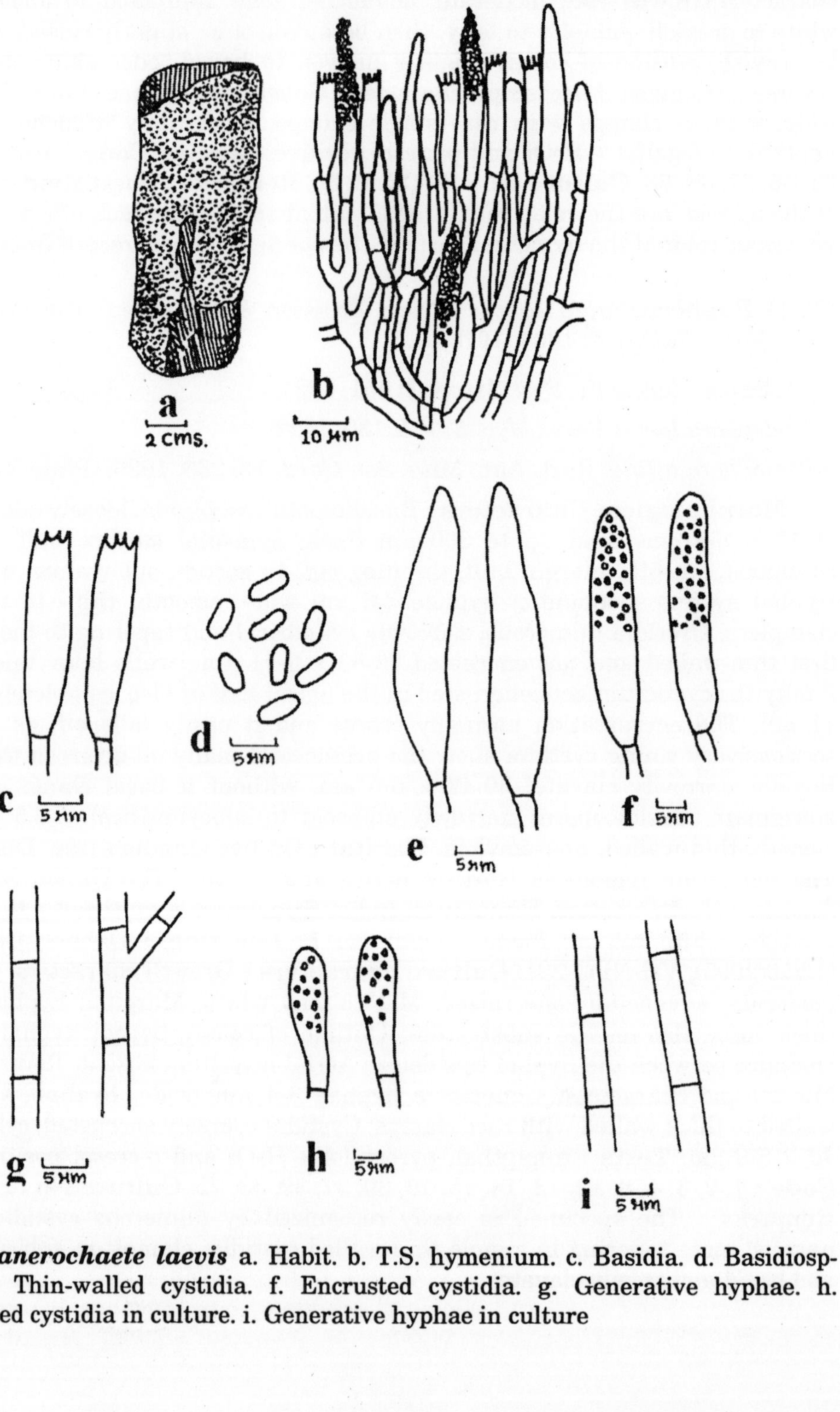

Fig. 57. *Phanerochaete laevis* a. Habit. b. T.S. hymenium. c. Basidia. d. Basidiospores. e. Thin-walled cystidia. f. Encrusted cystidia. g. Generative hyphae. h. Encrusted cystidia in culture. i. Generative hyphae in culture

(22-5). *Phanerochaete sanguinea* (Fries) Pouzar. *Ceská Mykol.* **27 (1)** : 26, 1973.

=*Thelephora sanguinea* Fr., *Elench. fung.* (Greifswald) **1:** 203, 1828. (Plate 5, 10 Fig. 58)

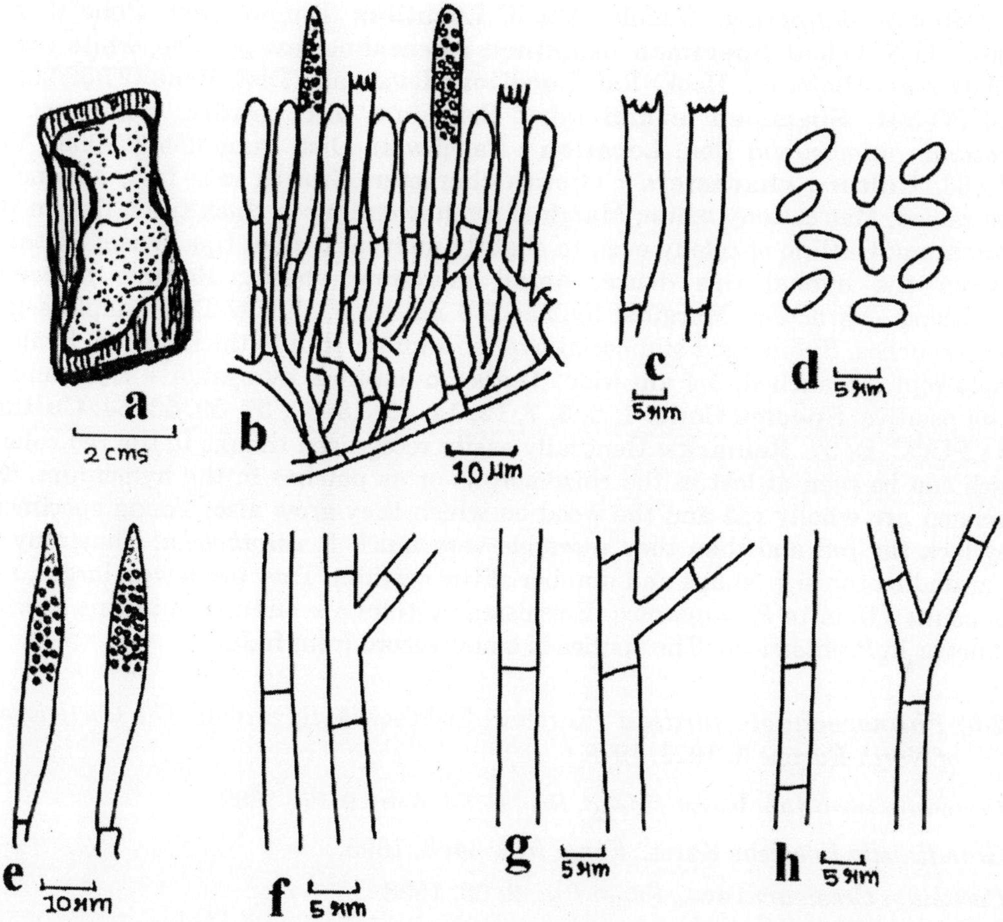

Fig. 58 *Phanerochaete sanguinea* a. Habit. b. T.S. hymenium. c. Basidia. d. Basidiospores. e. Encrusted cystidia. f. Generative hyphae. g. Marginal hyphae in culture. h. Submerge and aerial hyphae in culture

Morphological Characters : Basidiomata resupinate, loosely adnate, effused, membranous, 3-4 × 1-2 cms. and up to 500 μm thick; hymenial surface straw, buff, salmon, saffron to orange, smooth, sometimes cracked; margin concolorous with the hymenial surface, indeterminate, fimbriate with hyphal strands. In section subhyaline to pale red, pellicular to submembranous. Hyphal system monomitic; hyphae 3.5-6 μm wide, smooth, clampless, thin- to thick-walled. Cystidia (leptocystidia) cylindrical to subfusiform, 30-50 × 4.5-6 μm, smooth, thin- to slightly thick-walled, projecting up to 30 μm beyond the hymenial surface. Basidia narrowly clavate, 20-33 × 4.5-5.5 μm, without a basal clamp, producing 4 sterigmata. Basidiospores ellipsoid to subcylindrical, 5-6 × 2.5-3 μm, smooth, thin-walled, non-amyloid. **Habitat :** On live standing tree. **Distribution :** China, Japan, Taiwan, Europe, North America. **Specimen examined :** Benzidine test positive, white rot on: *Mallotus philippensis* (Lamk.) Muell. **Location:** Tamini, Dist. Pune (27/08/2006), VH-NPO-530. **Specimen examined :** Benzidine test positive, white rot on *Delonix regia* (Bojer ex. Hook) Raf. **Location:** Lonawala, Dist. Pune (27/08/2006), VH-NPO-504. **Specimen examined :** Benzidine test positive, white rot on *Jacaranda mimosifolia* Don. **Location :** Panchwati, Dist. Pune (24/09/2005), VH-NPO-63. **Culture characters :** Growth characters:Growth rate fast, advancing zone raised; Mat cottony white; Marginal hyphae ending in thick tufts, flat on the substratum; Outline of colony even to slightly uneven; hyphal tips dense; Distance between the hyphal tips dense; Aerial mycelium absent; Reverse bleached. Microscopic characters: Marginal hyphae 5-6 μm wide, thin walled, simple septa, often branches. Submerged and aerial hyphae similar, thin to thick-walled, hyaline, simple septa, branched, 3-4 μm wide. **Tests :** α-napthal, pyrogallol + H_2O_2 and p-cresol positive. **Species Code:** 1, 2, 3, 7, 12, 14, 15, 16, 30, 37, 39, 53, 54. **Culture No :** PUCC. B: 37. **Remarks:** Generally easily recognized thanks to the red colour, which can be seen at lest in the rhizomorphs or as patches in the hymenium. Old specimen are wholly red and the wood on which they grow also. Young specimens may lack the red and then they resemble very much *P. calothricha*. They may be separated by the size, shape and number of the cystidia. They are fewer, larger, and less pointed than in *P. sanguinea*. Encrusted cystidia are often seen in this species, but never in *P. calotricha*. The species is a new record from India.

(22-6) *Phanerochaete sordida* (Karsten) Eriksson & Ryvarden. *The Corticiaceae of North Europe* **5:** 1023, 1978.

= *Corticium sordidum* Karst., *Medd. Soc. F. F1. fenn.* **9:** 65, 1882.

= *Grandiniella livescens* Karst., *Fungi frid.* **34:** 8, 1895.

= *Corticium Cremeum* Bres., *Fungi frid.* **2:** 63, 1898.

= *Corticium eichlerinum* Bres. *Ann. Mycol.* **1:** 95, 1903.

= *Peniophora eichleriana* Sensu Bourdot and Galzin, Hym. De France: 306, 1928. (Plate 5, 10, Fig. 59)

Morphological Characters : Basidiomata resupinate, adnate, effused, membranous, 2. 5-3.5 × 0.5-1 cm. and 0.2-0. 5 mm. thick; hymenial surface yellowish white, buff, vinaceous buff to ochreous, sometimes becoming pale brown in 3% KOH, smooth, sometimes cracked; margin concolorous with the hymenial surface, thinning out, indeterminate, without hyphal strands. In section subhyaline, pellicular to sub-membranous. Hyphal system monomitic, hyphae 5-7 μm wide, smooth, clampless, thin-to thick-walled; cystidia (leptocystidia) fusiform to subcylindrical, 40-70 × 6-10 μm, without a basal clamp, smooth, thin- to thick-walled, encrusted over the upper half with large subhyaline crystals; basidia narrowly clavate to subcylindrical, 25-35 × 4.5-5.5 μm, without a basal clamp, producing 4 sterigmata; basidiospores ellipsoid to subcylindrical, 5-7.5 × 2.5-3.5 μm, smooth, thin-walled, non-amyloid. **Habitat :** On dead standing tree **Distribution :** Brazil, Canada, France, England, BritishColombia, India. **Specimen examined :** Benzidine test positive, white rot on: *Mallotus philippensis* (Lamk.) Muell. **Location :** Pune University, Dist. Pune (07/12/2006), VH-NPO-500.

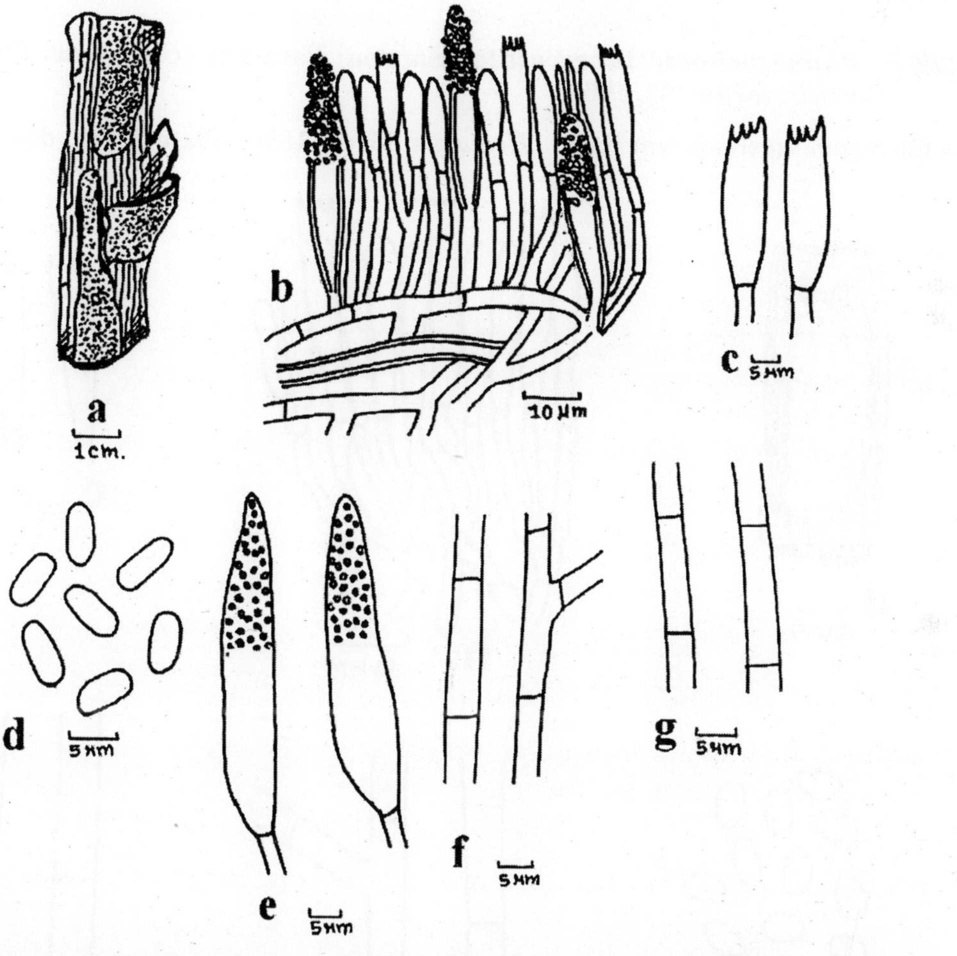

Fig. 59 ***Phanerochaete sordida*** a. Habit. b. T.S. hymenium. c. Basidia. d. Basidiospores. e. Encrusted cystidia. f. Generative hyphae. g. Generative hyphae in culture

Culture characters: Growth characters: Growth rate moderate; advancing zone raised; Mat cottony white, giving out hyphae in fan like structures; Marginal hyphae ending in thick hyphal tufts, but flat on surface; Marginal hyphae thick; Outline of colony bayed; hyphal tips dense and interwoven in tufts; Reverse completely bleached; Aerial mycelium absent. Microscopic characters: Generative hyphae thin-walled, 3-6 μm. wide, smooth, hyaline. **Tests :** α-napthal, pyrogallol +

H_2O_2 and p-cresol positive. **Species Code :** 1, 2, 3, 4, 8, 12, 13, 14, 15, 16, 30, 37, 53, 54, 72. **Culture No :** PUCC. B: 38. **Remarks** : The species shows a considerable variation among the specimen collected. The species is recognized by more or less thick-walled subicular hyphae, branched, at right angles and running in all directions, forming a well intertwined but open context. The cystidia are variable in shape, size, and number, thin-walled and little encrusted, which are characteristic of the species.

(22-7). *Phanerochaete tuberculata* (Karsten) Parmasto. *Conspectus Systematis Corticiacearum:* 83, 1968.

= *Corticium tuberculatum* Karst., *Hedwigia* **35:** 45, 1896. (Plate 5, Fig. 60)

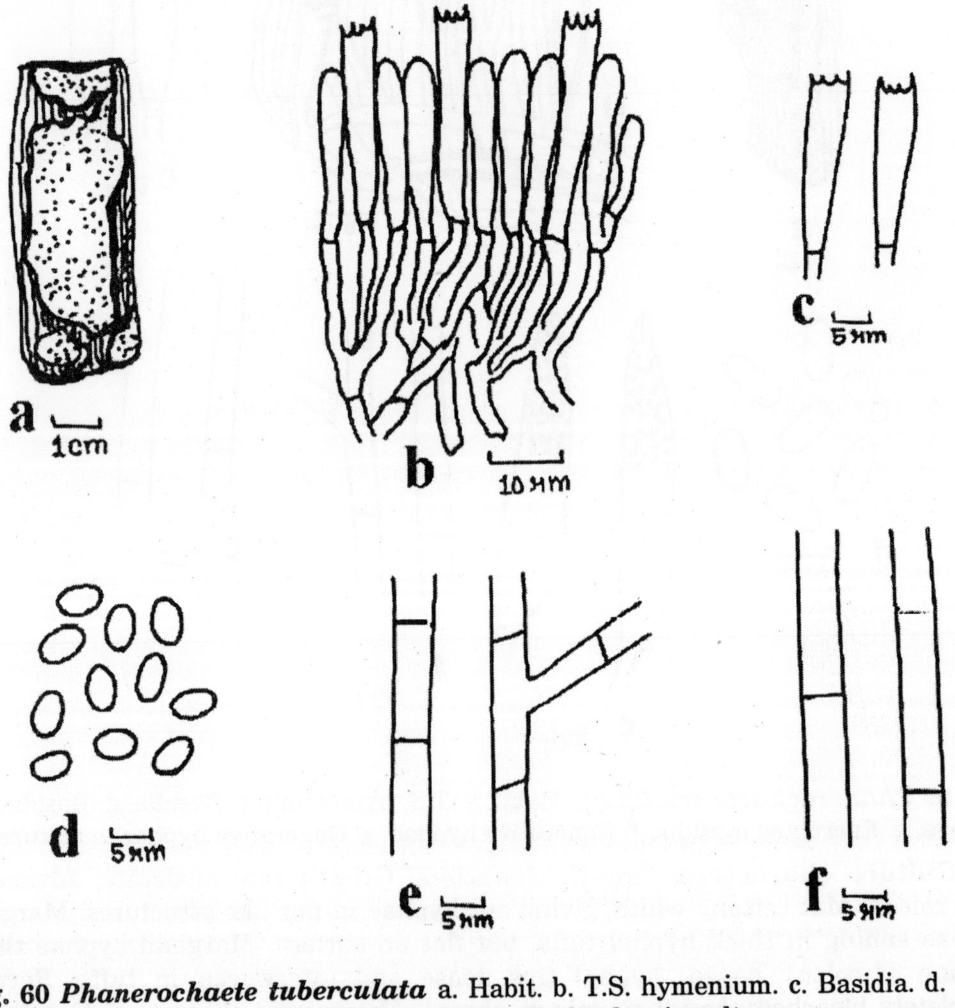

Fig. 60 *Phanerochaete tuberculata* a. Habit. b. T.S. hymenium. c. Basidia. d. Basidiospores. e. Generative hyphae. f. Generative hyphae in culture

Morphological Characters: Basidiocarps becoming broadly effused, adnate, 3-4 × 1-2 cms., membranous; hymenial surface tuberculate, deeply cracked when dry, cinnamon-buff to clay color; margin thinning out, white, arachnoid. Hyphal system monomitic, hyphae thin-walled, hyaline, simple-septate, 3.5-5 μm diam. Cystidia none. Basidia clavate, with 4 sterigmate, simple-septate at base, 25-35 × 5-6 μm. Basidiospores ellipsoid, 6-9 ×3.5-5 μm, smooth, non-amyloid. **Habitat :** On live standing tree. **Distribution:** France, U.S.A. **Specimen examined :** Benzidine test positive, white rot on: *Ficus bengamina* L. **Location :** Pune university, Dist. Pune (26/08/2006), VH-NPO-524. **Culture characters :** Growth characters: Growth rate fast; advancing zone raised; Mat white yellowish; Outline of colony diffuse; hyphal tips dense; Distance between the hyphal tips dense; Aerial mycelium absent; Odor none, Reverse unchanged. **Microscopic characters:** Generative hyphae 3-5 μm wide, thin walled, simple septate. **Tests :** α-napthal, pyrogallol + H_2O_2 and p-cresol positive. **Species Code :** 1, 2, 3, 8, 12, 14, 16, 31, 37, 53. **Culture No :** PUCC. B: 39. **Remarks :** *P. tuberculata* is characteristic and easily distinguished. Even if it varies with age and growth condition, it is more uniform and constant than the other species of the genus. Though lacking cystidia altogether there seem to be good reason for placing it in *Phanerochaete.* The species is a new record from India.

(23) *PHLEBIA* Fries. Systema Mycologicum 1: 426, 1821.

The genus *Phlebia* was proposed by Fries in 1821, with type species *Phlebia radiata* Schrad. The genus is cosmopolitan in distribution with 180 species recorded. Many species are tropical to subtropical in distribution. The genus is characterized by resupinate, effuso-reflexed or dimidiate, cartilaginous, subgelatinous or ceraceous basidiomata; hymenial surface smooth, tuberculate, odontoid, phlebioid or merulioid. In section compact, tightly agglutinated in subiculum, sometimes dimorphic with loosely intertwined hyphae near substrate or reflexed abhymenial layer composed of loosely inter-twined hyphae;hyphal system monomitic; hyphae smooth, thin- to thick-walled, usually with clamp, with or without a gelatinous matrix; cystidia present or lacking; basidiospores ellipsoid to allantoid, smooth, thin-walled, non-amyloid, acyanophilous. The specimen with characteristic reticulate folded to merulioid, reddish hymenophore, present of cystidia; narrowly ellipsoid to cylindrical, non-amyloid basidiospores was identified as *P. rufa* (Persoon) Christiansen. The specimen is recognized by pale yellow to cream hymenophore and absence of cystidia the specimen was identified as *P. subcretacea* (Litschauer) Christiansen.

KEY TO THE SPECIES OF *PHLEBIA*

1. Hymenophore reticulately folded to merulioid; reddish; Cystidia present; basidiospores 3-4 × 1-2 µm .... ***P. rufa***

1´. Hymenophore folded to merulioid; pale yellow to cream; Cystidia absent; basidiospores 4-6 × 2.5-3 µm ....***P. subceracea***

(23-1). *Phlebia rufa* (Persoon) Christiansen. *Dansk Bot. Ark.* **19:** 164, 1960.

=*Merulius rufus* Pers., *Syn. meth. fung.* (Göttingen) **2:** 498, 1801.

=*Serpula rufa* (Pers.) P. Karst., *Hedwigia* **35:** 45, 1896. (Plate 6, 11, Fig. 61)

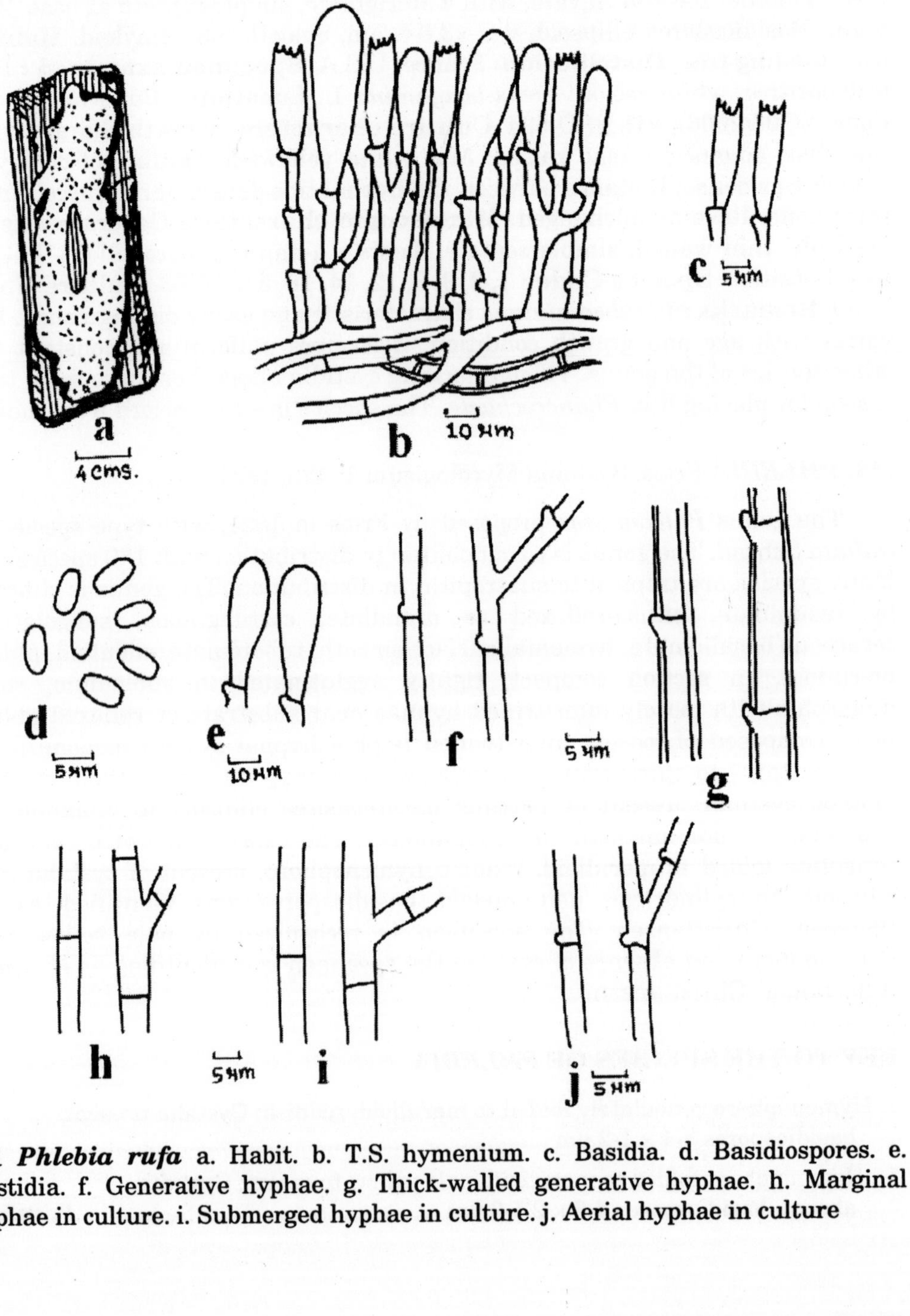

Fig. 61 *Phlebia rufa* a. Habit. b. T.S. hymenium. c. Basidia. d. Basidiospores. e. Cystidia. f. Generative hyphae. g. Thick-walled generative hyphae. h. Marginal hyphae in culture. i. Submerged hyphae in culture. j. Aerial hyphae in culture

Morphological Characters: Basidiomata resupinate, adnate, 17-20 × 5-7 cms. and up to 350 μm thick; hymenial surface variable in color and shape, usually white, rosy buff, ochreous umber to fawn, tuberculate, raduloid; margin concolorous with the hymenial surface, thinning out, often reflexed when dry. In section subhyaline, ceraceous, with a basal layer. Hyphal system monomitic; hyphae 3-5 μm wide, smooth, thin- to thick-walled, with clamp. Cystidia cylindrical to clavate, 50-80 × 8-12 μm, with a basal clamp. smooth. thin-walled. Basidia narrowly clavate 25-35 × 4-4.5 μm. with a basal clamp. producing 4 sterigmata. Basidiospores narrowly ellipsoid to cylindrical, 4.5-6 × 2-2.5 μm, smooth, thin-walled, non-amyloid. **Habitat** : On dead standing tree. **Distribution:** Chin, Iran, Japan, Europe, North and South America, India. **Specimen Examined**: Benzidine test positive, white rot on *Memecylon umbellatum* Burm. **Location:** Koyna: Dist: Satara (11/02/2007), VH-NPO-600. **Culture characters:** Growth characters: Growth rate fast; advancing zone appressed to submerged; Mats white, downy to woolly, sometimes developing cottony mounds at mat periphery, slightly raised, felty to tomentose, then becoming raised, thick, felty to woolly toward margins, outline of colony even; Distance between hyphal tips distant; odor absent or sour; Reverse unchanged or partially bleached. Microscopic characters: Marginal hyphae 4-7 μm diam, thin walled to slightly thick walled at first, becoming thick walled, simple septate, sparsely to moderately branched. Submerged hyphae broad, 6-8 μm diam, slightly thick walled to thick walled, simple septate, sparsely branched; Aerial hyphae 3-5 μm wide, thin walled to slightly thick walled with some becoming roughwalled, clamped. **Tests:** α-napthal and pyrogallol + H_2O_2 positive; p-cresol negative. **Species Code:** 1, 2, 3, 7, 13, 14, 15, 16, 30, 36, 37, 53, 54, (55). **Culture No:** PUCC. B: 40. **Remarks:** The species is recognized by reddish hymenophore and present of cystidia. The first record from Maharashtra in India.

(23-2) *Phlebia subcretacea* (Litsch.) Christ. *Dansk Bot. Ark.* **19:**165, 1960.

= *Corticium subcretaceum* Litsch., *Öst. bot. Z.* **88:** 110, 1939.

=*Jacksonomyces subcretaceus* (Litsch.) Jülich, *Persoonia* **11(4) :** 427, 1982.

=*Phlebia aerugineolivida* (Romell ex S. Lundell) Donk, *Fungus, Wageningen* **27:** 12, 1957.

=*Phlebia lichenoides* Parmasto, *Eesti NSV Tead. Akad. Toim.,* Biol. seer **16(4)** : 391, 1967. (Plate: 6; Fig. 62).

Morphological Characters : Basidiomata annual, resupinate, 3. 5 × 2.5 cms. and up to 200 μm thick, firmly attached, thin layered, soft and subgelatinous when fresh, hard and rigid when dry; hymenophore light yellow to cream, smooth to tuberculate or toothed, margin thin to abrupt, concolorous with the hymenial surface. Context thin or almost absent. Hyphal system monomitic;generative hyphae thin walled, septate, clamped, branched, hyaline, 2-4 μm wide, compactly arranged. Cystidia absent. *Basidia clavate*, thin walled, up to 20-25 × 5-7 μm; with 4 sterigmata. Basidiospores hyaline, cylindrical to suballantoid, thin walled, smooth, guttulate, non-amyloid, acyanophilous, 4-6 × 2.5-3 μm; basidioles clavate, guttulate, up to 15-20 × 5 μm. **Habitat :** On dead standing tree. **Distribution :** India-Tamilnado. **Specimen examined :** Benzidine test positive, white rot on *Pithecolobium saman* Benth. **Location :** Symboisis road, Dist. Pune (16/09/2005), VH-NPO-27. **Remarks :** Specific features of this species is the pale yellow to cream hymenophore and absence of cystidia. The culture was not obtained besides repeated trials. The species is a new record from Maharashtra.

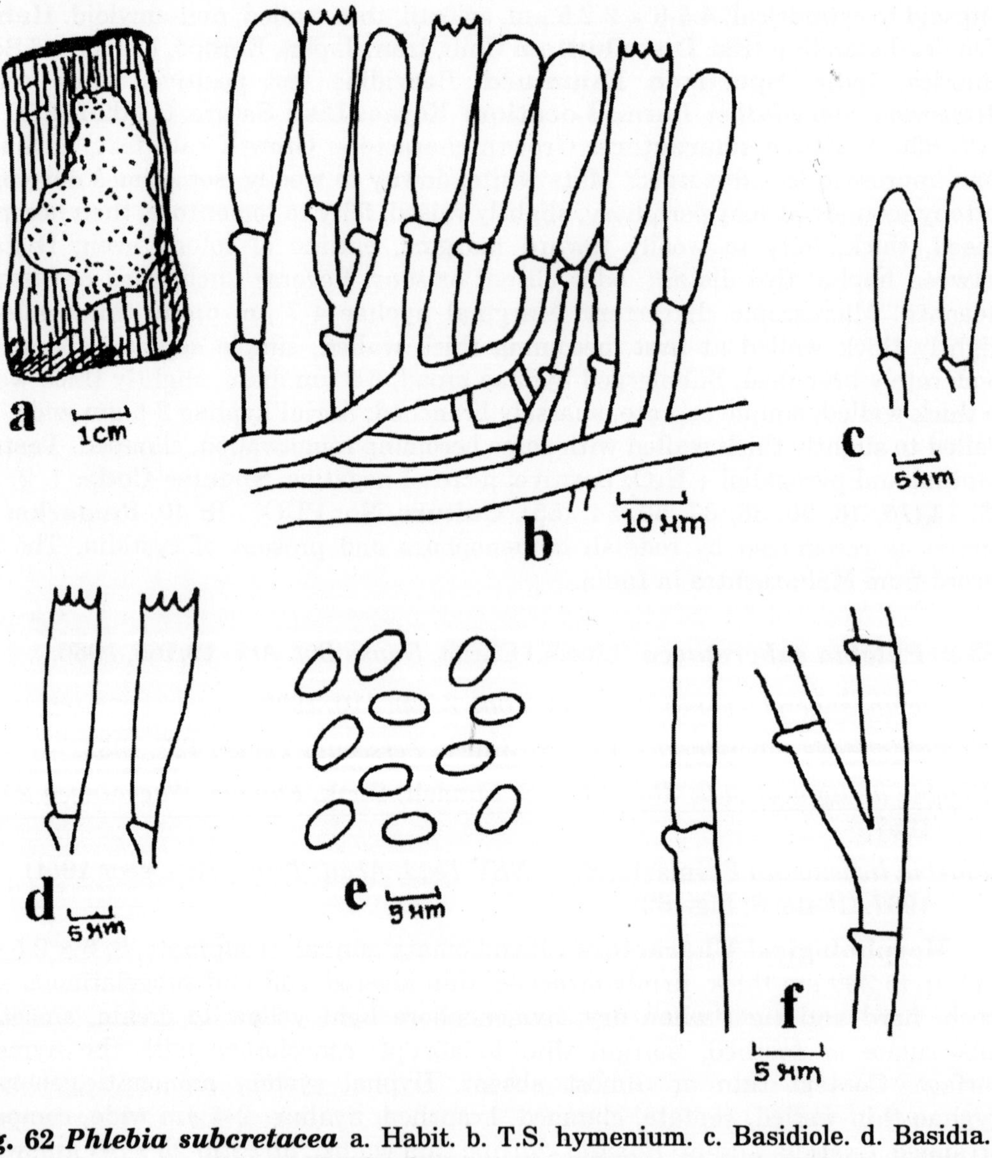

Fig. 62 ***Phlebia subcretacea*** a. Habit. b. T.S. hymenium. c. Basidiole. d. Basidia. e. Basidiospores. f. Generative hyphae

(24) *PHLEBIOPSIS* Jülich. *Persoonia* **10 (1) :** 137, 1978.

The genus was created by Julich in 1978, with type species *Phlebiopsis gigantea* (Fr.) Jülich. Ten species have been recorded under this genus.

The genus is characterized by resupinate, adnate; effused basidiomata; hymenial surface smooth or tuberculate; in section subhyaline to pale brown.; hyphal system monomitic; hyphae mostly clampless; cystidia (lamprocystidia) numerous, conical, thick-walled, heavily encrusted over the upper part; basidiospores narrowly ellipsoid to subcylindrical, thin-walled, non-amyloid, non-cyanophilous. From the investigation area four specimens were collected. The specimen with characteristic resupinate, effuse, adnate fruitbodies; hymenium smooth, light ochraceous, subiculum creamish and easily observed, tough; without clamps was identified as *P. galochroa* (Bresadola) Hjortstam & Ryvarden. The specimen is recognized by grayish white, "Buff" to "Pale Luteous" hymenial surface, basidiospores narrowly ellipsoid to sub-cylindrical the specimen was identified as *P. gigantea* (Fries) Jülich. The specimen with characteristic smooth, minutely farinose (pale violaceous gray) hymenial surface; pale vinaceous brown subiculum.; abundant, imbedded throughout subiculum and also projecting from hymenium, conical, heavily encrusted with coarse crystals cystidia; oblong basidiospores was identified as *P. peniophoroides* Gilbertson & Adaskaveg. The specimen is identified as *P. roumeguerei* (Bresadola) Jülich & Stalpers, due to the presence of hymenial surface whitish-pale ochraceous buff, avellaneus, wood brown- uneven; basidiospores oval.

KEY TO THE SPECIES OF *PHLEBIOPSIS*

1. Spores oblong or oval 2
1. Spores ellipsoid to sub-cylendrical 3
2. Spores oblong, 4-5 × 2-2.5 µm, cystidia 40 × 7-10 µm .... ***P. peniophoroides***
2´. Spores oval, 3.5-5 × 2-3 µm, cystidia 40-50 × 14-20 µm .... ***P. roumeguerei***
3. Spores ellipsoid 5-6 × 4 µm, cystidia 50-60 × 7-10 µm .... ***P. galochroa***
3´. Spores ellipsoid to sub-cylendrical, 5-6.5 × 2.5-3.5 µm, cystidia 45-90 × 8-15 µm .... ***P. gigantea***

(24-1) *Phlebiopsis galochroa* (Bresadola) Hjortstam & Ryvarden. *Mycotaxon* **10 (2) :** 285, 1980.

=*Peniophora galochroa* Bres., *Hedwigia* **35:** 290, 1896. (Plate 6, Fig. 63)

Morphological Characters : Fruitbody resupinate, effuse, adnate, 11-13 × 1-2 cms., hymenium smooth, light ochraceous, subiculum creamish and easily observed, tough, margin not especially differentiated. Hyphal system monomitic. Hyphae in KOH and Melzer's reagent lightrefracting, in cotton blue with the walls easily distingguishable but not cyanophilous; hyphae smooth 3-3.5 μm wide, non-amyloid or dextrinoid. Subhymenial hyphae branched, less light-refracting, usually thin-walled or thickened, all hyphae without clamps. Cystidia more or less imbedded, thick-walled and strongly encrusted (Metuloids), 50-60 × 7-10 μm. Basidia terminal, clavate or subcylindrical, 15-20 × 3.5-4.5 μm, with four sterigmata. Spores thin-walled, ellipsoid, usually 5-6 × 4 μm, non-amyloid or dextrinoid. **Habitat :** On fallen branches. **Distribution :** Brazil. **Specimen examined :** Benzidine test positive, white rot on *Memecylon umbellatum* Burm. **Location :** Amboli, Dist. Satara (19/02/2007), VH-NPO-806.

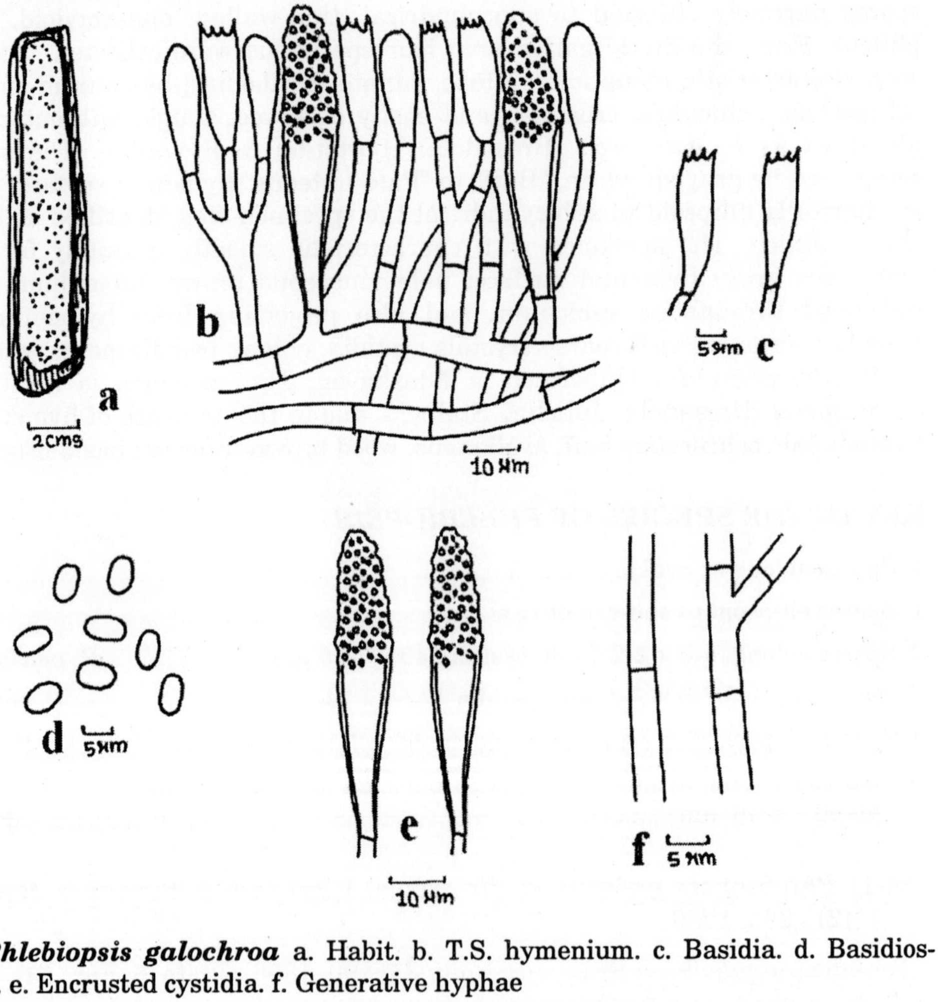

Fig. 63 ***Phlebiopsis galochroa*** a. Habit. b. T.S. hymenium. c. Basidia. d. Basidiospores. e. Encrusted cystidia. f. Generative hyphae

(24-2). ***Phlebiopsis gigantea*** (Fries) Jülich. *Persoonia* **10 (1)** : 137, 1978.

=*Corticium giganteum* (Fr.) Fr., *Epicr. syst. mycol.* (Upsaliae) : 559, 1838.

=*Kneiffia gigantea* (Fr.) Bres., *Annls mycol.* **1 (1/2)** : 99, 1903. (Plate 6, 11, Fig. 64)

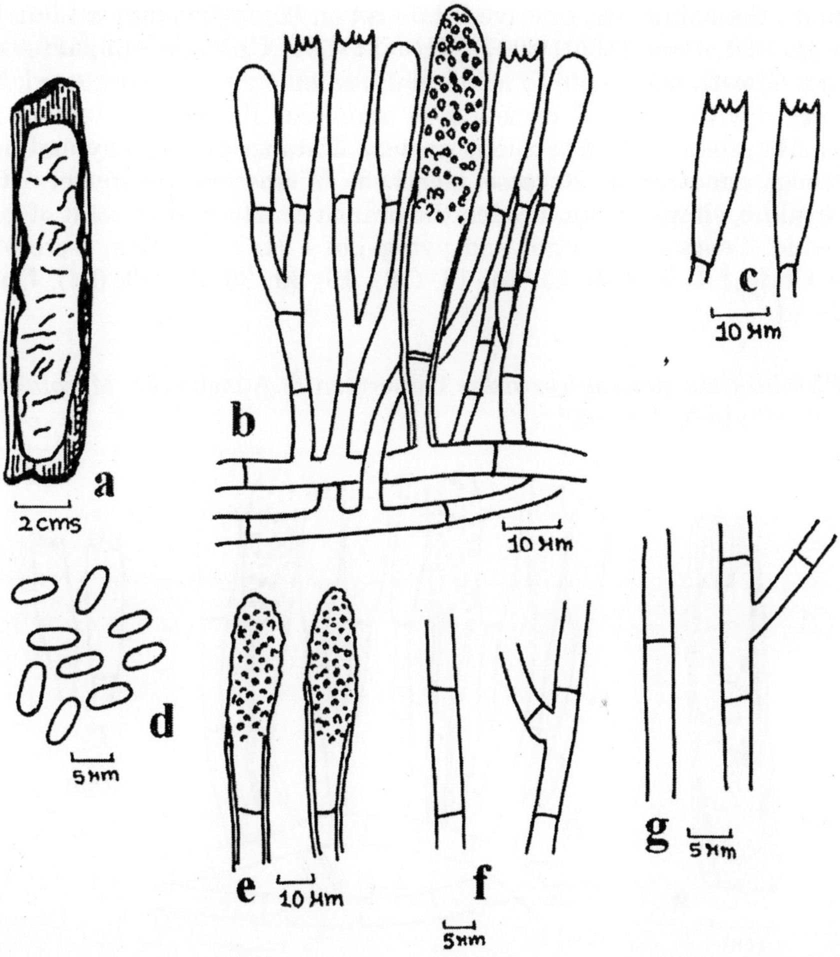

Fig. 64 ***Phlebiopsis gigantea*** a. Habit. b. T.S. hymenium. c. Basidia. d. Basidiospores. e. Encrusted cystidia. f. Generative hyphae. g. Generative hyphae in culture

Morphological Characters : Basidiomata resupinate, effused, adnate, 9-11 × 1-2 cms. and 200-600 μm thick; hymenial surface grayish white, buff to pale luteous, smooth, when dry cracked and sometimes rolled up from the subiculum; margin concolorous with the hymenial surface, thinning out, indeterminate. In section subhyaline, subpellicular to membranous. Hyphal system monomitic, hyphae 3-5 μm wide, smooth, clampless, thin- to slightly thick-walled. Cystidia (lamprocystidia) conical to subfusiform, 45-90 × 8-15 μm, smooth, thick-walled, heavily encrusted with subhyaline crystalloid materials over upper two-third of length. Basidia narrowly clavate to subcylindrical, 20-30 × 4-5 μm. producing 4 sterigmata. Basidiospores narrowly ellipsoid to sub-cylindrical, 5-6.5 × 2.5-3.5 μm, smooth, thin-walled, non-amyloid. **Habitat :** On fallen branches. **Distribution :** Widespread and apparently common throughout the boreal forests of the Northern, India. **Specimen**

Examined : Benzidine test positive, white rot on *Mangifera indica* Linn. **Location:** Baneshwar, Dist. Pune (13/01/2007), VH-NPO-591. **Culture characters :** Growth characters: Growth rate rapid to moderately; advancing zone appressed; Mat white, all submerged with no aerial mycelium or appressed floccose-farinaceous in more or less extensive areas; outline of colony uneven; Distance between hyphal tips distant; Odor strong, penetrating Reverse bleached. Microscopic characters: Generative hyphae hyaline, simple septa, branch the branch constricted at point of attachment, 3-6 μm wide. **Tests :** α-napthal and pyrogallol + H_2O_2 positive; p-cresol negative. **Species Code :** 1, 2, 3, 6, 13, 15, 16, (18), 19, 30, 36, 37, 53, (54). **Culture No:** PUCC. B: 41.

(24-3) *Phlebiopsis peniophoroides* Gilbertson & Adaskaveg. *Mycotaxon* **49:** 388, 1993. (Plate 6, Fig. 65)

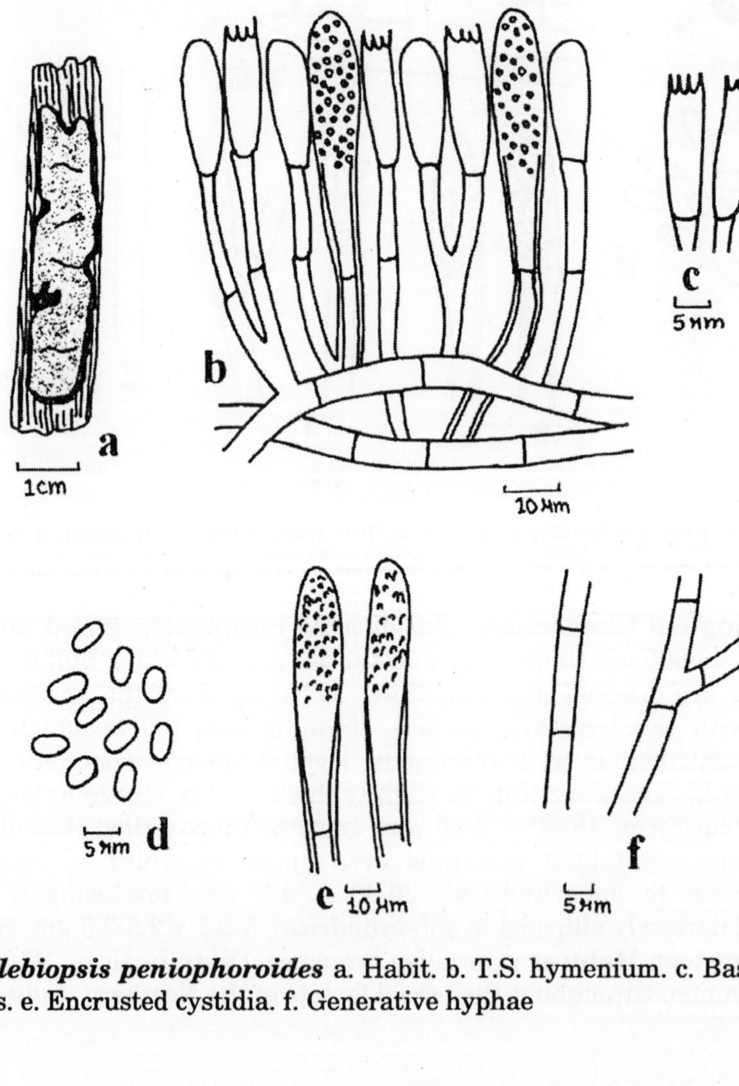

Fig. 65 *Phlebiopsis peniophoroides* a. Habit. b. T.S. hymenium. c. Basidia. d. Basidiospores. e. Encrusted cystidia. f. Generative hyphae

Morphological Characters : Basidiocarps annual, resupinate, adherent, effused, 5-8 × 2-4 cms.; hymenial surface, smooth, pale violaceous gray, margin thinning out, concolorous to whitish; subiculum very thin, pale vinaceous brown. Hyphal system monomitic; hyphae simple-septate, hyaline in KOH, thin-walled, with frequent branching, 3-5 μm wide, agglutinated in the subiculum; subhymenial hyphae branching repeatedly to form candelabrums of basidia. Cystidia abundant, conical, heavily encrusted with coarse crystals, 35-50 × 7-10 μm, hyphal base firm-walled. Gloeocystidia absent. Basidia in a dense palisade mixed with cystidia, narrowly clavate, 4-sterigmate, 20-25 × 3.5-5 μm, simple-septate at the base. Basidiospores oblong, hyaline, smooth, thin-walled, non-amyloid, 4-5 × 2-2.5 μm. **Habitat :** On fallen branches. **Distribution :** USA, Hawaii, India. **Specimen Examined** :Benzidine test positive, white rot on *Samanea saman* (Jacq.) Merr. **Location :** Sawantwadi, Dist. Satara (19/02/2007), VH-783.

(24-4) *Phlebiopsis roumeguerei* (Bresadola) Jülich & Stalpers. Verh. Kon. Ned. Akad. Wetensch., Afd. Natuurk., **II 74:** 190, 1980.

= *Corticium roumeguerei* Bres. [as 'roumeguerii'], *Fung. trident.* **2 (8-10) :** 36, 1892.

= *Kneiffia roumeguerei* (Bres.) Bres. [as '*roumeguerii*'], *Annls mycol.* **1 (1/2) :** 103, 1903.

= *Metulodontia roumeguerei* (Bres.) Parmasto, *Consp. System. Corticiac. (Tartu)*: 118, 1968.

= *Peniophora roumeguerei* (Bres.) Bres. [as 'roumeguerii'], in *Burt, Ann. Mo. bot. Gdn* **12:** 270, 1925.

= *Phlebia roumeguerei* (Bres.) Donk [as '*roumeguerii*'], *Fungus, Wageningen* **27:** 12, 1957. (Plate 6, Fig. 66)

Morphological Characters : Sporophores coriaceous, effused, adnate, 2-4 × 1-1.5 cms. and 0.5 mm thick; hymenial surface whitish-pale ochraceous buff, avellaneus, wood brown-uneven, minutely tuberculate; margin thinning out, concolorous or pallied, adnate. Context hyaline to subhyaline or light cream coloured in thicker sections, up to 450 μm thick, composed of a horizontal basal layer of compactly arranged hyphae and an intermediate layer of erect and interwoven hyphae with numerous cystidia arranged in distinct or overlapping parallel rows. Hyphal system monomitic, generative hyphae hyaline, branched, septate without clamp connections, thick-walled, 2-4 μm wide. Hymenium up to 40 μm deep, composed of basidia and cystidia. Cystidia abundant, hyaline, arising in the context and subhymenial layer, present in all portions except basal layer, conical with acute or obtuse apices, thick-walled with heavy encrustations, 40-65 × 12-15 μm, embedded or projecting up to 35 μm, beyond the hymenium. Basidia subclavate to clavate, 10-15 × 3-4.5 μm, 4 sterigmata. Basidiospores hyaline, oval, thin-walled, smooth, 4-5 × 2-3 μm. **Habitat :** On live standing tree. **Distribution :** U.S.A., England, France, Italy, India. **Specimen Examined**: Benzidine test positive, white rot on *Mangifera indica* Linn. **Location:** Dabhol; Dist:Ratnagiri (12/10/2005), VH-NPO-207.

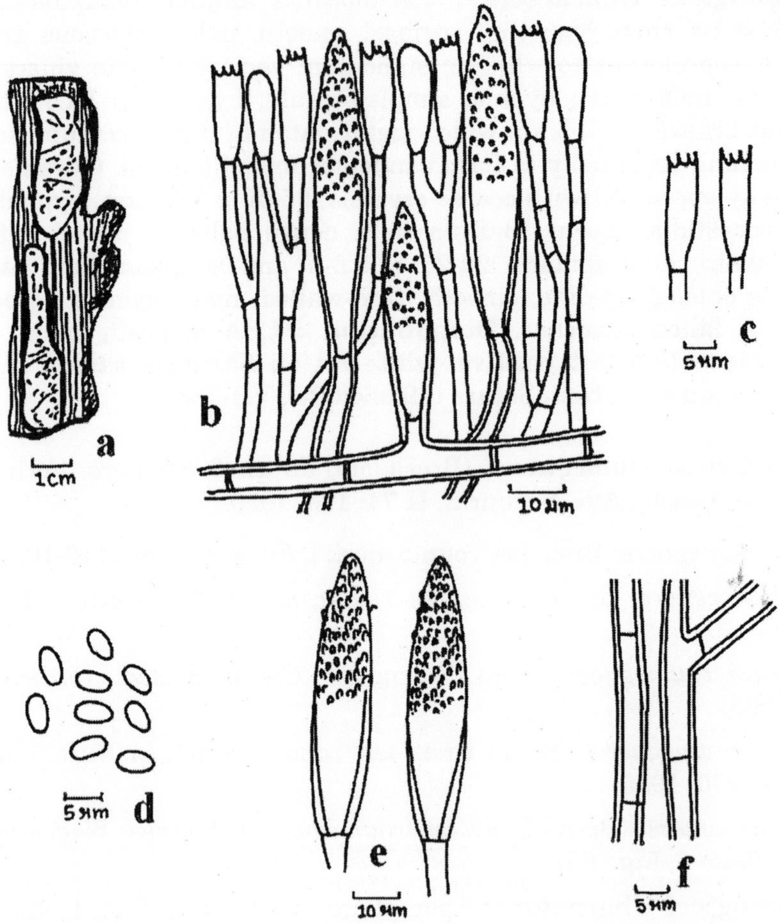

Fig. 66 ***Phlebiopsis roumeguerei*** a. Habit. b. T.S. hymenium. c. Basidia. d. Basidiospores. e. Encrusted cystidia. f. Generative hyphae

(25) ***PSEUDOXENASMA*** Larsson & Hjortstam. *Mycotaxon* **4(1)** : 307, 1976.

The genus was erected by Larsson & Hjortstam in 1976. The genus is with only one species. The genus is characterized by resupinate, effused, very thin, ceraceous fruitbodies, hymenium normally smooth; hyphal system monomitic; hyphae with clamps, thin-walled, densely interwoven; sulfocystidia always present, often with apical, globular schizopapilles and filled with grainy protoplasmatic contents, light-refracting in KOH (phase); spores broadly ellipsoid-subglobose, thick-walled, verrucose, strongly amyloid. The specimen with characteristic broadly ellipsoid to subglobose, thickwalled; verrucose and with strongly amyloid reaction basidiospores was identified as *P. verrucisporum* Larsson & Hjortstam.

(25-1) *Pseudoxenasma verrucisporum* Larsson & Hjortstam. *Mycotaxon* **4 (1)** : 307, 1976. (Plate 6, Fig. 67)

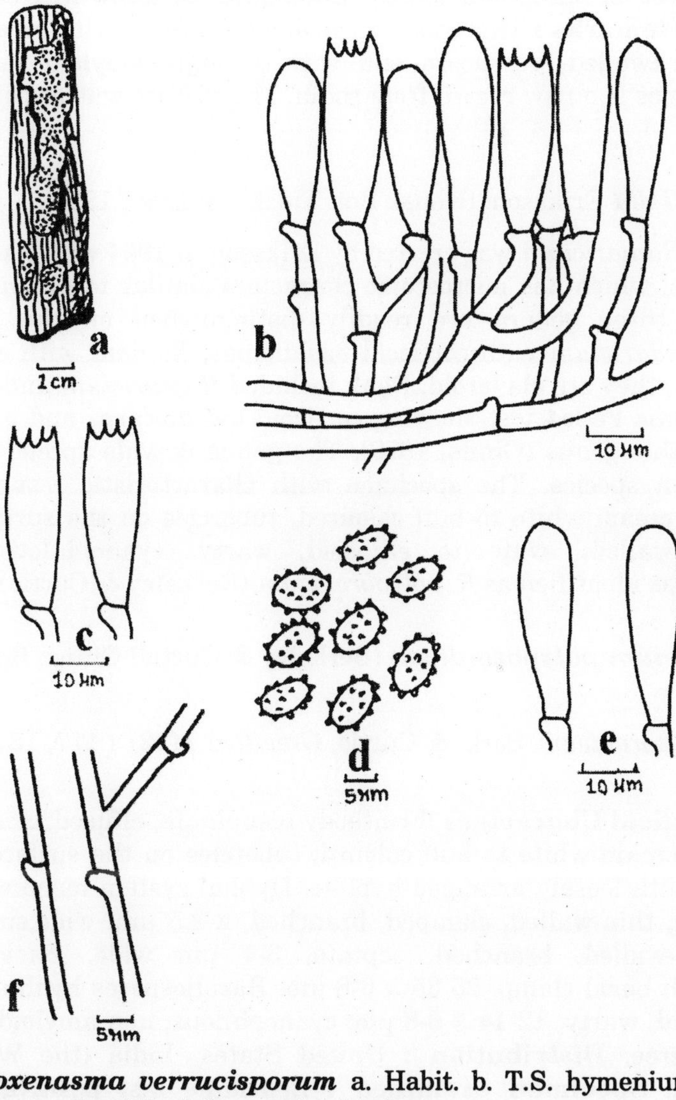

Fig. 67 *Pseudoxenasma verrucisporum* a. Habit. b. T.S. hymenium c. Basidia. d. Basidiospores. e. Cystidia f. Generative hyphae

Morphological Characters : Fruitbody resupinate, effused, 4-6 × 1-2 cms., closely attached to the substrate, ceraceous, thin, whitish to greyish white, hymenium pruinose under the lens or slightly furfuraceous, in dried condition almost hard, margin indistinctly thinning out. Hyphal system monomitic, hyphae narrow, thin-walled, 2-3 μm wide, with clamps, branched. Sulphocystidia rather frequent, 20-35 × 5-9 μm. Basidia clavate, 25-35 × 7-9 μm, with four sterigmata. Basidiospores broadly ellipsoid to subglobose, thickwalled; verrucose 8-10 × 7-8 μm, and with strongly amyloid reaction. **Habitat :** On live standing tree. **Distribution :** Sweden, Halland, Norway, Dalsland. **Specimem Examined:** Benzidine test positive, white rot on *Mangifera indica*. **Location:** Baneshwar, Dist. Pune (13/01/2006), VH-546. **Remarks :** This species is easily recognized with broadly ellipsoid to subglobose, thickwalled, verrucose and with strongly amyloid reaction basidiospores. The species is a new record from India. The culture was not obtained besides repeated trials.

(26) *RAMARICIUM* Eriksson. Svensk Bot. Tidskr. 48: 189, 1954.

The genus *Ramaricium* was erected by Eriksson in 1954 with its type species *R. occultum* Erikss, which has microscopic characters similar to *Kavinia* and some of *Ramaria*. The three genera are readily distinguished on the basis of gross morphology, *Ramaricium* with corticoid basidiomes, *Kavinia* with effused-hydnoid and in *Ramaria*, they are clavaroid. Donk included *Ramaricium* under Gomphaceae with characteristic FeSo4 test showing corticoid basidiocarps and only five species are known for this genus (Ginns, 1979). The genus is wide spread in distribution with five known species. The specimen with characteristic resupinate, effused, closely adnate, cream white to buff coloured, tubercles on the surface fruitbodies; hyaline, thick-walled, ovate to ellipsoid, warty, cyanophilous, non-amyloid basidiospores was identified as *R. polyporoideum* (Berkeley & Curtis) Ginns.

(26-1) *Ramaricium polyporoideum* (Berkeley & Curtis) Ginns. Bot. Not. 132: 98, 1979.

=*Corticium polyporoideum* Berk. & Curtis, *Grevillea* **1(12) :** 177, 1873. (Plate 6, 11, Fig. 68)

Morphological Characters : Fruitbody resupinate, effused, closely adnate, 10-12 × 3-4 cms., cream white to buff colourd, tubercles on the surface, margin pale, context broad with loosely arranged hyphae. Hyphal system monomitic, generative hyphae hyaline, thin-walled, clamped, branched, 2-2.5 µm. wide;subicular hyphae typically thick-walled, branched, septate, 3-4 µm wide. Basidia clavate, 4 sterigmate, with basal clamp, 25-35 × 6-8 µm. Basidiospores hyaline, thick-walled, ovate to ellipsoid, warty, 12-14 × 6-8 µm; cyanophilous, non-amyloid. **Habitat :** On live standing tree. **Distribution :** United States, India (the Western Ghats), Canada, Brazil. **Specimen examined :** Benzidine test positive, white rot on *Mangifera indica* Linn. **Location :** Baneshwar, Dist. Pune (11/02/2007), VH-NPO-763. **Culture characters :** Growth characters: Growth rate moderate; advancing zone raised; Mat white to grayish, downy to downy-woolly and tough, margin even to uneven, submerged spares, translucate. Reverse bleached. Microscopic characters: Aerial hyphae hyaline to yellowish, thin-walled, 2.5-5 μm wide, with clamps, marginal hyphae 3-5 μm. wide, with clamps, branched, thin-walled. **Tests :** α-napthal and pyrogallol + H_2O_2 positive; p-cresol negative. **Species Code :** 1, 2, 3, 9, 12, 15, 17 (22), 30, 37, 39, 52, 53. **Culture No :** PUCC. B: 42. **Remarks :** This species is differentiated from *R. albo-ochreaceum* by cream white to buff coloured basidiocarp and large basidiospores.

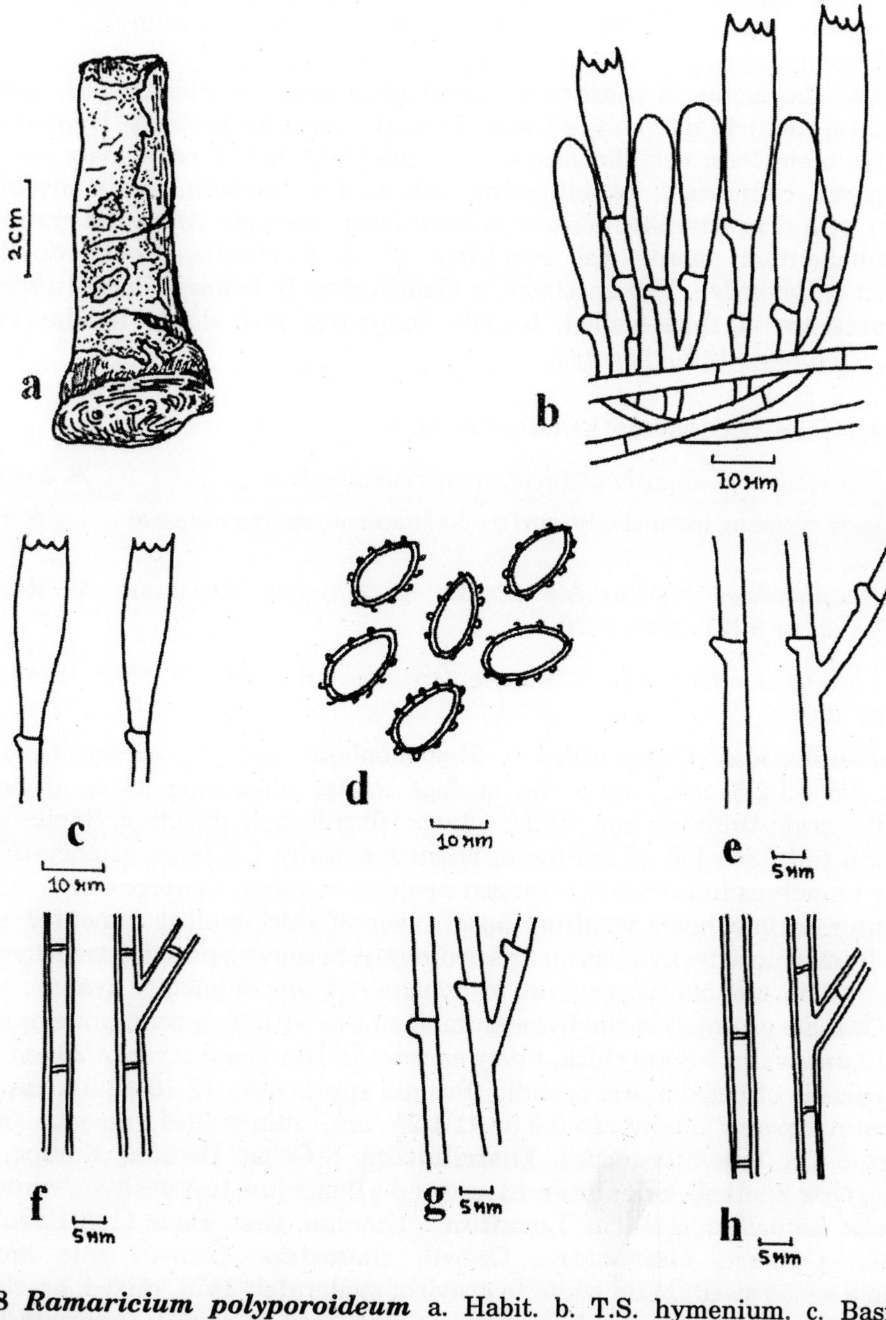

Fig. 68 ***Ramaricium polyporoideum*** a. Habit. b. T.S. hymenium. c. Basidia. d. Basidiospores. e. Generative hyphae. f. Thick-walled generative hyphae. g. Aerial hyphae in culture. h. Marginal hyphae in culture

(27) *SCOPULOIDES* (Massee) Höhnel & Litschauer. Wiesner Festschrift: 57, 58, 1908.

This genus erected by Höhnel & Litschauer in 1908, by studying the type of species *Scopuloides rimosa* (Cooke) Jülich. The genus has six species reported under this genus. The genus is characterized by resupinate, adnate, effused, ceraceous basidiomata; odontoid hymenial surface. In section subhyaline waxy. Hyphal system monomitic, clampless; cystidia numerous, conical, thick-walled heavily encrusted; basidiospores cylindrical to ellipsoid, thin-walled, smooth, non-amyloid. The specimen with characteristic dense hymenial layer, strongly encrusted cystidia and small suballantoid spores was identified as *S. hydnoides* (Cooke & Massee) Hjortstam & Ryvarden. The specimen is identified as *S. rimosa* (Cooke) Jülich, due to the presence of thick-walled, heavily encrusted with dense hyaline crystals cystidia, narrowly ellipsoid spores.

KEY TO THE SPECIES OF *SCOPULOIDES*

1. Fruitbody smooth to slightly hydnoid, spores suballantoid .... ***S. hydnoides***

1′. Fruitbody cracking extensively, surface hydnaceous, spores ellipsoid .... ***S. rimosa***

(27-1) *Scopuloides hydnoides* (Cooke & Massee) Hjortstam & Ryvarden. *Mycotaxon* **9 (2)** : 509, 1979.

=*Peniophora hydnoides* Cooke & Massee, *Grevillea* **16 (79)** : 77, 1888. (Plate 7, 11, Fig. 69)

Morphological Characters : Hymenophore annual, ceraceous, adnate, effused, 3-5 ×2-2.5 cms., hymenial surface white, remaining so on drying, not creviced; margin thinning out, white, adnate, fibrillose or pruinose. Spines conical, 90-130 μm tall, crowded or scattered, bearing apically 1-3 large septocystidia and laterally numerous imbricately arranged fusiform cystidia. Context white, 50-70 μm thick, intermediate layer wanting, basal layer of thick-walled cemented parallel hyphae from which the hymenium arises directly, becoming pseudoparenchymatous; Hyphal system monomitic, generative hyphae 5-7 μm diameter, hyaline, without clamp. Cystidia crowded in the hymenium, fusiform with long acuminate apices, 30-50 × 8-15 μm, walls 1-3 μm thick, finely encrusted. Hymenial layer to 20 μm deep, a dense palisade of basidia and cystidia. Basidia subclavate, 12-16 ×3.5-4 μm, with 4 sterigmata. Spores allantoid, 3-3.5 ×1-1.25 μm, thin-walled, smooth, hyaline. **Habitat :** On fallen branches. **Distribution :** Great Britain, Europe, North America, New Zealand. **Specimen examined :** Benzidine test positive, white rot on *Memecylon umbellatum* Burm. **Location :** Thamini, Dist. Pune (13/01/2007), VH-NPO-603. **Culture characters:** Growth characters: Growth rate moderate; advancing zone raised; Mats white to grayish, moderately thin, raised, arachnoid or silky to cottony; Outline of colony uneven, appressed to raised, fimbriate Reverse bleached. Microscopic characters: Generative hyphae thin–walled, hyaline, 5-8 μm., in wide, simple septate. **Tests:** α-napthal, pyrogallol + H_2O_2 and p-cresol positive. **Species Code :** 1, 2, 3, 8, 12, 14, 15, 30, 36, 37, 54, (55). **Culture No :** PUCC. B: 43. **Remarks :** Easily recognized species thanks to its dense hymenial layer, strongly encrusted cystidia and small suballantoid spores. The species is a new record from India.

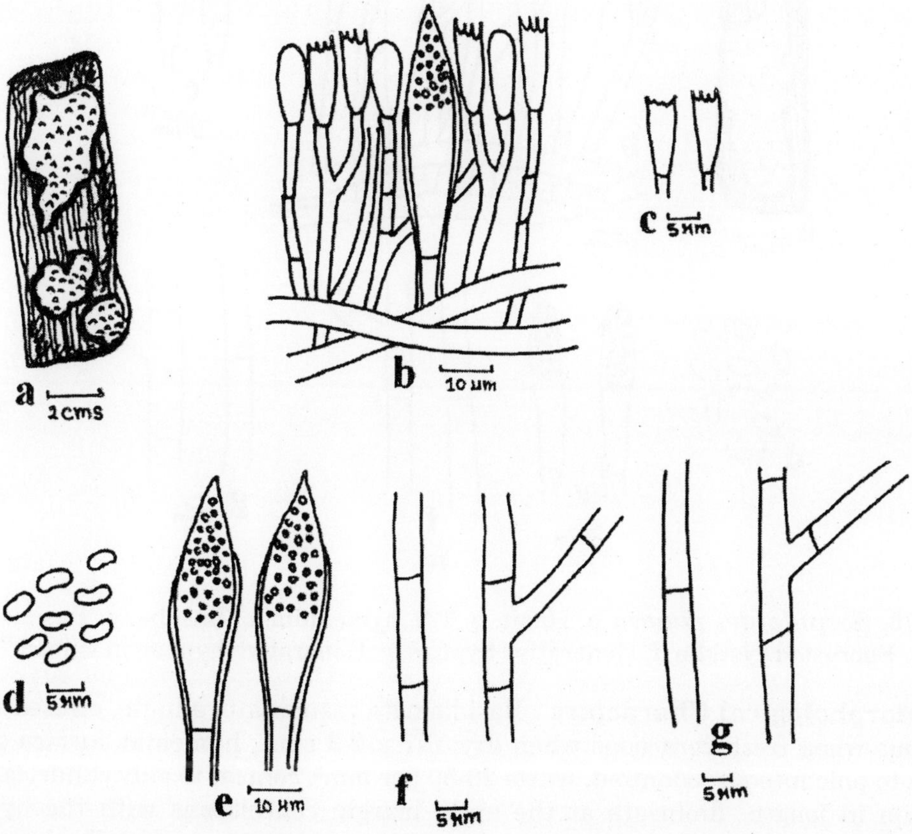

Fig. 69 ***Scopuloides hydnoides*** a. Habit. b. T.S. hymenium. c. Basidia. d. Basidiospores. e. Thick-walled encrusted cystidia. f. Generative hyphae. g. Generative hyphae in culture

(27-2). *Scopuloides rimosa* (Cooke) Jülich. Persoonia **11 (4) :** 422, 1982.

=*Odontia conspersa* Bres., *Atti Imp. Regia Accad. Rovereto, ser.* **3 3 (1) :** 10, 1897.

=*Peniophora rimosa* Cooke, *Grevillea* **9 (51) :** 94, 1881.

=*Peniophora terrestris* Massee, *Grevillea* **15 (76) :** 107, 1887.

=*Phanerochaete rimosa* (Cooke) Burds., *Mycol. Mem.* **10:** 107, 1985. (Plate 7, 11, Fig. 70)

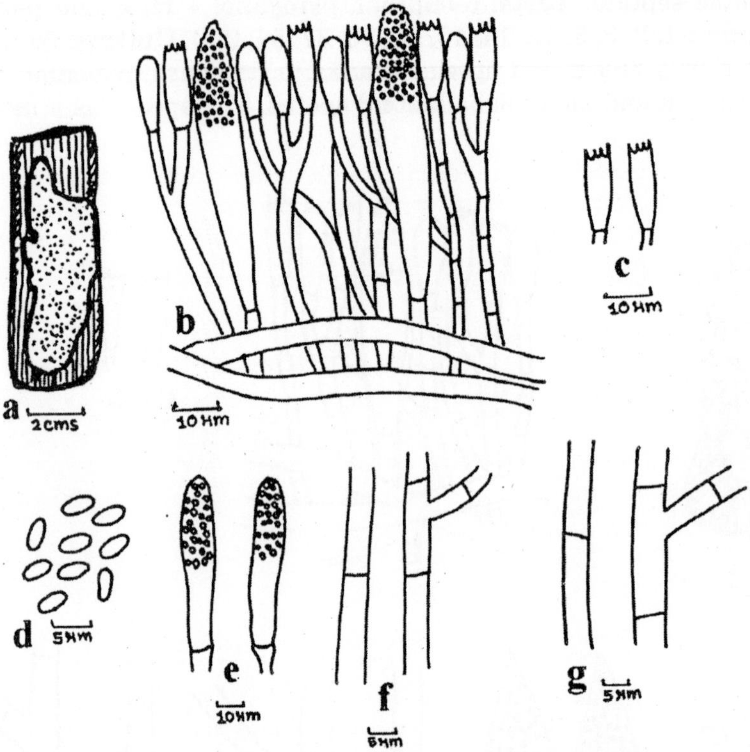

Fig. 70. ***Scopuloides rimosa*** a. Habit. b. T.S. hymenium. c. Basidia. d. Basidiospores. e. Encrusted cystidia. f. Generative hyphae. g. Generative hyphae in culture

Morphological Characters : Basidiomata resupinate, adnate, effused, subgelatinous when fresh ceraceous when dry; 6-7 × 2-3 cms., hymenial surface grayish white to pale luteous, odontoid, warts 20-50 per mm^2. conical to subcylindrical. up to 600 μm in length. fimbriate at the apex; margin concolorous with the hymenial surface. thinning out, indeterminate. In section subhyaline, waxy. Hyphal system monomitic: hyphae 3-4 μm wide, smooth, thin-walled, clampless, agglutinated in the subiculum. Lamprocystidia conical to fusoid, 50-80 × 7-12 μm. without a basal clamp, smooth, thick-walled, heavily encrusted. Basidia narrowly clavate to subcylindrical. 11-16 × 3-4.5 μm. without a basal clamp, producing 4 sterigmata. Basidiospores subcylindrical, 3.5-5 × 1.5- 2 μm, smooth, thin-walled, non-amyloid. **Habitat :** On fallen branches. **Distribution :** Iran, India, Nepal, Taiwan, Thailand, Europe, North and South America, Japan. **Specimen examined :** Benzidine test positive, white rot on *Artocarpus heterophyllus* Lam. **Location :** Panchwati, Dist. Pune (24/09/2005), VH-NPO-227. **Culture characters :** Growth characters: Growth rate moderate; advancing zone appressed to slightly raised; Mats white, thin to moderately thin, subfelty to arachnoid, then becoming slightly raised to raised, arachnoid to cottony toward margins, outline of colony even; Distance between hyphal tips distant; Aerial mycelium absent; odor absent;Reverse unchanged, bleached or stained yellow. Microscopic characters: Generative hyphae 4. 5-7μm wide, hyaline, initially thin walled, becoming slightly thick walled, simple speta, without clamps. **Tests :** α-napthal and pyrogallol + H_2O_2 positive; p-cresol negative. **Species Code :** 1, 2, 3, 7, (12), 13, 14, 15, 30, 36, 37, 53, 54. **Culture No :** PUCC. B: 44. **Remarks :** The species is recognized by thick-walled, heavily encrusted with dense hyaline crystals cystidia, narrowly ellipsoid spores.

(28). ***SCYTINOSTROMELLA*** Parmasto. Consp. Syst. Cort., p. 171, 1968.

This genus erected by Parmasto in 1968, by studying the type of species *Peniophora heterogenea* Bourdot & Galzin. The genus has 8 species reported under this genus. The genus is characterized by effuse basidiomata, typically several cm long and broad, thin to 0. 3 mm thick, membranous to pellicular, brittle, yellow, pale yellow, buff, pallid to tan, lignicolous or corticolous. Hymenial surface smooth, but conforming to the topography of the substrate. Subiculum pallid to white, thin, loosely cottony. Hyphal system dimitic. Generative hyphae with clamp, the walls hyaline to pale yellow, thin or moderately thickened, nonamyloid, acyanophilous. Skeletal hyphae aseptate, lacking branches, narrow, about 1-4 μm diam, the walls thick, pigmented, non-dextrinoid, non-amyloid, cyanophilous in one species. Cystidia in two species. Gloeocystidia in all species except one, fusoid, cylindrical or swollen basally, sinuous, the walls hyaline, thin; the contents oily to granular and pale yellow in KOH, sulfo-negative. Basidiospores ovoid, broadly ellipsoid, subglobose, amyloid, acyanophilous, verruculose. The specimen with characteristic mostly cylindrical or tubular, in KOH with light-refracting contents, no aldehyde reaction gloeocystidia and dimitic hyphal system was identified as *S. cerina* (Bresadola) Hjortstam & Ryvarden.

(28-1). ***Scytinostromella cerina*** (Bresadola) Hjortstam & Ryvarden. *Mycotaxon* **10 (2) :** 287, 1980.

=*Corticium cerinum* Bres., *Annls mycol.* **18 (1/3) :** 47, 1920. (Plate 7, Fig. 71)

Morphological Characters : Fruitbody resupinate, effuse, adnate, 8-10 × 2-3 cms., soft membranaceous, light yellowish to brownish, hymenium smooth, margin not especially differentiated or becoming slightly fibrillose. Hyphal system dimitic, generative hyphae ramified, thin-walled, 3-4 μm wide, with clamps, skeletal hyphae thick-walled, straight and uniform, indistinctly pigmented, 2-2. 5 μm wide, without clamps, non-dextrinoid but the walls apparently with weak cyanophilous reaction. Gloeocystidia present, mostly cylindrical or tubular, terminal, sometimes widened at the apex, in KOH with light-refracting contents, no aldehyde reaction, generally 80-100 ×10 μm. Basidia terminal, somewhat extended clavate, normally with four sterigmata, 40-50 × 5-7 μm. Spores thick-walled, ornamented (easily observed in Melzer's reagent), ellipsoid, 4. 5-5 × 2. 5-3 μm, amyloid. **Habitat :** On fallen branch. **Distribution :** United States, Canada. **Specimen examined :** Benzidine test positive, white rot on *Dalbergia melanoxylon* Guill. & Perr. **Location :** Baneshwar; Dist:Pune (13/01/2007), VH-NPO-660. **Remarks :** Easily recognized species owing to its gloeocystidia and dimitic hyphal system. The location to the genus *Scytinostromella* seems to be well established. For further information of this genus see Freeman and Petersen (1979). *Scytinostromella humifaciens* (Burt.) Freem. & Pet. seems to come close to *S. cerina* but has subulate and shorter cystidia, besides smaller spores. The species is a new record from India. The culture was not obtained besides repeated trials.

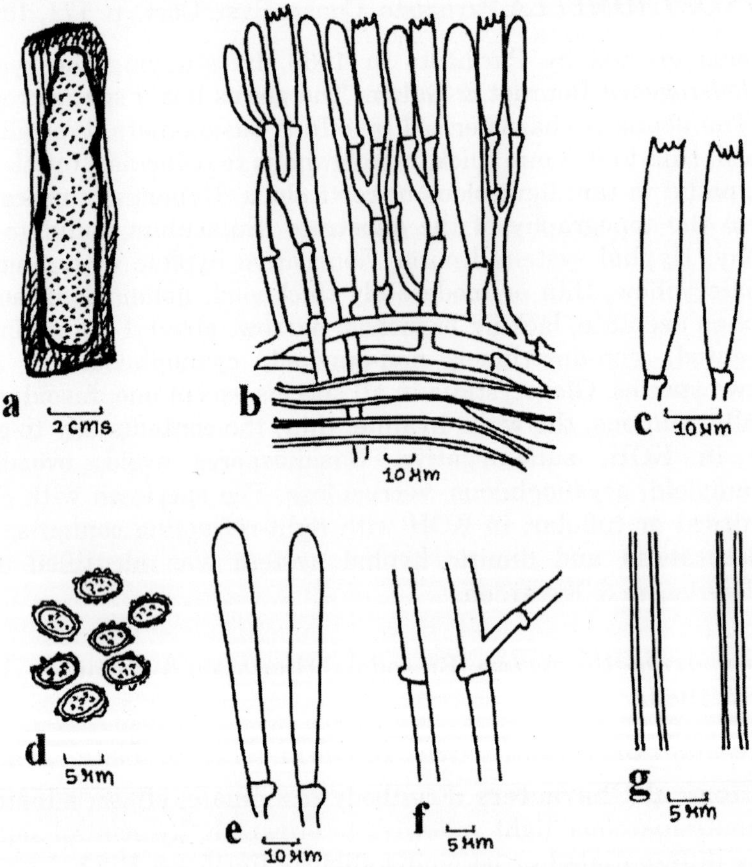

Fig. 71 ***Scytinostromella cerina*** a. Habit. b. T.S. hymenium. c. Basidia. d. Basidiospores. e. Gloeocystidia. f. Generative hyphae. g. Skeletal hyphae

(29). *STECCHERINUM* Gray. *Nat. Arrang. Brit.* Pl. **1:** 651, 1821.

The genus *Steccherinum* was erected by Gray in 1821with its type species *Steccherinum ochraceum* (Pers. : Fr.) Gray. The genus is wide spread in distribution with 73 known species. The genus is characterized by effused to effuso-reflexed, membraneous to coriaceous, loosely adnate basidiomata; hymenophore white when fresh, cream when dry, hydenoid, margin thin, mostly free, determinate, devoid of spines;separate spines in groups in the center. Context light brown, tough. Hyphal system dimitic; generative hyphae with clamps; cystidia, thick walled, encrusted, subhyaline; lamprocystidia tramal and subhymenial in origin; gloeocystidia absent. Basidiospores sub-globose, guttulate, non-amyloid. The specimen with characteristic the pubescent teeth, encrusted cystidia and ellipsoid to broadly cylindrical, slightly thickened, negative in Melzer's reagent spores was identified as *S. setulosum* (Berkeley & Curtis) Miller.

(29-1). *Steccherinum setulosum* (Berkeley & Curtis) Miller. *Mycologia* **27:** 362, 1935.

=*Hydnum setulosum* Berk. & Curtis, *Grevillea* **1 (7) :** 100, 1873. (Plate 7, Fig. 72).

Morphological Characters : Basidiomata annual, resupinate, adnate, up to 12-15 × 5-7 cms., moderately thin, up to 0.9 mm thick, ceraceous, hydnaceous but smooth between teeth, teeth velutinous, easily breaking off, sometimes with a few fine irregular cracks, with up to three layers, thin, subceraceous upper layer concolorous with hymenium and teeth, a middle layer of white, felty mycelium, and a lower layer next to substrate of ceraceous, brown mycelium; hymenial surface continuous, covering terete or slightly compressed teeth, teeth up to 10.0 ×0.5 mm, 1-3 teeth per mm, solitary, fused at base, or fused throughout length, gradually tapering to apex, apices entire and acute but in younger teeth obtuse and fimbriate, teeth and hymenial surface between teeth concolorous, sometimes teeth gradually becoming darker, occasionally paler, toward apex, pale yellow, light yellow, light orange, greyish orange, brownish orange, to yellowish brown, occasionally light brown, with fine white pubescence from projecting cystidia; margin up to 5 mm wide, smooth or verrucose, sterile but pubescent from projecting cystidia, gradually thinning out, fimbriate, adnate or slightly detached, smooth, tomentose to pilose, white to yellowish white or orange white, sometimes translucent. Hyphal system monomitic, 3-4 μm wide, with clamp, branched, somewhat agglutinated; walls hyaline, thin, smooth. Cystidia cylindrical to clavate, 60-100 ×7-10 μm, clamped at base, hyaline, smooth or lightly to heavily encrusted with fine to coarse, hyaline crystals, embedded cystidia often encrusted but if protuding then encrusted at apex only. Basidia clavate, 25-35 ×5-7 μm, with a basal clamp, 4-sterigmate; thin wall, smooth. Basidiospores ellipsoid to broadly cylindrical, 5-6.5 ×3-3.5 μm; walls hyaline, slightly thickened, smooth, non-amyloid. **Habitat :** On live standing tree. **Distribution :** Widely scattered in eastern United States, Canada. **Specimen examined :** Benzidine test positive, white rot on. *Peltophorum pterocarpum* (DC.) Baker ex K. Heyne. **Location :** NCL Ppark, Dist:Pune (19/09/2005) VH-49. **Specimen examined :** Benzidine test positive, white rot on *Azadriachata indica* Juss. **Location:** Film Institue; Dist: Pune (20/10/2005), VH-NPO-261. **Culture characters :** Growth characters: Growth rate moderate; advancing zone appressed to submerged; Mat white, thin, around inocula appressed to slightly raised, subfelty to subtomentose, then becoming thinner, subfelty toward margins; colony white;

outline of colony even to crenate; Distance between hyphal tips distant; Aerial mycelium absent; Reverse bleached. Microscopic characters: Generative hyphae branched, 3.5-6 μm, wide, thin walled, simple septate, sparsely branched; Cystidia clavate to ventricose-rostrate or spathulate, 30-50 × 5-7 μm, thin walled, terminal; Chlamydospores globose to limoniform, 6-15 ×6-10 μm, thin walled at first, then becoming slightly thick-walled. **Tests :** α-napthal and pyrogallol + H_2O_2 positive; p-cresol negative. **Species Code :** 1, 2, 3, 7, 13, 14, 15, 16, 30, 37, 53, 54, 72, 85. **Culture No :** PUCC. B: 45. **Remarks :** The species is easily recognized because of the pubescent teeth, encrusted cystidia and relatively large spores. The species is a new record from India.

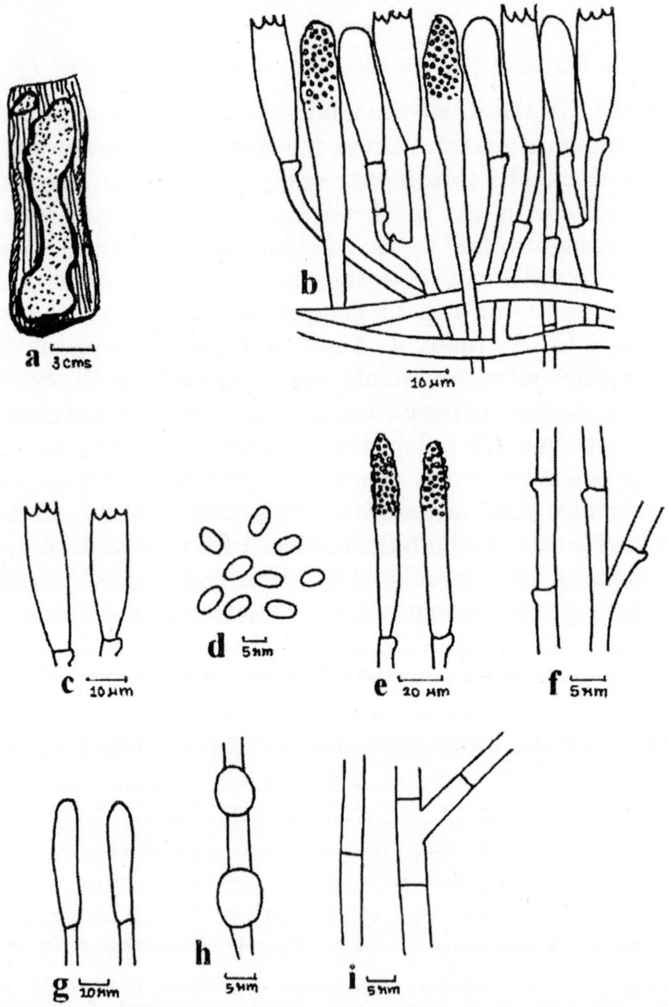

Fig. 72 ***Steccherinum setulosum*** a. Habit. b. T.S. hymenium. c. Basidia. d. Basidiospores. e. Encrusted cystidia. f. Generative hyphae. g. Cystidia. h. Chlamydospores in culture. i. Generative hyphae in culture

(30). *TUBULICRINIS* Donk. *Fungus* **26:** 13, 1956.

The genus was erected by Donk in 1956 by studying the type species *Peniophora gracillima* Ell. and Ev. The genus is widespread in distribution and 52 species have been reported under this genus. The genus is characterized by resupinate, effused basidiomata; hymenial surface reticulate, porulose to continuous, smooth, hispidurous by projecting lyocystidia under the lens (x 20); margin thinning out. In section subhyaline, subpellicular, membranous or ceraceous. Hyphal system monomitic; hyphae thin- to thick-walled, smooth, with clamped; cystidia (lyocystidia) abundant, projecting beyond the hymenial surface, sometimes encrusted, usually amyloid; basidiospores globose, ellipsoid, cylindrical or allantoid, smooth, thin-walled, non-amyloid. The specimen with characteristic the features of the lyocystidia, ellipsoid spores was identified as *T. ellipsoideus* Rajchanberg.

(30-1). *Tubulicrinis ellipsoideus* Rajchenberg. *Mycotaxon* **81:** 223, 2002. (Plate 7, 11, Fig. 73)

Morphological Characters : Basidiocarp resupinate, adnate, 7-10 × 2-3 cms., thin, openly to more or less densely felty, margin indeterminate, cobwebby to somewhat farinose. Hymenial surface smooth, under the lens porulose to close and dense, hispid because of the cystidia, white, cream to light beige, cracking with age or not. Hyphal system monomitic, generative hyphae clamped, hyaline, thickened walls, 2-3 μm. Cystidia cylindric, slightly expanded, non-amyloid, with homogeneously thickened walls, 50-90 × 12 μm. Basidia clavate, 20-25 × 7-10 μm, with 4 sterigmata. Basidiospores ellipsoid, 8-10 × 5-6 μm, thin-walled, non-amyloid. **Habitat :** On live standing tree. **Distribution :** Argentina. **Specimen examined :** Benzidine test positive, white rot on *Albizzia procera* (Roxb.) Bah. **Location :** Baneshwar, Dist. Pune (13/01/2007), VH-NPO-609. **Culture characters :** Growth characters: Growth rate slow, advancing zone appressed; mat at first cottony white; mat cottony to wooly, mycelium white, Marginal hyphae dense; Outline even, Reverse bleached with dark brown patches below the darkened mat. Microscopic characters: Generative hyphae, thin-walled, hyaline 3-3.5 μm., wide clamped, branched; Cystidia subulate to sub-cylindrical, thick-walled, bearing crystals 50-80 × 8-7 μm. **Tests :** α-napthal, pyrogallol + H_2O_2 and KOH positive; p-cresol negative. **Species Code :** 1, 2, 3, 4, 7, 13, 14, 15, 30 (34), 37 (38), 39, 53, 72. **Culture No :** PUCC. B: B: 46. **Remarks :** The species is distinguished on account of its rather large, wide and ellipsoid spores, the features of the lyocystidia. Other species with ellipsoid spores have either smaller or narrower spores and lyocystidia are different in shape. The species is a new record from India.

(31). *VARARIA* Karsten. Kritisk Öfversigt af Finlands Basidsvampar Tillagg **3:** 32, 1898.

The genus *Vararia* was originally published by Karsten in 1898 as the subgenus of *Xerocarpus* for the single species *X. alutarius*. The species was later found to be synonymous with that of *Radulum investiens* which was originally published from Pensylvania in 1832. Karsten in 1903 later on raised *Vararia* to genera rank with its type species *Vararia investiens.* Later on many species were described from temperate, tropical and subtropival areas. This lead to the total 84 species which are known from all over the world and genus is known to be cosmopolitan.

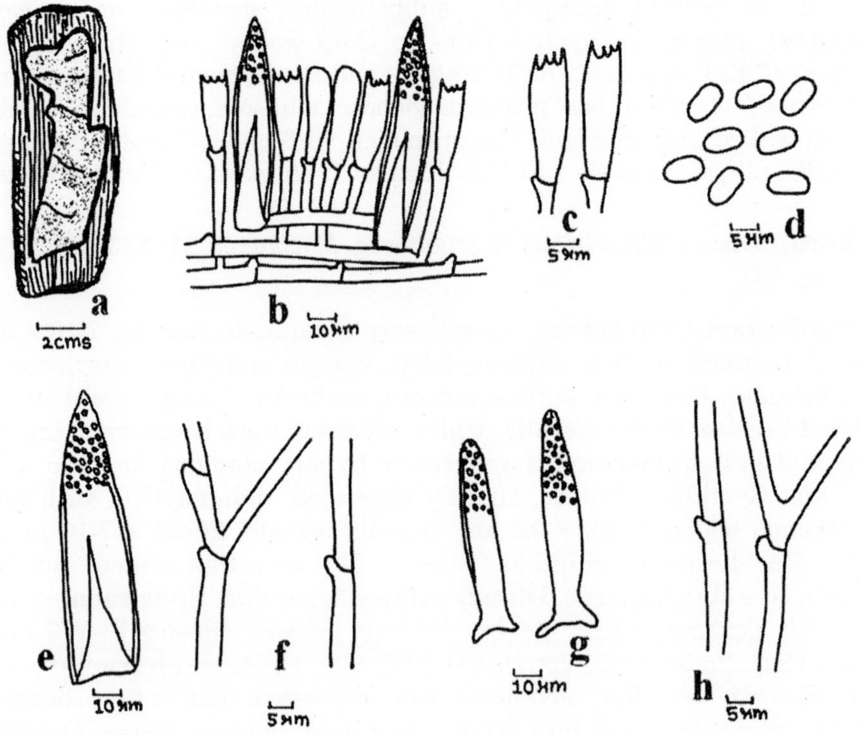

Fig. 73 ***Tubulicrinis ellipsoideus*** a. Habit. b. T.S. hymenium. c. Basidia. d. Basidiospores. e. Encrusted cystidia. f. Generative hyphae. g. Encrusted cystidia in culture. h. Generative hyphae in culture

The genus is characterized by annual, sometimes layered (perennial), effuse to effuse-reflexed, membranaceous to coriaceous basidiomata. Hymenium typically a catahymenium. Hymenial surface even to warted, with cream, yellowish, ochraceous or brownish tinges. Trama sometimes darkening in KOH. Hyphal system monomitic or dimitic with branched skeletal hyphae. Branched skeletal hyphae hyaline to yellowish or brown, thick-walled to solid, cyanophilous at least when young, dextrinoid. Dichohyphidia hyaline and cyanophilous when young, yellowish to brown when older, dextrinoid, repeatedly dichotomously branched, ultimate branches in the basal layer often long (and then often called dextrinoid binding hyphae or branched skeletal hyphae), in the (sub) hymenium relatively short, sometimes very short and coralloid. Unbranched skeletal hyphae absent. Subicular generative hyphae hyaline, thin- to somewhat thick-walled, with or without clamps. Gloeocystidia hyaline, thin- to thick-walled, ventricose or fusoid to subulate; old gloeocystidia often display contracted globular or even crystalline contents. Basidiospores hyaline, globose, ellipsoid, ovoid, cylindrical, navicular or fusoid, thin-walled, smooth or warted to echinulate, amyloid, not amyloid or with amyloid suprahilar plage. Substratum, parasitic or saprophytic on wood of angiosperms and gymnosperms, rarely on litter or soil, producing a white rot. The specimen with characteristic abundant throughout subiculum and hymenial region dichohyphidia, thick-walled and repeated dichotomous branching was identified as *Vararia sphaericospora* Gilbertson.

(31-1). *Vararia sphaericospora* Gilbertson. *Pap. Mich. Acad. Sci.* **50:** 176, 1965. (Plate 7, 11, Fig. 74)

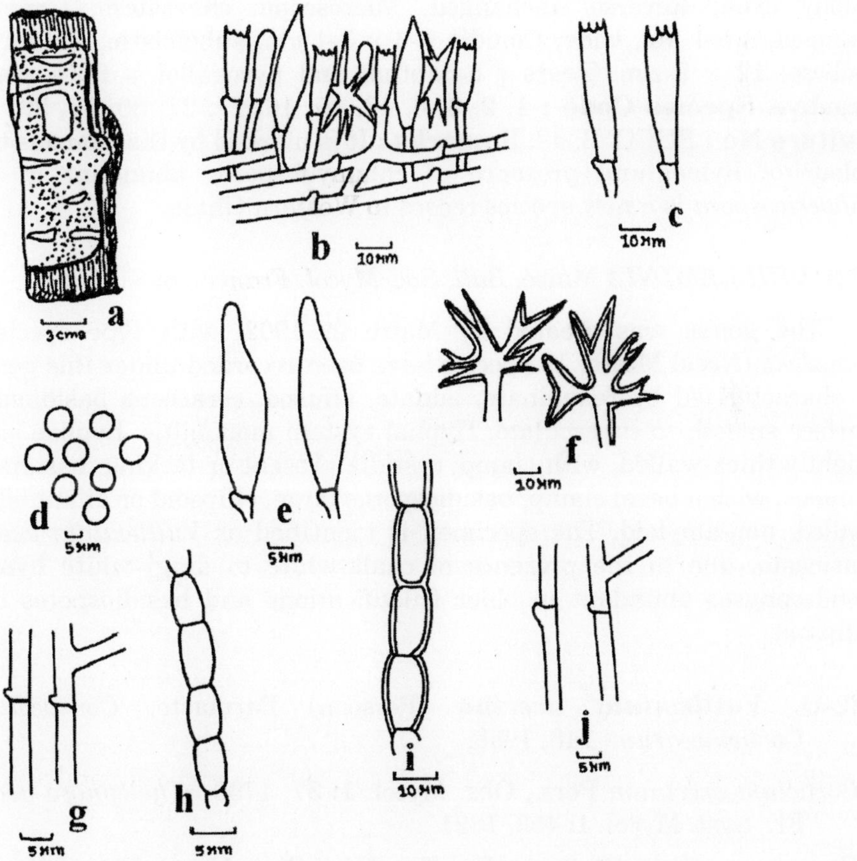

Fig. 74 *Vararia sphearicospora* a. Habit. b. T.S. hymenium. c. Basidia. d. Basidiospores. e. Gloeocystidia. f. Dichohyphidia g. Generative hyphae. h. Conidia in culture. i. Chlamydospores in culture. j. Generative hyphae in culture.

Morphological Characters : Basidiocarps annual, resupinate, effused, 15-18 × 6-9 cms., adherent but soft and easily scraped from the substratum with a needle; hymenial surface chamois to honey yellow, smooth; subiculum soft, concolorous, very thin. Hyphal system dimitic; subicular generative hyphae thin-walled, hyaline, non-amyloid, with clamps, 2-3 μmwide. Dichohyphidia abundant throughout subiculum and hymenial region, thickwalled, with well-defined and repeated dichotomous branching. Gloeocystidia thin-walled, 40-50 × 8-12 μm. Basidia elongated, cylindric-clavate, 4-sterigmate, 40-45 × 6-7.5 μm, with a basal clamp. Basidiospores spherical, hyaline, smooth, thin-walled, non-amyloid, 7-8 μm wide. **Habitat :** On dead standing tree. **Distribution :** USA, Gabon. **Specimen examined :** Benzidine test negative, brown rot on *Ficus religiosa* L. **Location :** Pune University; Dist: Pune (17/08/2006), VH-NPO-507. **Culture characters :** Growth characters: Growth rate moderate; advancing zone appressed; Mat downy to low cottony becoming cottony-wooly, white to cream, locally pinkish, outline of colony even; Reverse unchanged. Microscopic characters: Generative hyphae clamped, 1.5-4 μm, wide, Conidia 4-6 × 3-4 μm, subglobose, Chlamydospores subglobose, 12 × 9 μm. **Tests :** α-napthal and pyrogallol + H_2O_2 positive; p-cresol negative. **Species Code :** 1, 2, 3, 6, 13, 15, 16, 17, 21, 30, 37, 39, 52, 53, 84, 85. **Culture No :** PUCC. B: 47. **Remarks :** It is marked by the chamois to honey yellow colour of hymenium, presence of chlamydospores abundant in subiculum. *V. sphaericospora* is a new species record to Western Ghats.

(32). *VUILLEMINIA* Maire. *Bull. Soc. Mycol. France* **18:** 81, 1902.

The genus was created by Maire in 1902, with type species *Vuilleminia comedens* (Nees) Maire. 19 species have been recorded under this genus. The genus is characterized by resupinate, adnate, effused, ceraceous basidiomata; hymenial surface smooth to tuberculate. Hyphal system monomitic; hyphae smooth, thin- to slightly thick-walled, with clamp; cystidia present or lacking; basidia tubular, long, sinuous, with a basal clamp; basidiospores large, ellipsoid or allantoid, smooth, thin-walled, non-amyloid. The specimen is identified as *Vuilleminia acerina* (Persoon) Parmasto, due to the presence of chalk-white to dingy-white hymenial surface; dendrophyses abundant in older fructifications and basidiospores ovoid to ovoid-ellipsoid

(32-1). *Vuilleminia acerina* (Persoon) Parmasto. *Conspectus Systematis Corticiacearum*: 148, 1968.

=*Corticium acerinum* Pers., Ohs. Mycol. **1:** 37, 1796 =*Thelephora acerina* Pers. per Fr., Syst. Mycol. **1:** 453. 1821.

=*Stereum acerinum* (Pers. per Fr.) Fr., Epicr. Syst. Mycol. 554, 1838.

=*Hypochnus acerinus* (Pers. per Fr.) Pat., Rev. Mycol. **11:** 166, 1889.

=*Corticium acerinum* (Pers. per Fr.) Romell, Bot. Nat. **1895:** 71, 1895. (Plate 7, Fig. 75)

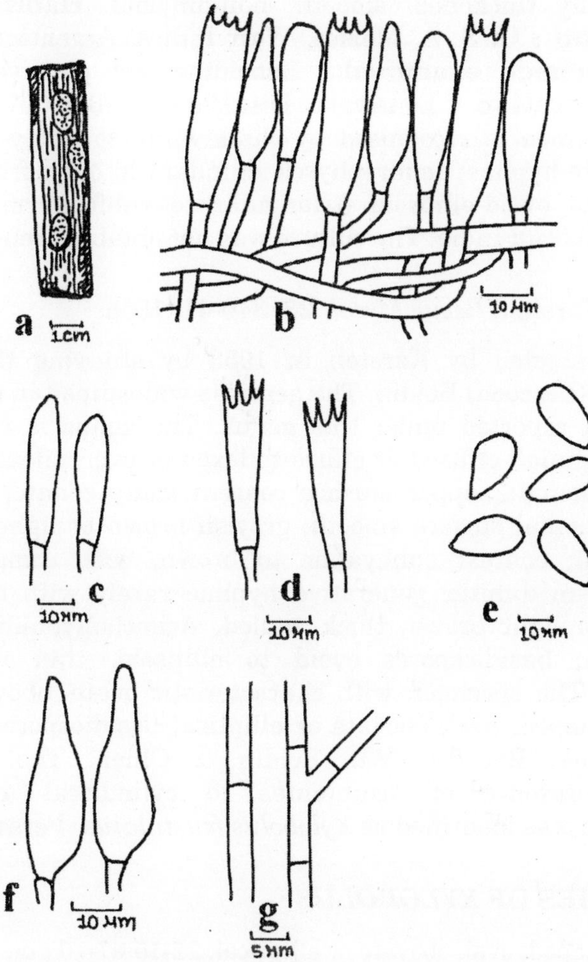

Fig. 75 ***Vuilleminia acerina*** a. Habit. b. T.S. hymenium. c. Basidiole d. Basidia. e. Basidiospores. f. Pseudocystidia. g. Generative hyphae

Morphological Characters : Basidiocarp corticioid, forming scattered colonies, 1-1. 5 × 0. 5-1 cms., confluent but not widely effused; margins determinate, adnate. Fructification 80-100 (-150) μm thick in cross section; texture pruinose to pruinose-pulverulent. Hymenial surface chalk-white to dingy-white or pallid; continuous to rimose on drying. Hyphal system monomitic, branched, thin-walled, irregular, clamless, 1. 5-2 μm. wide. Pseudocystidia scattered to absent, 30-40 × 8-12 μm, fusiform to clavate-ampulliform with an apical digitiform outgrowth, contents hyaline, staining lightly in phloxine, not darkening in sulphobenzaldehyde. Basidioles obovate to broadly clavate, 30-40 × 5-8 μm. Basidia subclavate to suburniform, 40-50 × 7-10 μm; at maturity bearing four sterigmata. Basidiospores ovoid to ovoid-ellipsoid, 10-13 × 7-10 μm, slightly flattened adaxially, apiculate, walls thin to partially thickened, smooth, non-amyloid. **Habitat :** On fallen branches. **Distribution :** U. S. A, Mexico, West Indies, Argentina, Europe, New Zealand, Japan. **Specimen examined :** Benzidine test positive, white rot on *Pongamia pinnata*. **Location :** Lonawala; Dist:Pune (13/01/2007), VH-NPO-556. **Remarks :** The specimen is recognised by chalk-white to dingy-white hymenial surface. Simple-septate hyphae; dendrophyses abundant in older fructifications and basidiospores ovoid to ovoid-ellipsoid (triangular to subfusiform in form). The species is a new record from India. The culture was not obtained repeated trials.

(33). *XYLOBOLUS* Karsten. *Revue Mycol.* **23:** 340-41, 1958.

The genus was erected by Karsten in 1958 by studying the type species *Xylobolus frustulatus* (Persoon) Boidin. The genus is widespread in distribution and 13 species have been reported under this genus. The genus is characterized by annual or mostly perennial, effused or effuso-reflexed or even pileate, hard, with or without cracks basidiomata; upper surface concentrically zonate, dark brown or shades of brown; hymenial surface smooth, greyish brown or light brown; margin thin or thick, abrupt; context subhyaline to brown, with compactly arranged hyphae. Hyphal system dimitic; generative hyphae rarely with clamps; skeletal hyphae subhyaline to light brown, thick walled. Acanthohyphidia, cystidia and gloeocystidia present; basidiospores ovoid to ellipsoid, thin walled, smooth, subhyaline, amyloid. The specimen with characteristic ovate, obovate or fusiform acanthophyses; pip-shaped, oval, obovate or elliptical basidiospores was identified as *Xylobolus apricans* (Bourdot) Wu, Boidin & Chien. The specimen with characteristic the presence of acuminates to cylindrical acanthohyphidia; basidiospores ellipsoid was identified as *Xylobolus frustulatus* (Persoon) Boidin.

KEY TO THE SPECIES OF *XYLOBOLUS*

1. Basidiospores pip-shaped, oval, obovate or elliptical, 12-15 ×7-10 μm, smooth to finely verruculose .... ***X. apricans***

1´. Basidiospores ellipsoid; hyaline; smooth; 3.5-6 (-7) × 2.5-3 (-3.5) μm .... ***X. frustulatus***

(33-1). *Xylobolus apricans* (Bourdot) Wu, Boidin & Chien. *Mycotaxon* **76:** 159, 2000.

= *Aleurodiscus apricans* Bourdot, *Revue scient. Bourbon. Cent.* Fr. **27:** 5, 1910. (Plate 8, Fig. 76)

Morphological Characters : Basidiocarp corticioid, effused, 6-9 × 1-2 cms., margin arachnoid, adherent, concolorous with the white to cream hymenophore. Hyphal system monomitic, with simple septate hyphae, 2.5-4 μm diam. Acanthophyses ovate, obovate or fusiform, 15-25 × 8-12 μm, upper part covered with numerous protuberances up to 4 μm long. Gloeocystidia yellow with a grainy content, 30-60 × 8-12 μm. Paraphyses subclavate or oval, 20-40 × 6-8 μm, some looking almost like cystidia. Basidia cylindrical or subclavate, 20-25 × 8-10 μm, with 4 sterigmata. Basidiospores oval, obovate or elliptical, 12-15 × 7-10 μm, smooth. **Habitat :** On dead standing tree **Distribution :** France, British, Spain, New Zealand. **Specimen Examined:** Benzidine test positive, white rot on *Grevilla robusta* A. Cunn. **Location :** NCL park; Dist:Pune (24/09/2005), VH-NPO-73. **Remarks :** Bourdot (1910) stated that both smooth and asperulate basidiospores were present and this was confirmed by Boidin *et al.* (1968). It is the only species in the genus with both smooth and ornamented basidiospores. The species is a new record from India. The cultures could not be obtained, besides repeated trials.

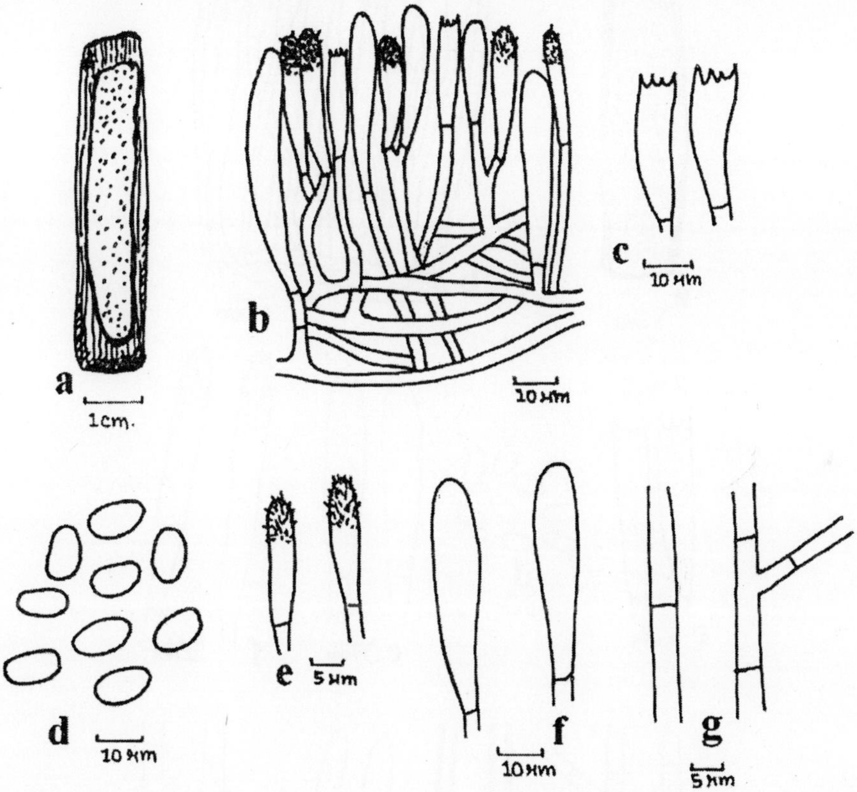

Fig. 76 ***Xylobolus apricans*** a. Habit. b. T.S. hymenium. c. Basidia. d. Basidiospores. e. Acanthophyses. f. Gloeocystidia. g. Generative hyphae

(33-2). ***Xylobolus frustulatus*** (Persoon) Boidin. *Revue Mycol. (Paris)* **23:** 341, 1958.

=*Thelephora frustulata* Pers., *Syn. meth. fung. (Göttingen)* **2:** 577, 1801.

=*Xerocarpus frustulosus* (Fr.) P. Karst., *Bidr. Känn. Finl. Nat. Folk* **37:** 134, 1882.

=*Xylobolus frustulatus* (Pers: Fr.) P. Karsten, *Acta Soc. Fauna Fl. Fenn.* **2**:40. 1881 (as "frustulosus"); superfluous combination by Boidin, Rev. *mycol.* (Paris) **23**:341. 1958.

=*Thelephora frustulata* Pers., *Syn. meth. fung.*, p. 577. 1801; Fr., Syst. mycol. **1**:445. 1821.

= *Stereum frustulatum* (Pers. : Fr.) Fr., Epicr. *syst. mycol.*, p. 552. 1838 (as "*frustulosus*"). (Plate 8, Fig. 77)

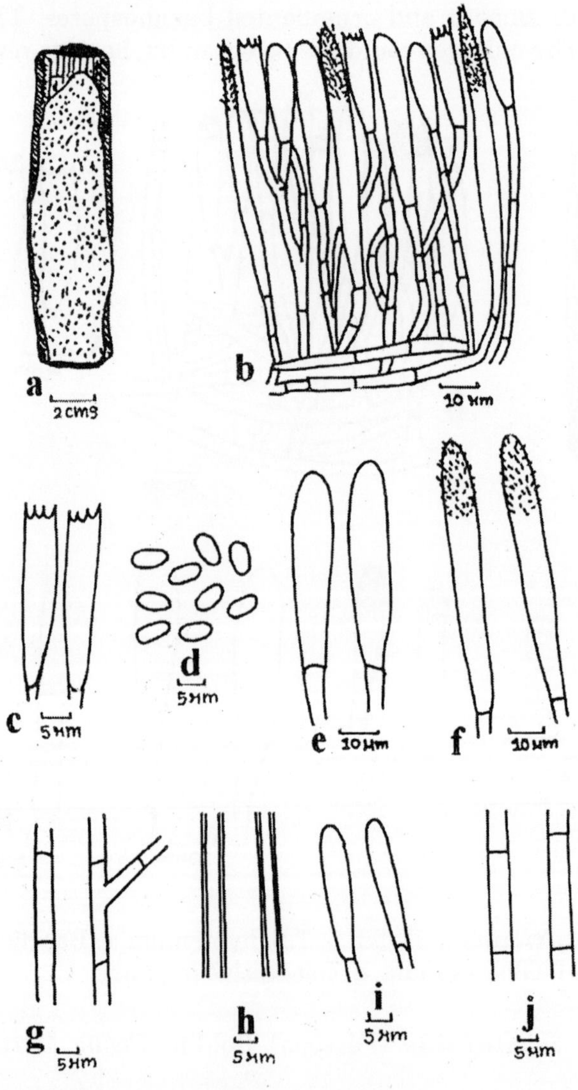

Fig. 77 ***Xylobolus frustulatus*** a. Habit. b. T.S. hymenium. c. Basidia. d. Basidiospores. e. Pseudocystidia. f. Acanthohyphidia. g. Generative hyphae. h. Skeletal hyphae. i. Cystidiole in culture. j. Generative hyphae in culture

Morphological Characters : Basidiomata perennial; resupinate, tuberculiform-frustulate, rarely appearing narrowly reflexed, 9-12 × 3-4 cms; sides of frustules dark brown to black, zonate; hymenial surface even to rugose or tuberculate; grayish white, pale orange, brownish orange, to grayish brown, sometimes with grayish violet tints when fresh. In section up to 5 mm thick, seldom thicker. Cutis brown; distinct when young, indistinct when old. Hymenium stratose, usually fewer than 40 μm. strata; youngest layer pale-colored, progressively darker toward the base. Hyphal system dimitic, generative hyphae thin-walled hyaline clampless, 3-4 μm. wide, skeletal hyphae thick-walled, 4-5 μm. wide. Pseudocystidia thick-walled; with pale brown contents or empty; walls hyaline to brown, smooth; 50-70×7-10 μm. Acanthohyphidia acuminate to cylindrical; hyaline, thickened brown walls; 80-100 × 8-10 μm. Basidia clavate to subcylindrical; with four sterigmate, basal clamp absent; 20-30 × 4-5 μm. Basidiospores ellipsoid, hyaline, smooth, thin-walled, amyloid, 3.5-6 × 2.5-3 μm. **Habitat :** On live standing tree. **Distribution :** U.S.A, Europe, Asia, Australasia. **Speciemen Examined:** Benzidine test positive, white rot on *Calicopteria floribunda* Lam. **Location :** Koyna; Dist:Satara (10/02/2007), VH-NPO-768. **Culture characters :** Growth characters: Growth rate moderate; advancing zone appressed to slightly raise; Mats Orange to Orange-Buff, moderately thick to thick, slightly raised and tomentose to raised and cottony to woolly; outline of colony even to slightly uneven; Reverse darkened. Microscopic characters: Generative hyphae 3-5 μm wide, thin walled, simple septate, sparsely branched, cystidiol thin-walled, hyaline, smooth, 20-30×5-7 μm. **Tests :** α-napthal and pyrogallol + H_2O_2 positive; p-cresol negative. **Species Code :** 1, 2, 3, 7, 13, 14, 15, 21, 22, 32, 38, 52, 53, 72. **Culture No :** PUCC. B: 48. **Remarks :** *Xylobolus frustulatus*, although differing from *X. subpileatus* in its peculiar tuberculate-frustulate growth form, is similar to the latter species microscopically. *Xylobolus subpileatus* is truly effuso-reflexed, retains a distinct context and cutis, and possesses encrusted pseudocystidia and acanthohyphidia. The species is a new record from India.

B. KEYS TO THE GENERA OF POROID APHILLOPHORALES

1. Basidiomata yellowish brown to dark brown, context xanthochroic, setae often present .... ***Phellinus*** (14)

1´. Basidiomata hyaline or coloured, context non-xanthochroic, setae usually Absent. 2

2. Hymenophore hydnoid, lamellate, daedaleoid 7

2´. Hymenophore with angular to round pores, sometimes slightly split and dentste in the dissepiments 3

3. Spores smooth 4

3´. Spores ornamented 11

4. Spores, cystidia or hyphae amyloid or dextrinoid 12

4´. Spores, cystidia or hyphae non-amyloid or dextrinoid 5

5. Generative hypha with simple septa.......... 15

5´. Generative hypha with simple clamps.......... 6

6. Cystidia present in hymenium or context.......... 19

6´. Cystidia absent from hymenium or context.......... 20

7. Hyphal system monomitic or dimitic.......... 8

7´. Hyphal system trimitic.......... 9

8. Basidiocarp resupinate, hymenophore as elongated pores, hyphal system monomitic, basidia lining also the bottom of the tubes .... ***Hymenogramme*** (10)

8´. Basidiocarp pilieate toresupinate, hymenophore hydnoid, lenzitoid to Dentate. .... ***Schizopora*** (17)

9. Pore surface pale brownish violet when fresh, fading to pale brown when old and dry, encrusted cystidia present in hymenium.......... ***Trichaptum*** (18)

9´. Pore surface white to grayish, encrusted cystidia absent from hymenium.......... 10

10. Pileus hirsute to velutinate and with a thin black zone between tomentum and context .... ***Cerrena*** (5)

10´. Pileus smooth and glabrous, not black zone in context.......... ***Earliella*** (7)

11. Skeletal hyphae more or less dextrinoid .... ***Wrightoporia*** (21)

11´. Skeletal hyphae non-dextrinoid.......... ***Amylosporus*** (2)

12. Spores amyloid.......... ***Anomoporia*** (3)

12´. Spores dextrinoid or negative in Melzers reagent.......... 13

13. Cystidia present.......... ***Amylocystis*** (1)

13´. Cystidia absent 14

14. Spores longer than 15µm, basidiocarps resupinate, hyphal pegs abundantly present.......... ***Grammothelopsis*** (9)

14´. Spores shorter than 15µm, basidiocarps resupinate to pileate, hyphal pegs rarely present. .... ***Perenniporia*** (13)

15. Cystidia smooth and thin wall 16

15´. Cystidia metuloid and encrusted.......... ***Oxyporus*** (12)

16. Basidiocarps orange to reddish or dark brown .... ***Pycnoporellus*** (15)

16´. Basidiocarps white to light coloured 17

17. Spores cylindrical .... ***Flavodon*** (8)

17´. Spores globose.......... 18

18. Pore surface vivid orange red or pinkish fading to reddish brown... ***Rigidoporus*** (16)

18´. Pore surface white to pale ochraceous.......... ***Wolfiporia*** (20)

19. Cystidia large and ventricose, skeletal hyphae few restricted to the context, pore surface yellow.......... ***Auriporia*** (4)

19′. Cystidia club shaped, formed by the outer encrusted parts of skeletal hyphae, skeletal hyphae dominating, pore surface cocoa-brown, pinkish to pale straw-coloured .... ***Junghuhnia*** (11)

20. Basidiocarp pileate, sessile to dimidiate, soft and warty when fresh mostly white to sordid brown when old or dried .... ***Tyromyces*** (19)

20′. Basidiocarp resupinate to effused-reflexed or pileate and then with a narrow pileus, pileus glabrous to velutinate .... ***Diplomitoporus*** (6)

DESCRIPTION OF NON-POROID GENERA AND SPECIES

(1). *AMYLOCYSTIS* Bondartsev & Singer ex Singer. *Mycologia* 36 (1) : 66, 1944.

The genus was created by Bondarew and Singer ex. Siner in 1944 by studying the type species *Polyporus lapponicus* Rom. *Amylocystis* is a well defined genus and four species are known. The genus is characterized by sessile or effused-reflexed basidiocarps, upper surface tomentose to hispid; pore surface and context white, bruising or drying reddish brown; monomitic hyphal system;with clamps; cystidia thick-walled, often incrusted at apex; basidiospores cylindrical, negative in Melzer 's reagent. The specimen is identified as *A. sericeomollis* (Romell) Teixeira, due to the presence of pores and spores are irregular in shape and size.

(1-1). *Amylocystis sericeomollis* (Romell) Teixeira. *Revta bras. Bot.* 15: 125, 1992.

=*Polyporus sericeomollis* Romell, *Arch. für Botanik* **11(3):** 22, 1911. (Plate 1; Fig. 78).

Morphological Characters : Hymenophore effused, to 5-7 ×3-4 cms., 2-3 mm. thick; white when fresh, becoming dingy cream; margin arachnoid, even or abrupt, not glancing, dissepiments not toothed. Pores not in strata, round, angular, concolorous in section, 1.5-2.5 mm. deep, 1-3 per/ mm., dissepiments equal, 100-200 μm thick, of woven hyphae with rounded ends. Context to 0.3 mm. thick, white, floccose. Hyphal system dimitic, generative hyphae 2-3 μm wide, thin-walled, branched, with clamps, skeletal hyphae 3-3.5 μm thick, aseptate, sparsely branched near ends. Basidia clavate, 16 × 5 μm. Spores oblong or elliptical, 3.5-6 × 2.5-3 μm, hyaline, smooth. **Habitat :** On live standing tree. **Distribution :** Europe, North America, New Zealand, Canada, Siberia. **Specimen Examined :** Benzidine test negative, brown rot on *Gliricidia sepium* (Jacquin) Kunth ex Walp. **Location :** Dongarwadi; Dist: Pune (08/10/2005), VH-PO-134. **Remarks :** Pores and spores are irregular in shape and size. The culture was not obtained besides repeated trials. The species is a new record from India.

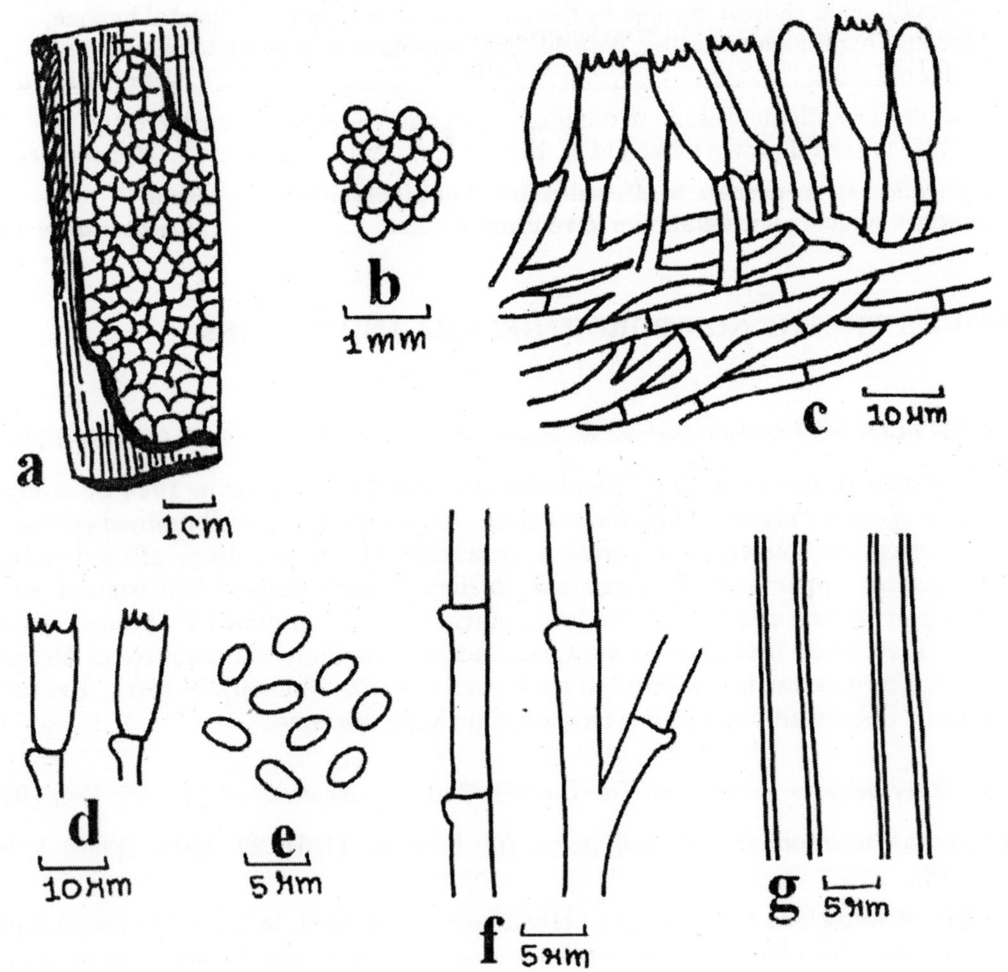

Fig. 78 ***Amylocystis sericeomollis*** a. Habit. b. Pores. c. T.S. hymenium. d. Basidia. e. Basidiospores. f. Generative hyphae. g. Skeletal hyphae

(2). ***AMYLOSPORUS*** Ryv. *Norw. J. Bot.* **20:** 1, 1973.

This genus erected by Ryvarden in 1973, by studying the type of species *Tyromyces graminicola* Murr. The distinctive characters of *Amylosporus* indicate a possible relationship to the family Hericiaceae rather than to other polpore genera. The genus has seven species reported under this genus. The genus is characterized by annual, terrestrial or lignicolous, sessile to stipitate basidiocarps; oileous surface cream to buff, finely tomentose, context white to pale brown, soft;pores regular, 2-4 per mm; hyphal system dimitic, generative hyphae with simple septa or single, double, or multiple clamps, gloeoplerous hyphae present, cystidia absent; basidiospores ellipsoid, smooth to finely ornamented, weakly to strongly amyloid in Melzers reagent. The specimen with characteristic widely effused fruitbodies, lilaceous pink pore surface was identified as *A. bracei* (Murrill) David & Rajchenberg.

(2-1). *Amylosporus bracei* (Murrill) David & Rajchenberg. *Mycotaxon* **22 (2) :** 288, 1985.

=*Poria bracei* Murrill, *Mycologia* **13 (2) :** 91, 1921.

=*Wrightoporia bracei* (Murrill) I. Lindblad & Ryvarden, *Mycotaxon* **71:** 357, 1999. (Plate 1; 8; Fig. 79).

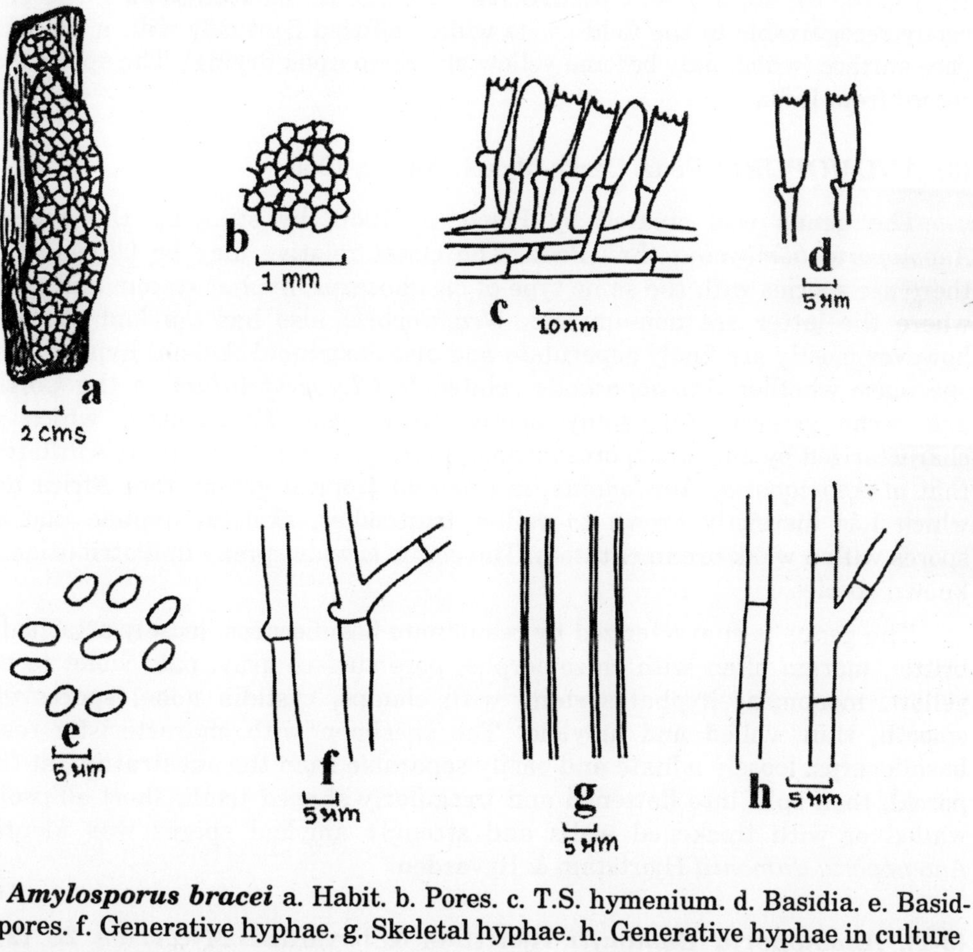

Fig. 79 *Amylosporus bracei* a. Habit. b. Pores. c. T.S. hymenium. d. Basidia. e. Basidiospores. f. Generative hyphae. g. Skeletal hyphae. h. Generative hyphae in culture

Morphological Characters : Fruit body widely effused, 15-18×3-4 cms., lilaceous pink (which may become yellowish cream upon drying), Pore rounded, 5-6 per/mm. Hyphal system dimitic, generative hyphae with clamps, thin wall, hyaline, 2-3 μm. wide; skeletal hyphae dextrinoid, 3-5 μm. wide, as well as the finely verruculose amyloid. Cystidia non. Basidia clavate, with four sterigmata, 15-18×5 μm. Spores ellipsoid, thin wall, hyaline, non-amyloid, 4-6 ×3 μm. **Habitat:** On dead standing tree. **Distribution :** Argentine. **Specimen examined:** Benzidine test negative, brown rot on *Bauhinia purpurea L.* **Location :** Baneshwar; Dist: Pune (13/01/2007), VH-PO-575. **Culture characters:** Growth characters: Growth rate moderate; advancing zone appressed to raised. Mat cottony white getting submerged in agar as colony advances giving the plate and hence the colony, a transparent look; colony white; outline of colony even slightly to uneven; Distance between hyphal tips distant; Aerial mycelium absent; Reverse bleached. Microscopic characters: Generative hyphae branched, 3-4 μm, wide, simple septate. **Tests :** α-napthal and pyrogallol+ H_2O_2 positive; p-cresol negative. **Species Code :** 1, 2, 3, 7, 12, 13, 14, 15, 30, 37, 53. **Culture No :** PUCC. B: 49. **Remarks :** The species is easily recognizable in the field by its widely effused fruitbody with a lilaceous pink pore surface (which may become yellowish cream upon drying). The species is a new record from India.

(3). *ANOMOPORIA* Pouz. Ceska Mykol. **20:**172, 1966.

The genus was created by Pouzr in 1966, by studying the type species *Anomoporia bombycina* (Fr.) Pouz. The closet relative may be *Oligoporus* where there are species with the same type of basidiocarps, hyphal system and spores, but where the latter are non-amyloid. *Wrightoporia* also has amyloid spores, which however mostly are finely asperulate and also dextrinoid skeletal hyphae. It may be speculate whether *Anomoporia* is related to *Gloeocystidillum* in the Corticiaceae and some genera with spiny spores within the Hericiaceae, which all are characterized by amyloid spores although the type of hymenophore is different from that of *Anomoporia, Amylonotus*, is a poroid tropical genus from Africa and Asia which has distinctly brown to yellow fruitbodies, skeletal hyphae and amyloid spores with a weak ornamentation. The genus is wide spread in distribution with 11 known species.

The genus is characterized by resupinate basidiocarps, loosely attached, soft to brittle, margin often with rhizomorphs, pore surface gray, pale violet to white or yellow, monomitic hyphal system, with clamps, cystidia none, spores ellipsoid, smooth, thin walled and amyloid. The specimen with characteristic resupinate basidiocarps, loosely adnate and easily separable from the substratum, at first sub-poroid, then split into flattened and irregularly shaped teeth, short-ellipsoid, thin-walled or with thickened walls and strongly amyloid spores was identified as *Anomoporia dumontii* Hjortstam & Ryvarden.

(3-1). *Anomoporia dumontii* Hjortstam & Ryvarden. *Mycotaxon* **28 (2) :** 554, 1987. (Plate 1;8, Fig. 80).

Morphological Characters : Basidiocarp annual, resupinate, 7-9 × 3-4 cms., loosely adnate and easily separable from the substratum, at first subporoid, then split into flattened and irregularly shaped teeth, 0.5-2 mm long, apically penicillate to fimbriate, at least when fully developed, cream-coloured to pale yellow, margin white, cottony, rhizomorphs absent. Hyphal system monomitic, generative hyphae thin-walled, smooth, 3-5 μm wide, hyaline, with clamps. Cystidia lacking. Basidia more or less clavate, 15-20 ×4-5 μm, with 4 sterigmata and a basal clamp. Spores short-ellipsoid, thin-walled or with thickened walls, hyaline, smooth, 3-4 ×3 μm, and strongly amyloid. **Habitat :** On live standing tree. **Distribution:** Venezuela. **Specimen examined :** Benzidine test negative, brown rot on *Samanea saman* (Jacq.) Merr. **Location :** Sambhaji park; Dist: Pune (21/10/2005), VH-PO-300. **Culture characters :** Growth characters: Growth is moderate, advancing zone appressed; Mat cottony white; outline of colony even; Reverse bleached slightly; Aerial mycelium absent. Microscopic characters: Generative hyphae branched, hyaline, thin-walled, with clamps, 3-4 μm. wide. **Test:** α-napthal and pyrogallol+ H_2O_2 positive; p-cresol negative. **Species Code**: 1, 2, 3, 7, 15, 16, 30, 37, 39, 53. **Culture No :** PUCC. B: 50. **Remarks :** The features of this species are essentially the same as described for *A. irpicoides* excepting the morphology of the spores and the width of the aculeal hyphae. Also the amyloid reaction is stronger in *A. dumontii*. The species is a new record from India.

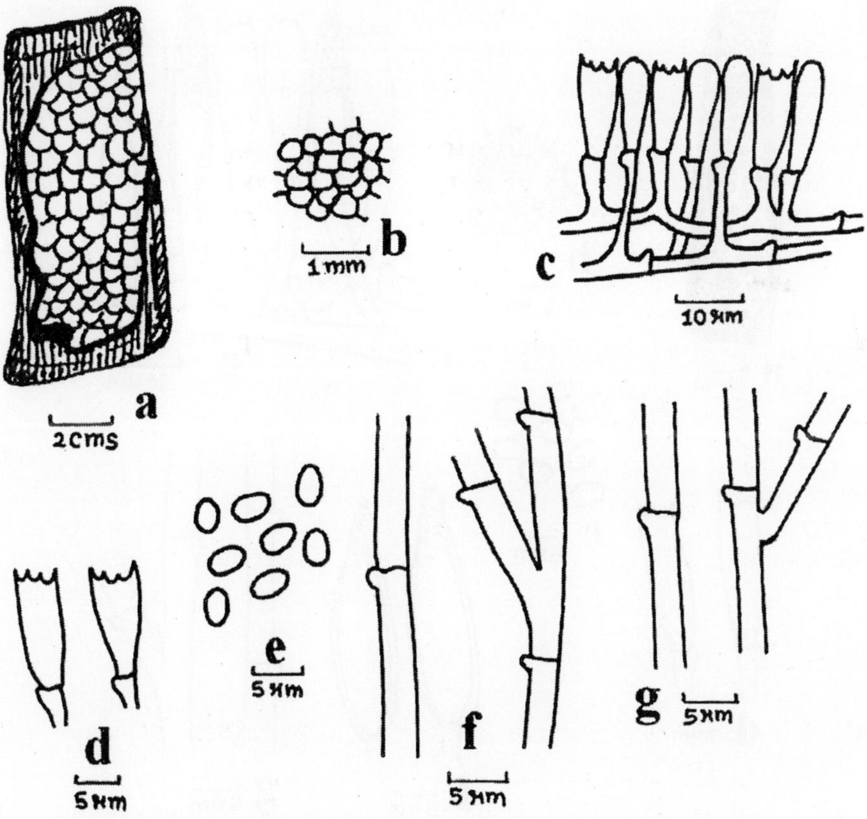

Fig. 80 ***Anomoporia dumontii*** a. Habit. b. Pores. c. T.S. hymenium. d. Basidia. e. Basidiospores. f. Generative hyphae. g. Generative hyphae in culture

(4). *AURIPORIA* Ryvarden. *Norweg. J. Bot.* **20:** 2, 1973.

The genus was created by Ryvarden in 1973, by studying the type species *Auriporia aurea* (Pk.) Ryv. The genus is wide spread in distribution with 11 known species. The genus is characterized by resupinate or pileate basidiomata. Pores pale luteous, luteous or orange; the colour of basidiomata may turn red in KOH. Hyphal system monomitic; generative hyphae clamped; numerous cystidia, inocyboid, thick-walled, fusiform, muricate, pedicellate, non-amyloid (and nondextrinoid). Spores hyaline, thin-walled, smooth, non-amyloid, noncyanophilous, ellipsoidal or allantoid (cylindrical). The specimen with characteristic resupinate basidiocarp with hymenophore poroid, deep yellow-orange in colour was identified as *A. aurulenta* David, Tortic & Jelic.

(4-1). *Auriporia aurulenta* David, Tortic & Jelic. *Bull. trimest. Soc. mycol. Fr.* **90:** 364, 1975. (Plate 1, Fig. 81).

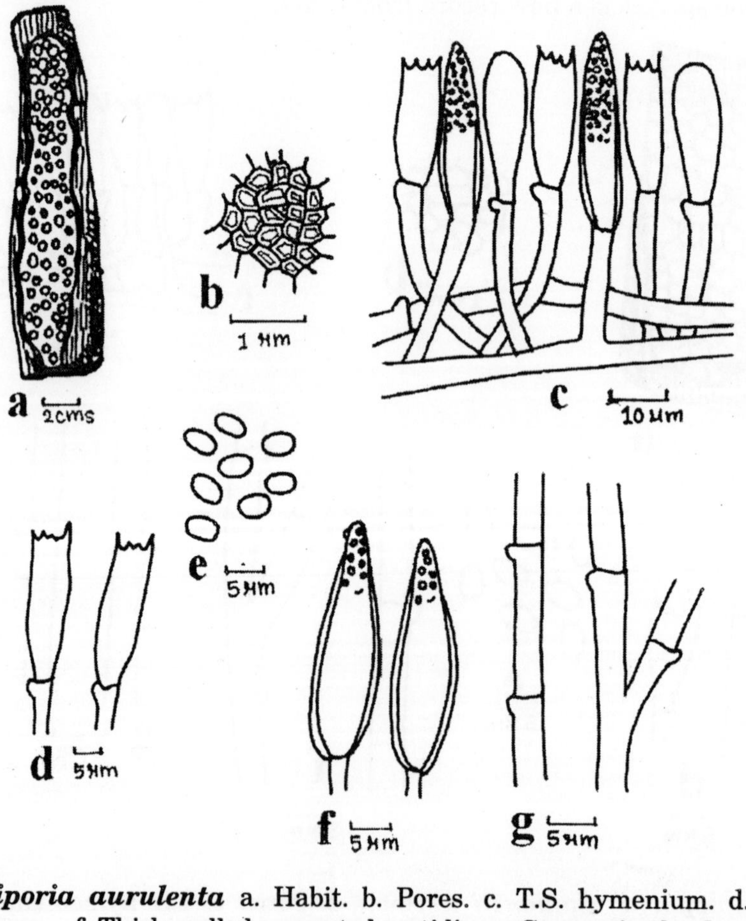

Fig. 81 *Auriporia aurulenta* a. Habit. b. Pores. c. T.S. hymenium. d. Basidia. e. Basidiospores. f. Thick-walled encrusted cystidia. g. Generative hyphae

Morphological Characters : Basidiocarp resupinate, 10-13 × 2-3 cms., with hymenophore poroid, pores angular, 4 pre /mm, deep yellow-orange in colour, but turning ochre with age; hyphal system monomitic, hyphae with clamps, 3-5 μm. wide. Cystidia ventricose, more or less fusiform, thick-walled, 20-35 × 8-12 μm, with grainy crystals. Basidia clavat, with 4 sterigmata, 25-30 × 5-7μm. Basidiospores ellipsoid, thin wall, hyaline, 4.5-6 × 2-3 μm., non-amyloid, non-cyanophilous. **Habitat :** On live standing tree. **Distribution :** Europe, Austria, Czechoslovakia, France, Yugoslavia. **Specimen examined :** Benzidine test negative, brown rot on *Cassine gluca* (Roxb.) Kuntze. **Location :** Dongerwadi; Dist: Pune (13/01/2007), VH-PO-561. **Remarks :** This species is easily recognized in the field by its basidiocarp resupinate with hymenophore poroid, deep yellow-orange in colour, but turning ochre with age. The species is a new record from India. The culture was not obtained besides repeated trials.

(5). *CERRENA* Gray. *A natural arrangement of British plants* **1:** 649, 1821.

The genus was erected by Gray in 1821 by studying the type species *Cerrena unicolor* (Fr.) Murr. The genus is widespread with twenty five species studied under the genus. The genus is recognized by its irregular hymenophore and the distinct black line between the tomentum and the context. Undoubtedly, it is closely related to *Trametes*, sharing the same hyphal system and type of spores. However, the characteristics mentioned above seem to justify its position, even if there are *Trametes* species with a similar dark zone, e.g. *T. versicolor* and in old specimens of *T. hirsuta* and *T. zonatella*. If they all should be put in one genus, then of course *Cerrena* has priority before genera like *Coriolus* and *Trametes*. The specimen with characteristic the hirsute pileus, the black line in the context and the labyrinthine hymenophore was identified as *C. unicolor* (Bulliard) Murrill.

(5-1). *Cerrena unicolor* (Bulliard) Murrill. *J. Mycol.* **9 (2) :** 91, 1903.

=*Daedalea unicolor* (Bull.) Fr., *Syst. mycol.* (Lundae) **1:** 336 (1821). (Plate 1, 8, Fig. 82).

Morphological Characters : Fruitbody perennial, pileate, dimidiate, effuso-reflexed to resupinate, 7-10 × 4-6 cms., tough to corky. Pileus flat, densely tomentose to hirsute, concentrically zoned in shades from white at the margin to greyish or even light brown towards the base, often with shades of green due to algae, at the base also often overgrown by mosses, margin acute and slightly lobed. Pore surface white to light cinerous or smoky, pores elongated to sinuous, partly split to labyrinthine, or even deeply split into flattened teeth, first 2-3 per/mm, measured tangentially, in old specimens even larger, tubes concolorous with hymenophore, up to 5 mm long. Context white and thin, up to 2 mm thick, separated from the tomentum on the pileus by thin, very distinct black zone. Hyphal system trimitic, generative hyphae with clamps, thin-walled, 2-3 μm. wide, skeletal hyphae hyaline, thick-walled to solid, dominating in the fruitbody, 3-6 μm wide, binding hyphae mostly rare and confined to the context, thick-walled to solid, hyaline and with relatively short branches, 3 - 5 μm wide. Cystidia thin wall, clavate, 30-40 × 5-7 μm. Basidia clavate, 35-40 × 7 μm, with 4 sterigmata, basal with clamps. Spores ellipsoid, hyaline, thin-walled, smooth and non-amyloid, 4.5-6. 5 ×3-3.5 μm. **Habitat :** On fallen branches. **Distribution :** Tanzania. **Specimen examined :** Benzidine test positive, white rot on *Artocarpus heterophyllus* Lam. **Location :** Amboli; Dist: Satara (19/02/2007), VH-PO-787. **Culture characters :** Growth characters: Growth rate fast; advancing zone raised; Mat white, raised, loosely arranged, cottony-floccose. Out line of colony even; Reverse bleached. Microscopic characters: Generative hyphae branched, 1. 5-4.5 μm, wide, with clamps. **Tests :** α-napthal and pyrogallol+ H_2O_2 positive; p-cresol negative. **Species Code :** 1, 2, 3, 5, 12, 15, 16, 30, 37, 39, 53. **Culture No :** PUCC. B: 51. **Remarks :** *Cerrena unicolor* is easy to recognize because of the hirsute pileus, the black line in the context and the labyrinthine hymenophore. The species is a new record from India. .

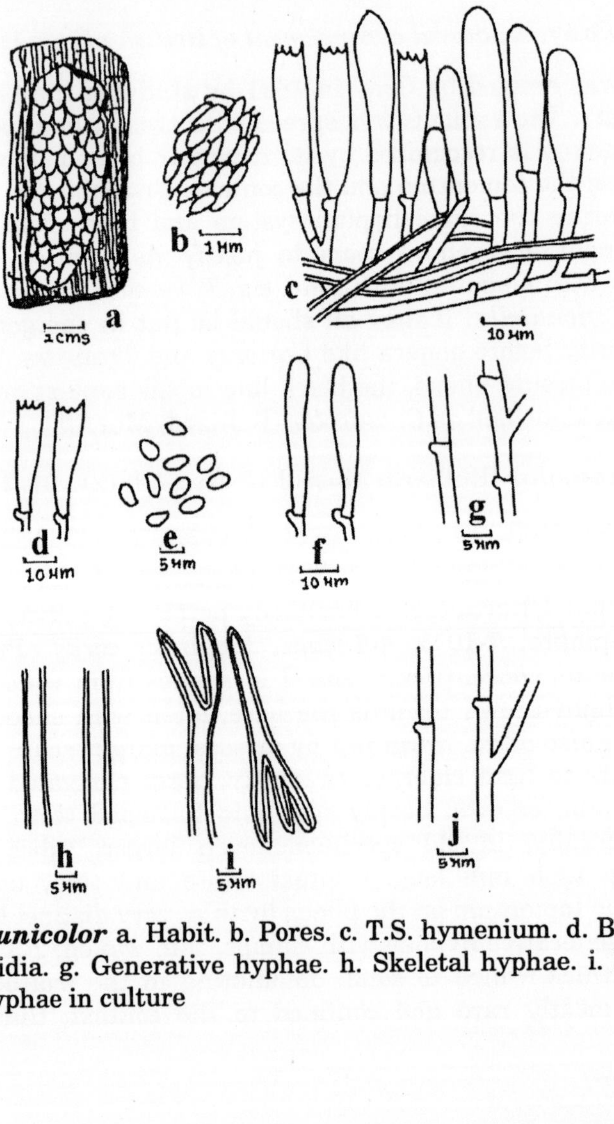

Fig. 82 ***Cerrena unicolor*** a. Habit. b. Pores. c. T.S. hymenium. d. Basidia. e. Basidiospores. f. Cystidia. g. Generative hyphae. h. Skeletal hyphae. i. Binding hyphae. j. Generative hyphae in culture

(6). *DIPLOMITOPORUS* Dom. *Acta Sco. Bot. Pol.* **39:** 191, 1970.

The genus was erected by Domanski in 1958 by studying the type species *Trametes flavescens* Bres. The genus is similar to *Antrodia* and is separated mainly because all *Antrodia* species have a brown rot. The borderline towards *Anterodiella* is vague as both genera include species with the same hyphal system, type of spores of *Antrodiella* are small and ellipsoid and the fruitbodies are usually dense and cartilaginous. The genus is widespread with fourteen species known. The genus is characterized by resupinate to effused reflexed basidiocarps, white to lightcoloured, pores round to angular, medium to small, hyphal system dimitic, generative hyphae with clamps, skeletal hyphae hyaline and thick-walled, non-amyloid to weakly amyloid, cystidia absent or present, spores allantoid to ellipsoid, thin-walled, smooth and non-amyloid. The specimen is recognized by white pore surface, pores angular, present dendrohyphidia, cystidia and other sterile hymenial elements absent the specimen was identified as *D. hondurensis* (Murrill) Ryvarden.

(6-1). *Diplomitoporus hondurensis* (Murrill) Ryvarden. *Mycotaxon* **74 (1) :** 121, 2000.

=*Grammothele hondurensis* (Murrill) Ryvarden, *Mycotaxon* **23:** 185, 1985.

=*Poria hondurensis* Murrill, *Mycologia* **12 (6) :** 303, 1920. (Plate 1; Fig. 83).

Morphological Characters : Basidiocarps resupinate, 5-7 × 2-3 cms. and up to 2 mm thick, adnate, brittle when dry, margin narrow, white to cream, pore surface white, pores angular, in parts irregular and slightly incised, 2-4 per mm, in parts with hyphal pegs, some as hydnoid protuberances, others as an initial development of partition walls, tube layer concolorous with pores, up to 2 mm thick, subiculum very thin and white. Hyphal system dimitic, generative hyphae hyaline, with clamps, 2-3 μm wide, skeletal hyphae predominant, solid to thick-walled, hyaline, non-amyloid, 2-3 μm. wide. Cystidia and other sterile hymenial elements absent. Dendrohyphidia present, both along the dissepiments where they are abundant and prominent, 12-15 × 4-6 μm. Basidia 10-15 × 5-6 μm with basal clamps, 4 sterigmata. Basidiospores oblong ellipsoid to cylindrical, thin-walled, non-amyloid, 5-8 × 3-3.5 μm. **Habitat**: On fallen branch. **Distribution** : Puerto Rico and Honduras. **Specimen Examined**: Benzidine test positive, white rot on *Mallotus philippensis*. **Location :** Baneshwar, Dist. Pune (13/01/2007), VH-PO-653. **Remarks:** The species is microscopically separated by the dendrohyphidia and larger basidiospores from the following species, *D. incisus*, which is macroscopically rather similar except that its pores are more irregularly incised. The species is a new record from India. The culture was not obtained besides repeated trials.

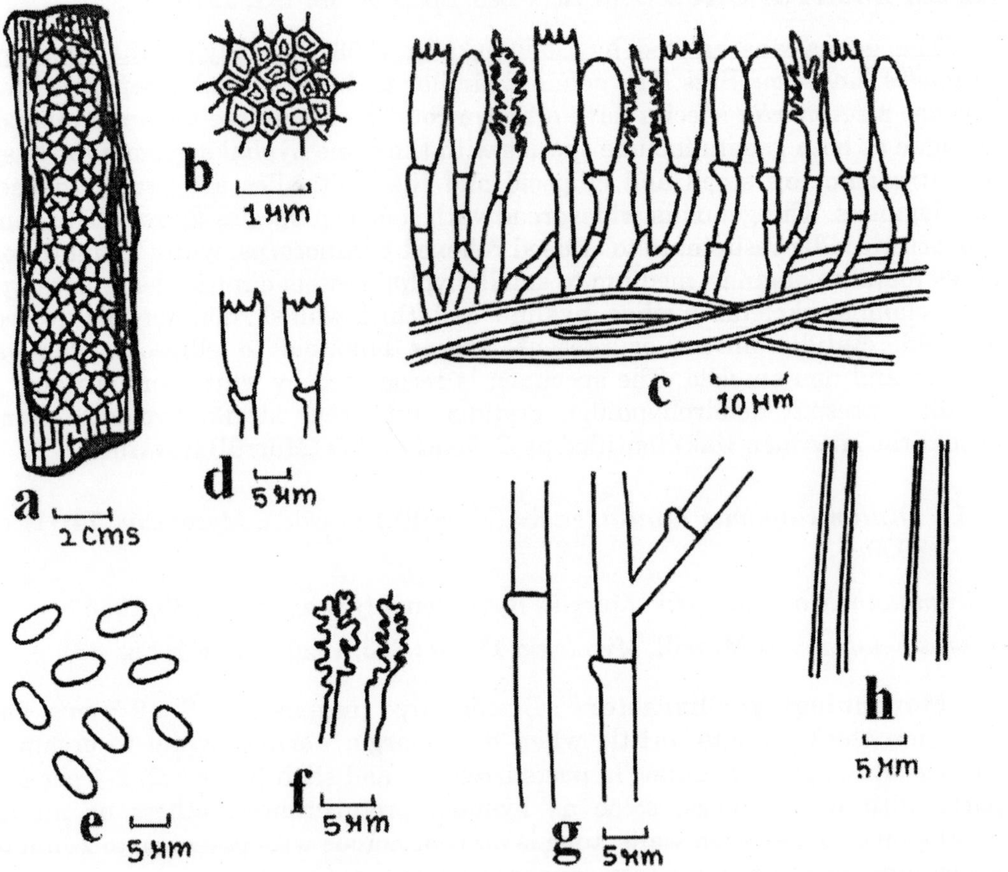

Fig. 83 ***Diplomitoporus hondurensis*** a. Habit. b. Pores. c. T.S. hymenium. d. Basidia. e. Basidiospores. f. Dendrohyphidia. g. Generative hyphae. h. Skeletal hyphae

(7). ***EARLIELLA*** Murrill. *Bull. Torrey Bot. Club* **32 (9) :** 478, 1905.

This genus erected by Murrill in 1905, by studying the type of species *Earliella cubensis* (Pres.) Murr. The genus has only three species reported under this genus. The genus is characterized by resupinate, effused-reflexed to pileate, annual to perennial, tough basidiocarps, upper surface when present, glabreous, first white to cream, then with a reddish cuticle spreading from the base; pore surface white to cork-colored, pores round to sinuous, context white to wood-colored, hyphal system trimitic, generative hyphae with clamps, skeletal hyphae and binding hyphae hyaline, cystidia none, basidiospores cylindrical to oblong-ellipsoid, hyaline and IKI negative. The specimen with characteristic effused tough basidiocarps with a reddish cuticle and somewhat irregular elongated and sinuous pores was identified as *E. scabrosa* (Persoon) Gilbertson & Ryvarden.

(7-1). *Earliella scabrosa* (Persoon) Gilbertson & Ryvarden. *Mycotaxon* **22 (2)** : 364, 1985.

=*Polyporus scabrosus* Pers., *Voy. Uranie. Bot.* **5:** 172 (1827).

=*Polyporus corrugatus* Pers., *Voy. Uranie. Bot.* **5:** 172 (1827).

=*Earliella corrugata* (Pers.) Murrill, *Bull. Torrey bot. Club* **34:** 468 (1907). (Plate 2, Fig. 84).

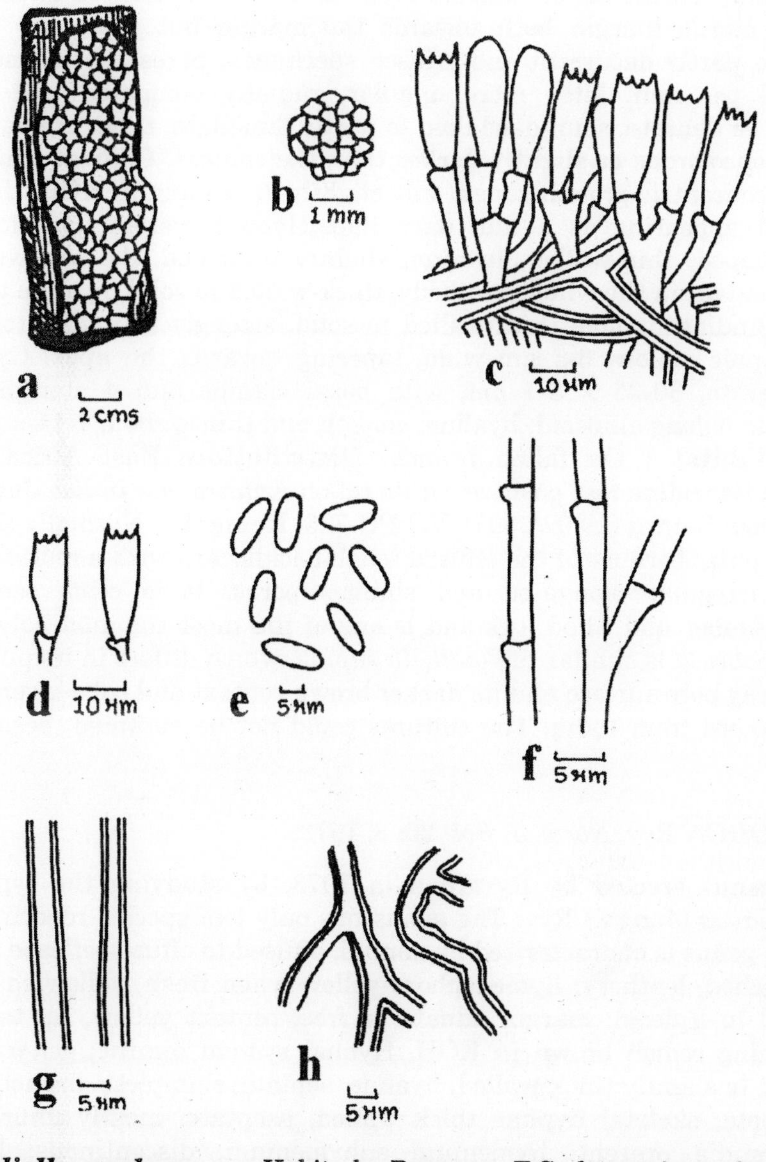

Fig. 84 *Earliella scabrosa* a. Habit. b. Pores. c. T.S. hymenium. d. Basidia. e. Basidiospores. f. Generative hyphae. g. Skeletal hyphae. h. Binding hyphae.

Morphological Characters : Fruitbody perennial, pileate, broadly attached, effuseo-reflexed or resupinate, 14-18 × 4-6 cms., solitary, imbricate or fused into horizontal rows, consistency corky to rigid when dry. Pileus glabrous, concentrically zonate, sulcate and furrowed, when dry somewhat radially wrinkled, colour cream, ochraceous, pale brown to brownish grey, when older covered with a reddish to bay or brown cuticle spreading from the base and later may cover most of the pileus, but may also lack completely, margin acute, thin, entire or undulate and lobed. Pore surface cream, ochraceous or luteous, even fulvous to discoloured in old specimens, often with sterile margin both towards the margin but especially towards the substrate in partly decurrent and effused specimens, pores first round and thick-walled, 2-3 per mm, later more angular, radially elongated and thin-walled, splitting to be dentate, semi-daedaloid to partly lamellate, tubes single layered, 1-3 mm long, concolorous or slightly darker than the context. Context white to cream, with weak concentric growth zones, only slightly darkening in KOH, the red-brown pilear crust appearing as a thin dark line. Hyphal system trimitic, generative hyphae clamped, thin-walled, hyaline, slightly branched, 3-5 μm wide. Skeletal hyphae abundant in the whole fruitbody, thick-walled to solid, hyaline to yellow, 4-6 μm wide. Binding hyphae thick-walled to solid, strongly to moderately branched, hyaline to pale yellow, 3-5 μm wide, tapering towards the apex. Cystidia none. Basidia clavate, 30-35 × 5-7 μm, with basal clamps and 4 sterigmata. Spores cylindrical to oblong ellipsoid, hyaline, smooth and thin-walled, 7-11 × 3-4 μm, non-amyloid. **Habitat :** On fallen branch. **Distribution:** East Africa. **Specimen examined:** Benzidine test positive, white rot on *Azadirachta indica* Juss. **Location :** Amboli; Dist: Satara (19/02/2007), VH-PO-726. **Remarks**: Normally this species is easy to recognize because of the effused tough basidiocarp with a reddish cuticle and somewhat irregular elongated and sinuous pores. It is often seen on poles, structural timber and filled logs and is one of the most common polypores in the tropic. *E. scabrosa* is similar to *Fomitella supine*, which differs in its purplish brown to smoky gray pore surface and its darker brown context and tube layer. The species is a new record from India. The cultures could not be obtained, besides repeated trials.

(8). *FLAVODON* Ryv. *Norw. J. Bot.* 20: 3, 1973.

This genus erected by Ryvarden in 1973, by studying the type of species *Flavodon flavus* (Jungh.) Ryv. The genus has only two species reported under this genus. The genus is characterized by annual, effused to effuso-reflexed basidiomata, firmly attached, leathery; hymenophore yellow when fresh, yellowish brown when dry, poroid to irpicoid; margin adnate or free; context yellow, up to 1mm thick; tissue turning redish brown in KOH. Hyphal system dimitic; generative hyphae thin walled to slightly thick walled, hyaline, septate, clampless, branched, branches near to septa; skeletal hyphae thick walled, aseptate, mostly unbranched, pale yellow; cystidia present; hymenium euhymenium, discontinous; basidiospores hyaline, thin walled, smooth, ellipsoid, non-amyloid. The specimen with characteristic yellowish colour and the hydnoid hymenophore, two types of cystidia abundantly present, terminally encrusted, thick walled true cystidia, arising from the basidial side branches and bent, skeletal hyphae (cystidial), arising from the trama, tissue turning radish brown in KOH was identified as *F. flavus* (Klotzsch.) Ryv.

(8-1). *Flavodon flavus* (Klotzsch.) Ryv. *Norw. J. Bot.* 20:3, 1973.

=*Irpex flavus* Klotzsch, *Linnaea* 8: 488, 1833. (Plate 2, 8, Fig. 85).

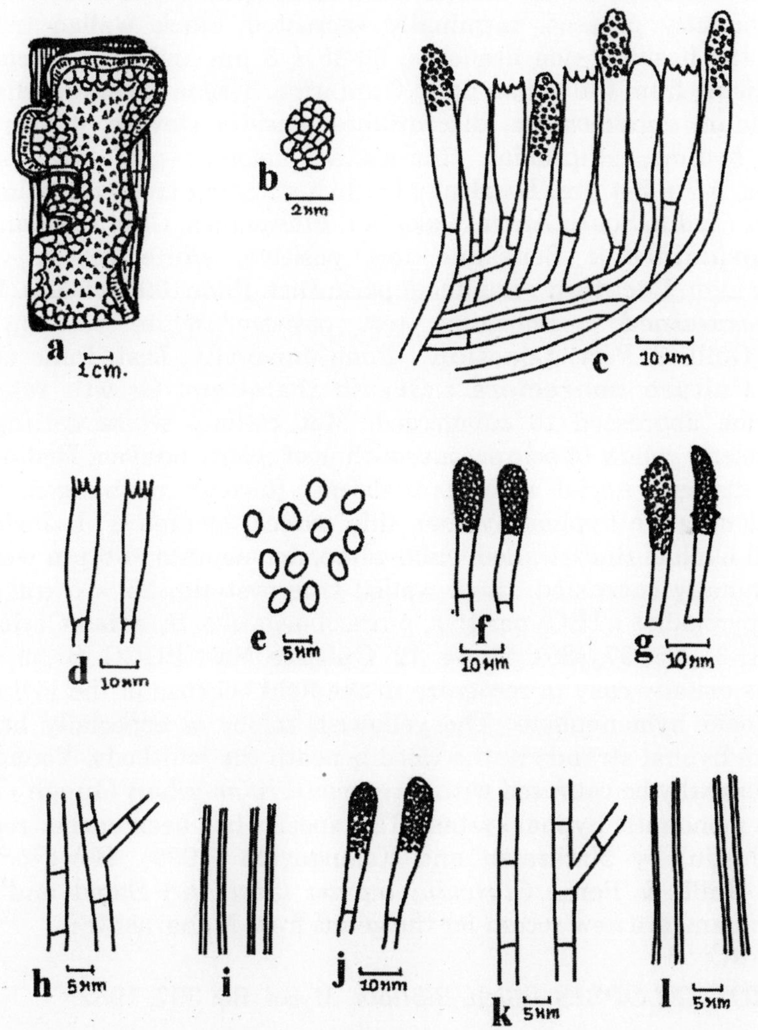

Fig. 85. *Flavodon flavus*. a. Habit. b. Pores c. T.S. hymenium. d. Basidia. e. Basidiospores. f. Encrusted cystidia. g. Terminally encrusted skeletal hyphae. h. Generative hyphae. i. Skeletal hyphae. j. Encrusted cystidia in culture. k. Generative hyphae in culture. l. Skeletal hyphae in culture

Morphological Characters : Basidiomata annual, effused to effuso-reflexed, firmly attached, leathery, 10-15 × 6-8 cms. and 2 mm thick; hymenophore yellow when fresh, pale yellow, when dry or even yellowish brown, poroid initially, later becoming irpicoid with irregular teeth, margin adnate or and reflexed, reflexed pileeus light brown to yellow, tomentose, without any zone. context yellow, up to 1mm thick, with compactly arranged hyphae in upper part. Hyphal system dimitic, generative hyphae hyaline, thin walled to slightly thick walled, septate, clamless, branched, branches near the septa, 2-4 μm wide, skeletal hyphae thick walled, subhyaline, non-septate, mostly unbranched, abundant, 3-5 μm wide. Two types of cystidia abundantly present, terminally encrusted, thick walled true cystidia, arising from the basidial side branches, 30-35 × 5 μm and bent, skeletal hyphae (cystidial), arising from the trama up to 6 μm wide. Tissue turning radish brown in KOH. Hymenium euhymenium, discontinues, basidia clavate, 4 spored, curved, basidiospore hyaline, ellipsoidal, thin-walled, smooth, gattulate, non-amyloid, acyanophilous, 5-7 × 3-4 μm. **Habitat :** On live standing tree and fallen branches. **Distribution :** India, South Africa, Pakistan, Philippines, China, Japan, Australia. **Specimen examined :** Benzidine test positive, white rot on: *Artocarpus heterophyllus* Lam. **Location :** Sambhaji park; Dist: Pune (09/10/2005), VH-PO-105. **Specimen examined :** Benzidine test positive, white rot on: *Dalbergia melanoxylon* Guill. & Perr. **Location :** Pune university; Dist: Pune (19/02/2007), VH-PO-531. **Culture characters :** Growth characters: Growth rate moderate; advancing zone appressed to submerged; Mat cottony white getting cream to ochraceous; colony yellow to ochraceous; outline of colony uneven; Distance between hyphal tips distant; Aerial mycelium absent; Reverse unchanged. Microscopic characters: Generative hyphae hyaline, thin walled, clampless, branched, 3-5 μm wide, skeletal hyphae thick walled, subhyaline, non-septate, 4-6 μm wide. Cystidia present, terminally encrusted, thick walled true cystidia, 25 × 5 μm. **Tests :** α-napthal and pyrogallol + H_2O_2 positive; p-cresol negative. **Species Code :** 1, 2, 3, 7, 13, 14, 15, 21, 31, 35, 37, (38), 53, 54, 72. **Culture No :** PUCC. B: 52. **Remarks :** The species is usually easy to recognize in the field because of the yellowish colour and the hydnoid hymenophore. The yellowish colour is especially bright in the context and in hyphal strands in the wood beneath the fruitbody. Young specimens may morphologically be confused with *Oxyporus cervino-gilvus* (Jungh.) Ryv. which, however, has monomitic hyphal system. This species has been earlier reported from India (Tamilnadu) by Natarajan and Kolandavelu (1998). However, *Dalbergia melanoxylon* Guill. & Perr., *Gliricidia sepium* (Jacquin.) Stend and *Artocarpus heterophyllus* Lam. are new record for the genus from Maharashtra.

(9). *GRAMMOTHELOPSIS* Jülich. *Biblioth. Mycol.* 85: 397, 1982.

This genus erected by Julich in 1982, by studying the type of species *Grammothelopsis macrospora* (Ryvarden) Jülich. The genus has only four species reported under this genus. The genus is characterized by resupinate, effused, adnate fruitbodies; pale brown pore surface; margin white to pale ochraceous, narrow to wide, pores angular to elongated, finely dentate, pore mouths and tube walls dotted with white hyphal pegs, partly as conical studs, partly as elongated short ridges, tubes pale brown. Hyphal system di-trimitic, generative hyphae hyaline and with clamps, trama and subhymenium dominated by skeletal hyphae, thick-walled to solid, mostly unbranched, non-dextrinoid. Dendrohyphidia richly present both in the hymenium and along the sterile pore-mouths. Basidia both in the bases of the tubes and on the tube walls. Spores broadly ellipsoid, thick-walled and strongly dextrinoid. The specimen with characteristic ochraceous pore surface, thick-walled and strongly dextrinoid basidiospores was identified as *G. puiggarii* (Spegazzini) Rajchenberg & Wright.

(9-1). *Grammothelopsis puiggarii* (Spegazzini) Rajchenberg & Wright. *Mycologia* **79 (2) :** 253, 1987.

=*Hymenogramme puiggarii* Speg., *Boletín de la Academia Nacional de Ciencias de Córdoba* **23:** 412, 1919. (Plate 3; Fig. 86)

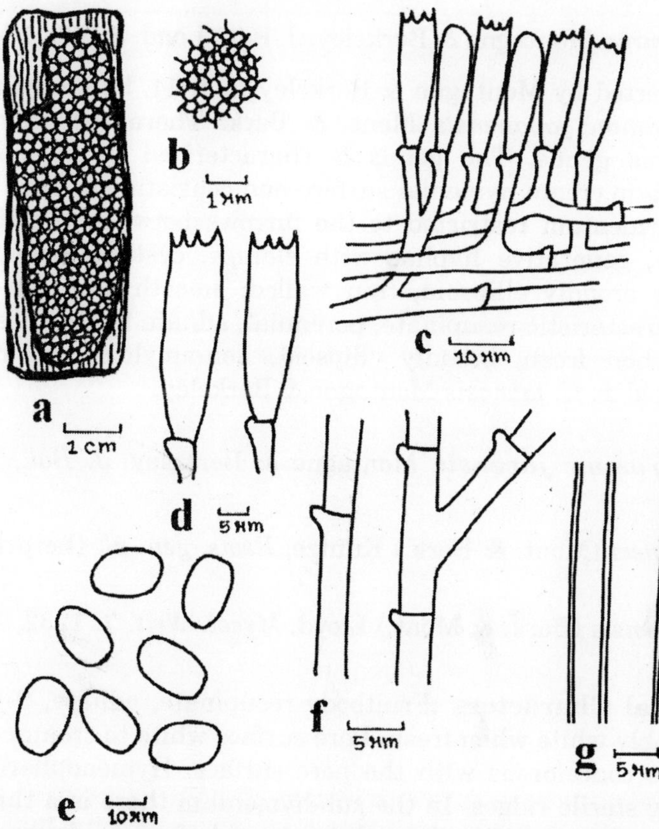

Fig. 86 *Grammothelopsis puiggarii.* a. Habit. b. Pores c. T.S. hymenium. d. Basidia. e. Basidiospores. f. Generative hyphae. g. Skeletal hyphae

Morphological Characters : Basidiocarp resupinate, effused, adnate, 3-6 × 2-3 cms., pore surface pale ochraceous, margin white, pores angular and regular, 2-3 per/mm, tubes up to 6 mm. deep, context 300 μm thick, white and cottony. Hyphal system dimitic, generative hyphae hyaline and with clamps, 2-3 μm. wide, skeletal hyphae thick-walled to solid, 4-6 μm wide, mostly unbranched, but in the pore mouths distinctly arboriform and may easily be interpreted as binding hyphae, strongly dextrinoid, especially in the pore mouths. Cystidia absent. Basidia clavate, 30-35 ×10-12 μm, with four sterigmata. Basidiospores broadly ellipsoid, hyaline, strongly dextrinoid, 15-20 ×10-12 μm. **Habitat :** On fallen branch. **Distribution :** Brazil. **Specimen examined :** Benzidine test positive, white rot on *Memecylon umbellatum Burm*. **Location :** Amboli; Dist: Satara (19/02/2007), VH-PO-769. **Remarks :** The specimen is recognized by ochraceous pore surface. Skeletal hyphae in the pore mouths distinctly arboriform and may easily be interpreted as binding hyphae, strongly dextrinoid, especially in the pore mouths. Basidiospores thick-walled and strongly dextrinoid. The species is a new record from India. The cultures could not be obtained, besides repeated trials.

(10). *Hymenogramme* Montagne & Berkeley. J. Bot., Lond. 3: 329, 1844.

This genus erected by Montagne & Berkeley in 1844, by studying the type of species *Hymenogramme javanensis* Mont. & Berk. There are only four species reported under this genus. The genus is characterized by resupinate, adnate fruitbodies, whitish to cream hymenial surface and consisting of long anastomosing ridges with the hymenium restricted to the furrows between the ridges. Hyphal system monomitic, generative hyphae with clamps, cystidiole abundant in the hymenium, spores broadly ellipsoid, thin walled, smooth and non-amyloid. The specimen with characteristic resupinate, perennial, adnate fruitbodies, light cream, probably white when fresh; broadly ellipsoid, non-amyloid and non-dextrinoid spores was identified as *H. javensis* Montagne & Berkeley.

(10-1). *Hymenogramme javensis* Montagne & Berkeley. *J. Bot., Lond.* 3: 330, 1844.

=*Aschersonia javensis* (Mont. & Berk.) Kuntze, *Revis. gen. pl.* (Leipzig) 3 (2) : 444, 1898.

=*Grammothele javensis* (Berk. & Mont.) Lloyd, *Mycol. Writ.* 7: 1232, 1923. (Plate 3, Fig. 87).

Morphological Characters : Fruitbody resupinate, adnate, 9-12 × 5-7 cms., light cream, probably white when fresh. Pore surface white to cream, pores angular, 1-2 per/mm, tubes concolorous with the pore surface. Hymenophore consisting of long anastomosing sterile ridges. In the sub-hymenium there is a thin brown zone, below is an intermittent light-coloured layer and then a dense, darker zone. Hymenium as a palisade on the bottom of the furrows defined by the ridges, 0-40 μm deep. Hyphal system monomitic, generative hyphae with clamp, hyaline, thin-walled to thick-walled, 3-4 μm wide. Cystidiole abundantly present in the hymenium, subulate with a rounded apex, mostly 20-30 × 6-7 μm, slightly projecting above the hymenium. Basidia clavate, 20-25 × 4-5 μm, with four sterigmata. Spores broadly ellipsoid, hyaline, thin-walled, smooth, non-amyloid and non-dextrinoid, 4-5 × 2.5-3 μm. **Habitat :** On fallen branch. **Distribution :** Philippines. **Specimen examined :** Benzidine test positive, white rot on *Livistona chinensis* Rbr. **Location :** Pune university; Dist: Pune (19/08/2006), VH-PO-514. **Remarks :** The unique hymenial configuration makes the species and the genus distinct. The species is a new record from India. The cultures could not be obtained, besides repeated trials.

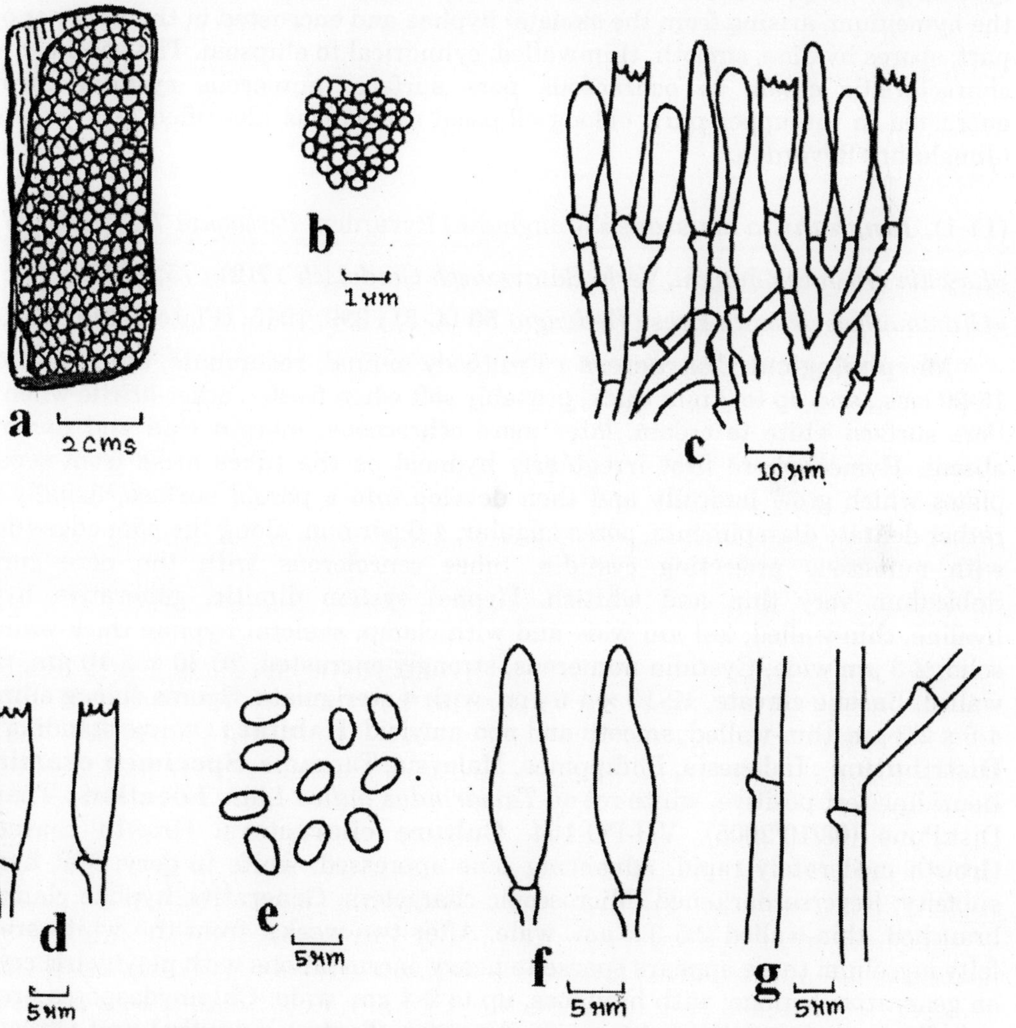

Fig. 87 ***Hymenogramme javensis*** a. Habit. b. Pores c. T.S. hymenium. d. Basidia. e. Basidiospores. f. Thin-walled cystidia. g. Generative hyphae

(11). *JUNGHUHNIA* Corda. Anleitung zum Studium der Mykologie: 195, 1842.

The genus was raised with the type species *Junghuhnia crustacean* (Junghuhn) Ryvarden. The genus is considerd as tropical in distribution with 27 species. The genus is characterized by annual to perennial, resupinate fruitbodies, pore surface white, cream or yellow-pink to cocoa-brown, pores mostly small and slightly irregular, context thin, white to ochraceous; hyphal system dimitic, generative hyphae thin-walled, hyaline, narrow and with clamps at the septa, the often difficult to find in old fruitbodies, skeletal hyphae dominate in the fruitbodies; cystidia present, both in the subhymenium and slightly projecting into and beyond the hymenium, arising from the skeletal hyphae and encrusted in the upper or outer part; spores hyaline, smooth, thin-walled, cylindrical to ellipsoid. The specimen with characteristic cream to ochraceous pore surface; numerous cystidia, strongly encrusted in the upper part, oblong ellipsoid spores was identified as *J. crustacea* (Junghuhn) Ryvarden.

(11-1). *Junghuhnia crustacea* (Junghuhn) Ryvarden. *Persoonia* **7(1) :** 18, 1972.

=*Laschia crustacea* Jungh., *Verh. Bataviaasch Gerdotsch* **17(2) :** 75, 1839.

=*Grammothele delicata* Bres., *Hedwigia* **56 (4, 5) :** 299, 1915. (Plate 4, Fig. 88).

Morphological Characters : Fruitbody annual, resupinate, effused, 30-35 × 15-20 cms., and up to 2 mm thick, probably soft when fresh, rather brittle when dry. Pore surface white to cream, later more ochraceous, margin thin and narrow to absent. Hymenophore first irregularly hydnoid as the tubes arise from separate plates which grow laterally and then develop into a poroid surface, usually with rather dentate dissepiments, pores angular, 4-6 per mm, along the pore edges dotted with numerous projecting cystidia, tubes concolorous with the pore surface. Subiculum very thin and whitish. Hyphal system dimitic, generative hyphae hyaline, thin-walled, 2-3 μm wide and with clamp, skeletal hyphae thick-walled to solid, 2-3 μm wide. Cystidia numerous, strongly encrusted, 20-40 × 5-10 μm, thick-walled. Basidia clavate, 12-18 × 4-6 μm., with 4 sterigmata. Spores oblong ellipsoid, 4-5 × 2-3 μm, thin-walled, smooth and non-amyloid. **Habitat :** On live standing tree. **Distribution** : Indonesia, Philippines, Malaysia, Tanzania. **Specimen examined :** Benzidine test positive, white rot on *Tamarindus indica* Linn. **Location :** Thamini; Dist:Pune (08/10/2005), VH-PO-134. **Culture characters:** Growth characters: Growth moderately rapid, advancing zone appressed, white to gery; Mat hyaline, subfelty; Reverse darkened. Microscopic characters: Generative hyphae clampless, branched, thin-walled 2.5-3.5 μm, wide. After two weeks, from the white areas of felty mycelium there appears sparse to heavy encrustations with polyhydric crystals on generative hyphae; with branches, up to 2-4 μm. wide. Chlamydospores present, apically in chain, globose, 4-6 μm in diameter. **Tests :** α-napthal and pyrogallol+ H_2O_2 positive; p-cresol negative. **Species Code :** 1, 2, 3, 7, 13, 18, 13, 24, 30, 52, 53, 57, 82, 85. **Culture No :** PUCC. B: 53. **Remarks :** The species is easily to recognize by its cream to ochraceous pore surface. The specimen shows hymenium both along the tube walls and the bases of the tubes. The species was reported by Ryvarden and Johansen (1980) from South East Asia and Africa on many deciduous woods.

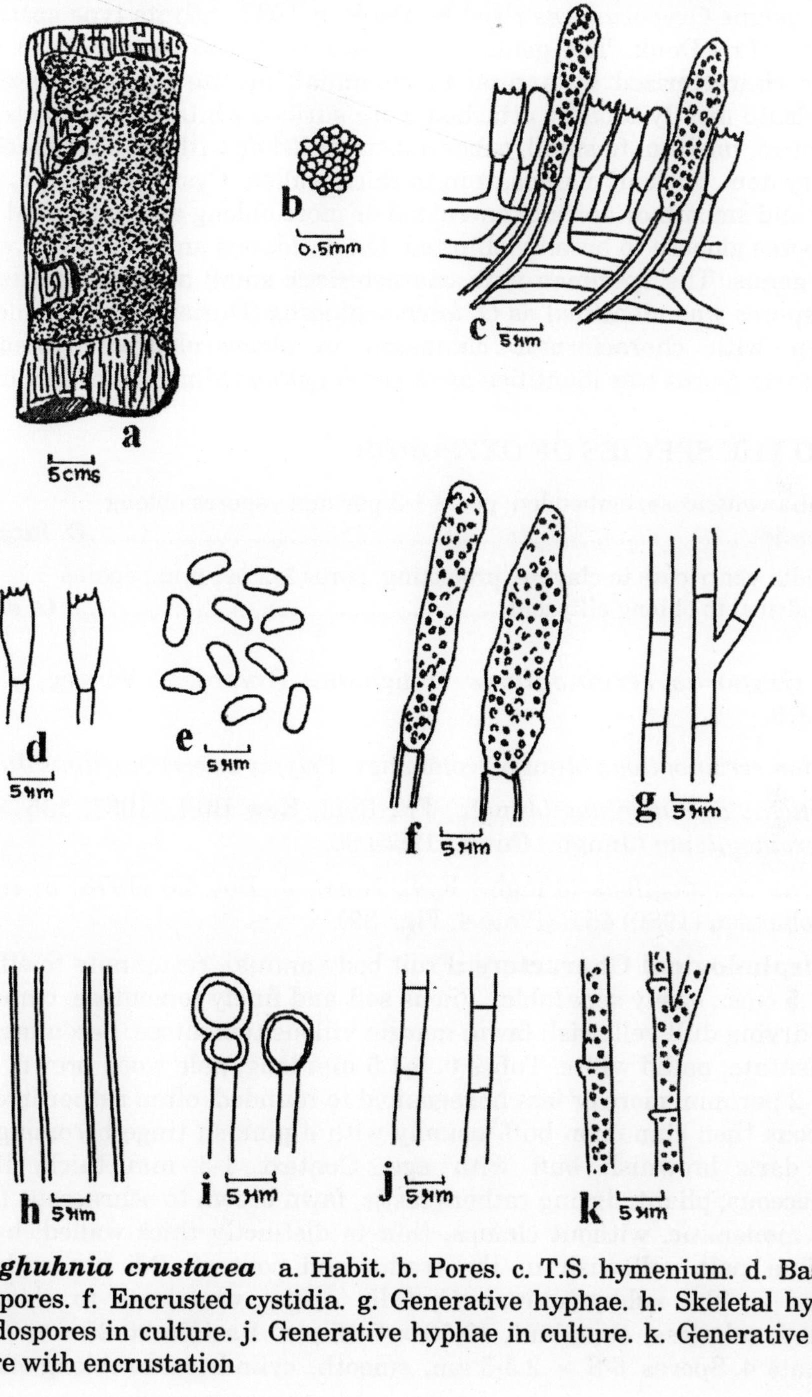

Fig. 88 ***Junghuhnia crustacea*** a. Habit. b. Pores. c. T.S. hymenium. d. Basidia. e. Basidiospores. f. Encrusted cystidia. g. Generative hyphae. h. Skeletal hyphae. i. Chlamydospores in culture. j. Generative hyphae in culture. k. Generative hyphae in culture with encrustation

(12). *OXYPOROUS* (Bourdot & Galzin) Donk. *Meded. Bot. Mus. Herb. Rijksuniv. Utrecht* **9:** 202, 1933.

The genus *Oxyporus* was rised by Donk in 1933 with its type species *Oxyporus populinus* (Fr.) Donk. The genus is represented by 35 cosmopolitan species. The genus is characterized by annual to perennial, pileate to resupinate fruitbodies, when pileate mostly broadly attached; pore surface white, cream to straw-coloured, pores entire, medium to small, tubes nonstratified or with distinct zones; monomitic hyphal system, without clamps, thin to thick-walled. Cystidia present, ventricose to tubular and smooth or apically encrusted or more oblong and encrusted in the upper part. Spores globose to broadly ellipsoid. On deciduous and coniferous wood. Cosmopolitan genus. The specimen with characteristic small apically encrusted cystidia, oblong spores was identified as *O. latemarginatus* (Durieu & Montagne) Donk. The specimen with characteristic cinnamon to straw-coloured hymenial surface, subcylindric spores was identified as *O. cervinogilvus* (Junghuhn) Ryvarden.

KEY TO THE SPECIES OF *OXYPORUS*

1. Cystidia ventricose, embedded, poers 1-3 per mm., spores oblong ellipsoid .... ***O. latemarginatus***

1´. Cystidia ventricose to clavate, projecting, pores 1-2 per mm., spores cylindrical to oblong ellipsoid .... ***O. cervinogilvus***

(12-1). *Oxyporus cervinogilvus* (Junghuhn) Ryvarden. *Norweg. J. Bot.* **20:** 3, 1973.

=*Flavodon cervinogilvus* (Jungh.) comb. nov. *Polyporus cervinogilvus Jungh.* (1838).

= *Polystictus cervinogilvus* (Jungh.) Fr., Reid, Kew Bull., 1957, 135. =*Trichaptum cervinogilvum* (Jungh.) Cunn. (1965) 96.

=*Oxyporus cervinogilvus* (Jungh.) Ryv., Norw. J. Bot. 20 (1973) 3; Ryvarden and Johansen (1980) 451. (Plate 4, Fig. 89).

Morphological Characters : Fruit body annual, resupinate to effuso-reflexed, 4-6×1-1.5 cms., easily separable,. Pileus soft and finely tomentose, cinnamon to pale brown, drying dull yellowish fawn; margin villous tomentose, becoming thin, entire, to subdentate, pallid white. Tubes 0. 5-1.5 mm long, pale wood brown, not stratose; pores 1-2 per/mm, more or less hexagonioid to rounded, often rather irregular, pallid ochraceous then cinnamon buff, usually with a pinkish tinge becoming ochraceous, finally dark brownish buff with age. Context 1-3 mm thick, floccosofelted, subcoriaceous, pliant, drying rather elastic, fawn brown to ochraceous fawn. Hyphal system monomitic, without clamps, thin to distinctly thick-walled, hyaline in the hymenium pale yellowish in the trama and context, 3-5 μm wide, frequently branched in the subhymenium. Cystidia acute, ventricose to clavate, coarsely encrusted, colorless, abundant, 40-50 × 7-15 μm. Basidia 20-25 × 6-7 μm, clavate; sterigmata 4. Spores 6-8 × 2.5-3 μm, smooth, cylindrical to oblong ellipsoid, thin-

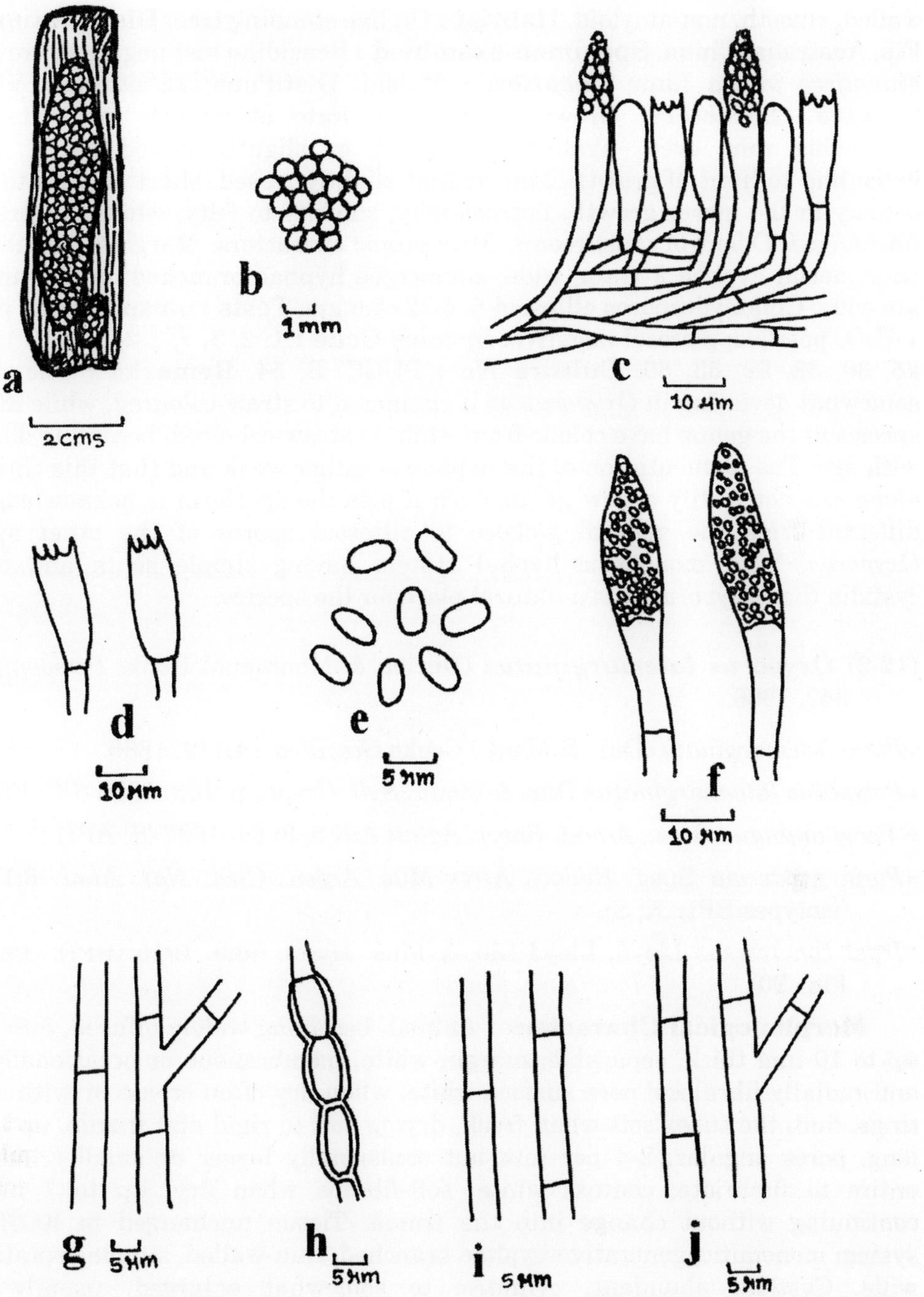

Fig. 89. *Oxyporus cervino-gilvus*. a. Habit. b. Pores. c. T.S. hymenium. d. Basidia. e. Basidiospores. f. Encrusted cystidia. g. Generative hyphae. h. Chlamydospores in culture. i. Marginal and aerial in culture. j. Submerged hyphae in culture

walled, smooth, non-amyloid. **Habitat :** On live standing tree. **Distribution :** India, Fiji, Australia, China. **Specimen examined :** Benzidine test negative, brown rot on *Mangifera indica* Linn. **Location :** Mulshi. **Dist:**Pune (12/09/2005), VH-PO-20. **Culture characters:** Growth characters: Rate of growth moderately rapid; Advancing zone even, hyaline, appressed, or slightly raised aerial mycelium extending to limit of growth. Mat at first slightly raised, short cottony to floccose-cottony in the newer growth, floccose-felty, subfelty to felty, white. Reverse mostly unchanged, Odour of mushrooms. Microscopic characters: Marginal hyphae 2-4 μm wide, aerial hyphae 2-4 μm, wide; submerged hyphae branched or unbranched 3-4 μm wide. Chlamydospores ellipsoid 5. 5-12×4-6 μm. **Tests :** α-napthal and pyrogallol + H_2O_2 positive; p-cresol negative. **Species Code :** 1, 2, 3, 7, 1 2, 13, 17, 19, 21, 24, 25, 30, 38, 52, 53, 85. **Culture No :** PUCC. B: 54. **Remarks :** The species is somewhat deviating in *Oxyporus* as it cinnamon to straw-coloured, while most other species in the genus have colour from white to straw-coloured, becoming darker only with age. The pigmentation of the hyphae is rather weak and that this chrematistic alone can not justify a new genus even if also the sporform is acknowledged to be different from the general globose to ellipsoid spores of the other species in *Oxyporus*. With monomitic hyphal system having simple septa and encrusted cystidia that *Oxyporus* was a natural place for the species.

(12-2) *Oxyporus latemarginatus* (Durieu & Montagne) Donk. *Persoonia* **4 (3) :** 342, 1966.

=*Poria latemarginata* (Dur. & Mont.) Cooke *Grevillea* **14:**112. 1886.

=*Polyporus late-marginatus* Dur. & Mont., *Syll. Crypt.*, p. 163. 1856 (PC; FH).

= *Poria ambigua* Bres., *Accad. Rover. Agiati Atti* **3, 3:** 84. 1897 (S; BPI).

=*Poria geoderma* Speg, *Buenos Aires Mus. Argen. Cien. Nat. Anal.* **6:**171. 1899 (isotypes-BPI;. K; S).

=*Irpex concrescens* Lloyd, Lloyd Lib. & Mus. *Mycol.* **60:**9. 1915 (BPI). (Plate 4, 9, Fig. 90).

Morphological Characters : Annual, becoming widely effused, 5-8×3-4 cms., up to 10 mm thick, separable; margin white, membranous or occasionally matted and radially fibrillose; pore surface white, when dry often cream or with a pinkish tinge, dull, the tubes soft when fresh, drying soft to rigid and fragile, up to 10 mm long, pores angular, 2-4 per mm but occasionally larger or smaller, edges thin, entire to fimbriate; context white, soft-fibrous when dry, up to 1 mm thick, continuing without change into the trama. Tissue unchanged in KOH; hyphal system monomitic, generative hyphae branched, thin-walled, simple-septate, 4-6 μm wide. Cystidia abundant, cylindric to somewhat enlarged, usually coarsely encrusted at the apex, 15-22 × 4-7 μm; basidia slender-clavate, 15-20 × 4-6 μm; spores hyaline, smooth, non-amyloid, oblong-ellipsoid, 4-7 × 3-4 μm. **Habitat :** On fallen branches. **Distribution :** East Africa, India, Kenya, Tanzania, Uganda. **Specimen examined :** Benzidine test negative, brown rot on *Tectona grandis* Linn.

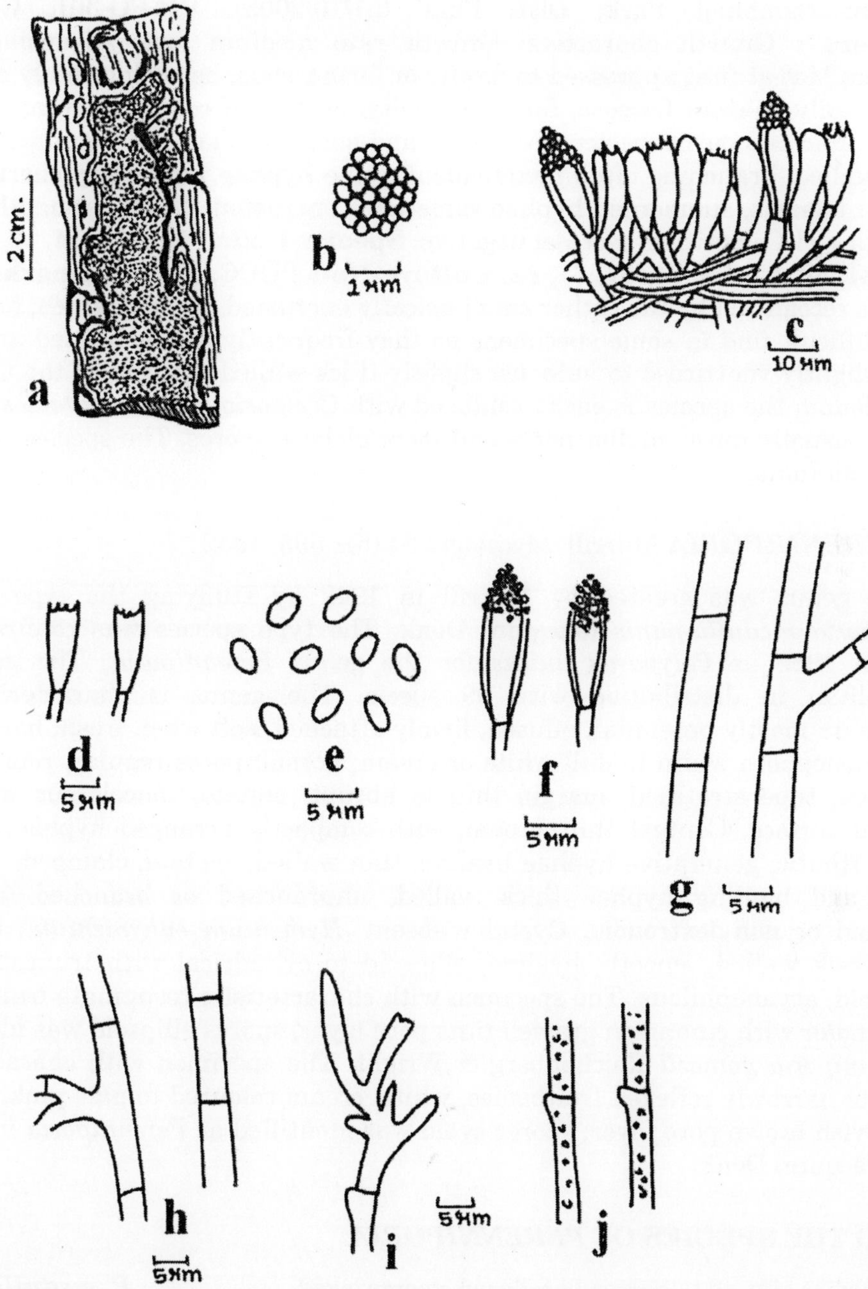

Fig. 90. ***Oxyporus latamarginatus*** a. Habit. b. Pores. c. T.S. hymenium. d. Basidia. e. Basidiospores. f. Encrusted cystidia. g. Generative hyphae. h. Marginal and aerial hyphae in culture. i. Much branched laterals hyphae in culture. j. Encrusted submerged hyphae in culture

Location: Sambhaji Park; Dist: Pune (21/10/2005), VH-PO-301. **Culture characters :** Growth characters: Growth rate medium fast; advancing zone appressed; Mat at first appressed to downy or farinaceous, becoming thinly cottony, cottony-woolly or even floccose, finally subfelty. outline of colony uneven; Reverse bleached. Microscopic characters: Marginal and aerial hyphae 1. 5-6 μm, wide, thin to firm walled, branching often inequivalent, some hyphae with many short, much branched laterals, submerged hyphae sometimes encrusted. **Tests :** α-napthal and pyrogallol + H_2O_2 positive; p-cresol negative. **Species Code :** 1, 6, 13, 14, 17, 18, 19, 21, 22, 24, 30, 52, 53, 54, 57, 61, 82. **Culture No :** PUCC. B: 55. **Remarks :** The species is recognized by the rather small apically encrusted cystidia which, however, can be difficult find in some specimens as they frequently are embedded and then appear slightly ventricose to bulbous, slightly thick-walled oranges. If the Cystidia are not found, the species is easily confused with *Ceriporia xylostromatoides* which, however, usually have smaller pores and more globose spores. The species is a new record from India.

(13). *PERENNIPORIA* Murrill. *Mycologia* 34 (5) : 595, 1942.

The genus was created by Murrill in 1942 by studying the type species *Perenniporia medulla-panis* (Jacquin) Donk. The type species was confirmed by Cooke in 1953 as *Polyporus unitus* for the genus *Perenniporia*. The genus is cosmopolitan in distribution with 99 specie. The genus is characterized by basidiomata mostly perennial, effused, firmly attached, soft when fresh, hard when dry; hymenophore white to dull white or cream; poroid; pores regular, round; pore wall thick, tube stratified; margin thin to abrupt, adnate, concolorous with the hymenial surface. Context thin, cream, with compactly arranged hyphae; hyphal system trimitic; generative hyphae hyaline, thin walled, septate, clamped; skeletal hyphae and binding hyphae thick walled, unbranched or branched, hyaline, dextronoid or non-dextronoid. Cystidia absent. *Hymenium euhymenium*; basidiospores thick walled, smooth, hyaline, ellipsoid to cylindrical with truncate apex, dextronoid, acyanophilous. The specimen with characteristic resupinate fruitbodies, brown and/or with cinnamon grayish tints pore layer, spores ellipsoid was identified as *Perenniporia gomezii* Rajchenberg & Wright. The specimen with characteristic effused to narrowly reflexed fruitbodies, white, cream coloured to pale pinkish-buff, pale greyish brown pore layer; spores ovoid was identified as *Perenniporia medulla-panis* (Jacquin) Donk.

KEY TO THE SPECIES OF *PERENNIPORIA*

1. Fruitbodies effused to narrowly reflexed, spores ovoid .... ***P. medulla-panis***

1´. Fruitbodies resupinate, spores ellipsoid .... ***P. gomezii***

(13-1) *Perenniporia gomezii* Rajchenberg & Wright. *Mycotaxon* 15: 306, 1982.

=*Poria gomezii* (Rajchenb. & Wright) Ginns, *Mycotaxon* 21: 331, 1984. (Plate: 5, Fig. 91).

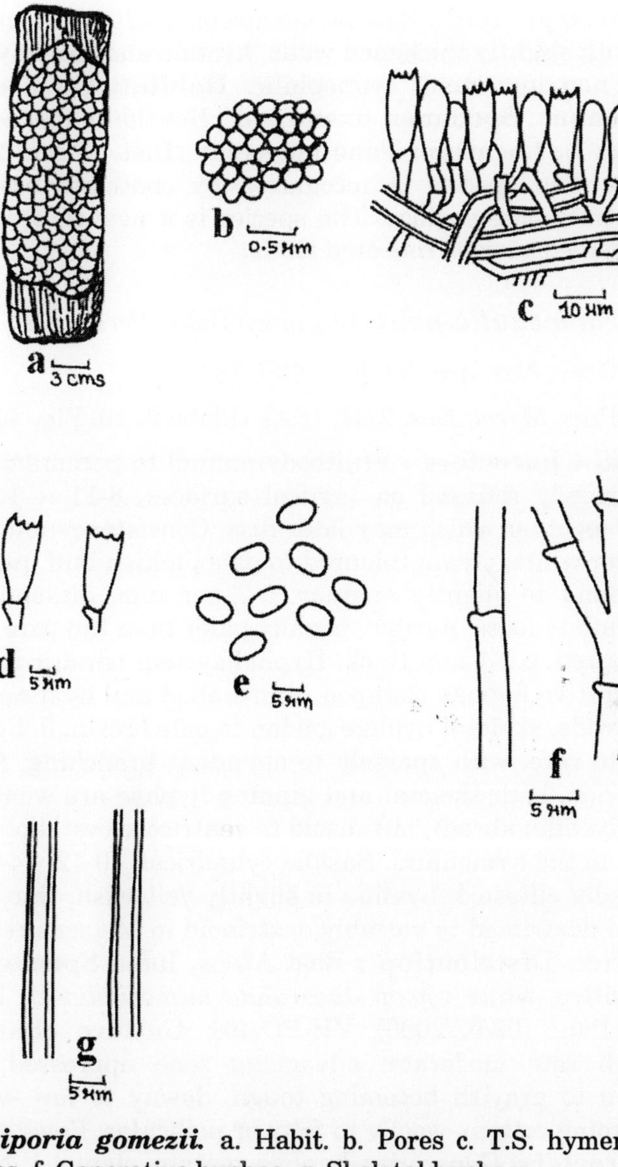

Fig. 91 *Perenniporia gomezii.* a. Habit. b. Pores c. T.S. hymenium. d. Basidia. e. Basidiospores. f. Generative hyphae. g. Skeletal hyphae

Morphological Characters: Basidiocarp annual, lignicolous, widely resupinate and totally attached to substrate, on vertical surfaces, 11-17 × 7-10 cms. Pore surface chocolate brown or with cinnamon greyish tints. Pores round or isodiametric, of variable size, 4-8 per mm, depending on the part and position of fruitbody. Context thin, wood-coloured, corky, up to 1 mm wide. Tubes concolorous with pore surface or violaceus grey, bony. Reaction with KOH not xanthochroic. Hyphal system dimitic. Generative hyphae clamped, hyaline, thin or faintly thickened walls, branched, 3-5 μm wide, skeletal hyphae 3-6 μm wide, with thick-wall, Hymenium 16-20 μm thick. Basidia claviform, 15-20 × 5-7 μm. Basidiospores oblong, ellipsoid, with slightly thickened walls, hyaline and rarely yellowish, 7 × 3-4 μm, non-amyloid, non-dextrinoid, cyanophilic. **Habitat:** On live standing tree. **Distribution:** Argentina. **Specimen examined :** Benzidine test positive, white rot on *Ficus bengamina* L. **Location:** Pune university, Dist. Pune (26/08/2006), VH-PO-525. **Remarks:** The species is recognized by chocolate brown and/or with cinnamon grayish tints pore surface. The species is a new record from India. The culture was not obtained besides repeated trials.

(13-1). ***Perenniporia medulla-panis*** (Jacquin) Donk. *Persoonia* 5 (1) : 76, 1967.

=*Poria medullaris* Gray, *Nat. Arr. Br. Pl.* 1:639, 1821.

=*Polyporus unitus* Pres. *Mycol. Eur.* 2:93, 1825. (Plate: 5, 10, Fig. 92).

Morphological Characters : Fruitbody annual to perennial, usually effused but sometimes narrowly reflexed on vertical surfaces, 8-11 × 1. 5-3 cms., often several fruitbodies together which may later fuse. Consistency tough-corky, to hard when dry. Pore layer white, cream coloured to pale pinkish-buff, pale greyish brown with age, pores round to slightly angular, 5-7 per mm, dissepiments thick and entire, margin adnate to loose, narrow, usually paler than the pore surface. Context cream to wood coloured, 0.2-1 mm thick. Hyphal system trimitic in the context and dissepiments, generative hyphae clamped, thin-walled and hyaline, with occasional branching, 3-4 μm wide, skeletal hyphae golden to pale brown, 3-4 μm wide, binding hyphae abundant to rare, with sparsely to abundant branching, tapering towards the ends, 2-3 μm wide, both skeletal and binding hyphae are weakly dextrinoid in Melzer's reagent. Cystidia absent, but fusoid to ventricose cystidiols 15-20 × 6-8 μm, sometimes present in the hymenium. Basidia cylindrical, 10-12 × 4-5 μm, with basal clamp. Spores broadly ellipsoid, hyaline to slightly yellowish, thin- to thick-walled, 4-7 × 3.5-6 μm, non-dextrinoid to variably dextrinoid in Melzer's reagent. **Habitat :** On live standing tree. **Distribution :** East Africa, India **Specimen examined :** Benzidine test positive, white rot on *Jacaranda mimosifolia* D. Don. **Location :** Kondhona, Dist. Pune (09/02/2006), VH-PO-491. **Culture characters:** Growth characters: Growth rate moderate; advancing zone appressed, hyphae rather distant, mat cream to grayish becoming tough, downy to low woolly or slightly floccose, soon becoming cottony-woolly to felty or pellicular; Reverse often bleached. **Microscopic characters:** Generative hyphae septate, clamped, branched 3-5 μm, wide. **Tests :** α-napthal and pyrogallol + H_2O_2 positive; p-cresol negative. **Species Code :** 1, 2, 3, 4, 6, (7), 13, 15, 17, 18, 22, 24, 25, 30, 37, 39, 52, 53. **Culture No :** PUCC. B: 56. **Remarks :** The species is rather variable in surface colour in the tropics. The surface colour varies from white to cream to pale pinkish buff and pale grayish brown with age. Vegetative hyphae are dextrinoid, while they are non-dextrinoid in European species. The species is recorded by Thind and Rattan in 1973 on stumps of *Cedrus deodora* and *Qurecus*, from Dalhousie and Mussoorie.

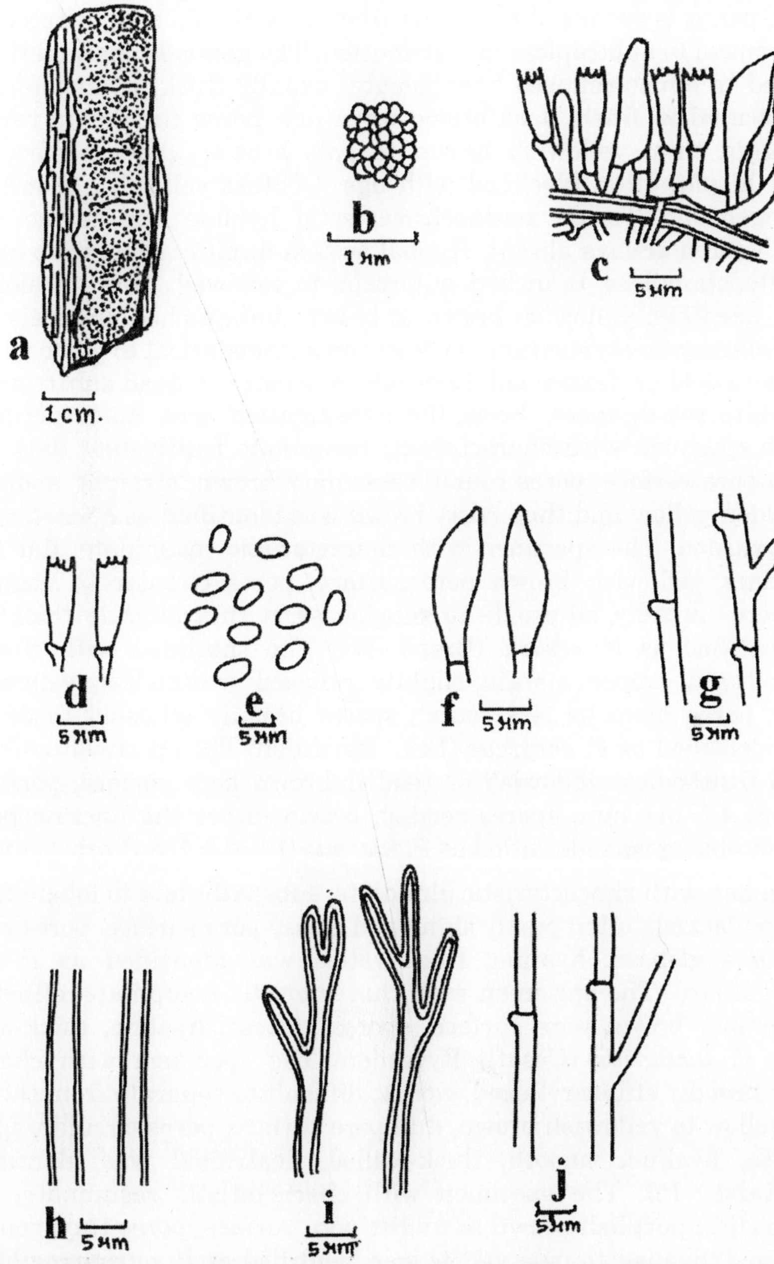

Fig. 92 *Perenniporia medulla-panis* a. Habit. b. Pores. c. T.S. hymenium. d. Basidia. e. Basidiospores. f. Cystidia. g. Generative hyphae. h. Skeletal hyphae. i. Binding hyphae. j. Generative hyphae in culture

(14). *PHELLINUS* Quel. Ench. Fung. 172, 1886.

The genus was proposed by Quelet in 1886, with type species *Phellinus rubriporus* Quel. The genus is cosmopolitan in distribution with 355 species recorded. Many species are tropical to subtropical in distribution. The genus is characterized by perennial, effused to effuso-reflexed basidiomata, usually thick; hymenophore poroid, yellowish brown when fresh, dark brown when dry, pores circular to various form, stratose; margin thin, yellowish brown to pale brown, upper surface glabrous, tomentose or hispid, often blackend with age. Context yellowish brown to redish brown, homogeneous. Tissue xanthochroic; setal hyphae and seatae present in hymenium. Cystidia always absent. Hyphal system dimitic; generative hyphae thin walled, septate, clampless, branched, subhyalin to yellowish brown;skeletal hyphae thick walled, aseptate, yellowish brown to brown, unbranched or rarely branched, abundant. *Hymenium euhymenium*; basidiospores cylenderical to globose, hyaline or coloured, nonamyloid or dextronoid. Lignicolous, mostly on dead substrates, capable of causing white rot diseases. From the investigation area nine specimens were collected. The specimen with characteristic resupinate fruitbodies; dark fulvous to umber brown pore surface; pores round; setae dark brown, straight; spores globose, hyaline to golden-yellow and then rusty brown was identified as *P. carteri* (Berkeley ex Cooke) Ryvarden. The specimen with characteristic resupinate; flat to concave fruitbodies; dark yellowish brown pore surface; pores circular to angular; setae yellowish; spores broadly ellipsoide to subglobose, with a slightly thickend yellow wall was identified as *P. cesatii* (Bres.) Ryv. The specimen with characteristic resupinate, effused, upper margin slightly reflexed fruitbodies; yellowish-brown pore surface; pores more or less round; spores broadly ellipsoid to drop-shaped; hyaline was identified as *P. chryseus* (Lév.) Ryvarden. The specimen with characteristic effused fruitbodies; yellowish or reddish-brown pore surface; pores round to angular; about 4-5 per mm, spores reddish brown under the microscope, smooth, broadly oval or oblong was identified as *P. inermis* (Ellis & Everhart) Cunningham.

The specimen with characteristic dimidiate, subspathulate to lobate fruitbodies; dark brown to blackish; often partly shiny and glossy pore surface, pores round, 8-10 per mm; spores globose, hyaline, thick-walled was identified as *P. lamaensis* (Murrill) Patouillard. The specimen with characteristic resupinate, effused, adnate fruitbodies, umber brown pore surface, spores globose, hyaline, thick-walled was identified as *P. luctuosus* (Cesati) Ryvarden. The specimen with characteristic resupinate to broadly effuse-reflexed, woody, difficult to separate from the substrate fruitbodies; yellow to yellowish brown, dull pore surface; pores circular; spores ovoid to sub-globose, hyaline, smooth, thick-walled, dextrinoid was identified as *P. punctatus* (Karst.) Pil. The specimen with characteristic resupinate, perennial, adnate fruitbodies; purplish-brown to umber pore surface; pores tiny; round; spores broadly ellipsoid, hyaline to pale yellow was identified as *P. purpureogilvus* (Petch) Ryvarden. The specimen with characteristic resupinate to slightly reflexed fruitbodies; golden brown pore surface; pores circular to angular; spores sub-globose, hyaline, thick-walled, smooth, dextrinoid was identified as *P. sonorae* Gilbertson.

KEY TO THE SPECIES OF *PHELLINUS*

1. Contextual or tramal setal hyphae or tramal setae present .... ***P. lamaensis***

1´. Contextual or tramal setal hyphae or tramal setae absent 2

2. Hymenial setae present 3

2´. Hymenial setae absent 8

3. Basidiospores remaining hyaline 4

3. Basidiospores eventually pigmented 6

4. Known from Borneo;hymenial setae 10-15×5-8 μm .... ***P. luctuosus***

4´. Not as above 5

5. Basidiospores globose to mostly subglobose .... ***P. sonorae***

5´. Basidiospores ellipsoid to mostly ovoid .... ***P. chryseus***

6. Basidiospores becoming brown .... ***P. carteri***

6´. Basidiospores becoming yellow 7

7. Hymenial setae 13-25×3. 5-7 μm, somewhat ventricose, hymeial tube up to 0. 5 mm long;pore surface dull grayish brown;basidiosores 3. 5-4. 5×3-4 μm .... ***P. cesatii***

7´. Hymenial setae (20-) 25-40×5. 5-8 μm, acuminate, subulate, hymeial tube up to 2mm long;pore surface purplish brown to umber;basidiosores 3. 5-4. 5×3 μm .... ***P. purpureogilvus***

8. Basidiospores remainig hyaline 9

8´. Basidiospores eventually pigmented .... ***P. inermis***

9. Basidiospores 5-8. 5μm long .... ***P. punctatus***

9. Basidiospores up to 5. 2 μm long .... ***P. cesatii***

(14-1) *Phellinus carteri* (Berkeley ex Cooke) Ryvarden. *Norweg. J. Bot.* **19:** 234, 1972.

=*Fuscoporia carteri (*Berk. ex Cooke) G. Cunn., *Bull. N.Z. Dept. Sci. Industr. Res., Pl. Dis. Div.* **73:** 11, 1948.

=*Poria carteri* Berk. ex Cooke, *Grevillea* **15 (73) :** 25, 1886. (Plate 5, Fig. 93).

Morphological Characters : Fruitbody resupinate, effused, perennial, woody hard, 10-13 × 5-8 cms. and up to 6 mm thick, margin narrow and finely velutinate. Pore surface dark fulvous to umber brown, pores round, small, 7-9 per mm, up to 3 mm deep, tubes concolorous up to 3 mm deep. Context very thin, 0. 1-0. 4 mm deep, fulvous to snuff brown. Hyphal system dimitic, generative hyphae hyaline, simple septate, 2-3 μm wide, skeletal hyphae yellow to rusty brown, 2-5 μm wide, thick-walled to almost solid. Setae 20-30 × 8-10 μm, dark brown, straight, slightly ventricose to acuminate, thick-walled. Basidia broadly clavate, 15-20 × 4-6 μm. Spores globose, 4.5-6.5 μm wide, hyaline to golden-yellow and then rusty brown.

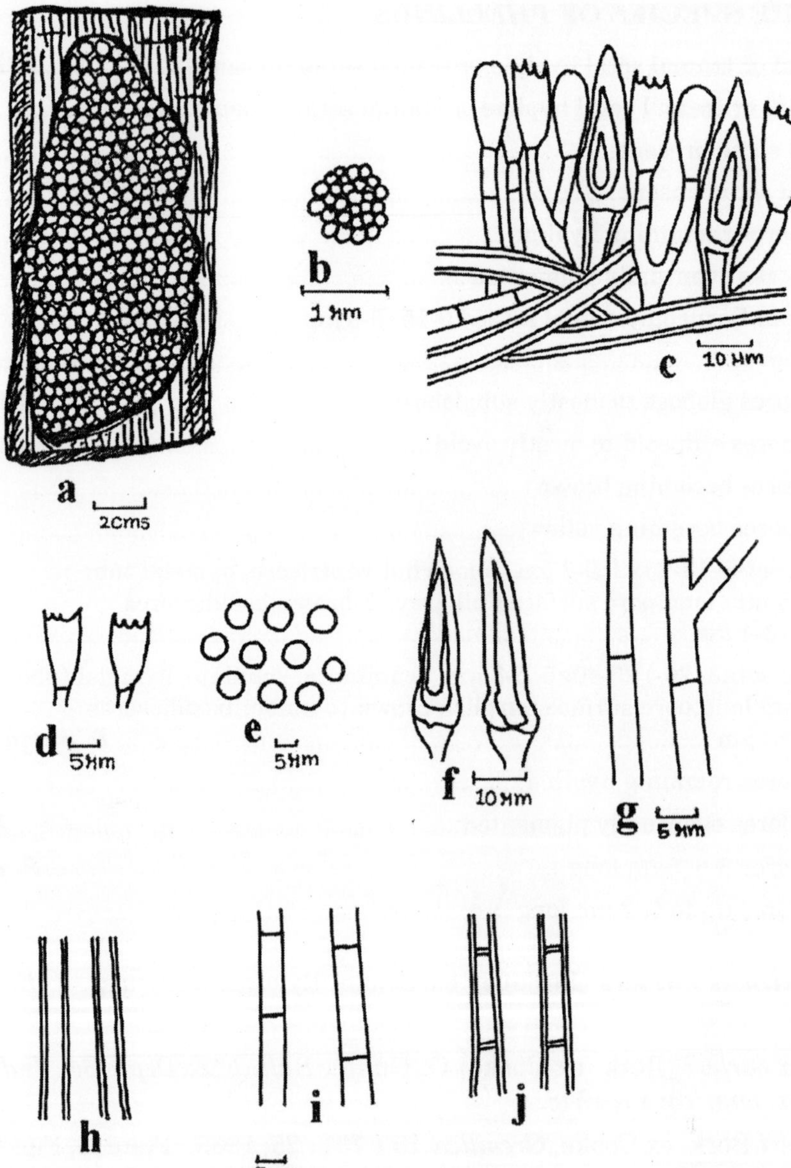

Fig. 93 ***Phellinus carteri.*** a. Habit. b. Pores c. T.S. hymenium. d. Basidia. e. Basidiospores. f. Setae. g. Generative hyphae. h. Skeletal hyphae. i. Advancing hyphae in culture. j. Aerial hyphae in culture

Habitat : On live standing tree. **Distribution :** India, Ghana. **Specimen examined :** Benzidine test positive, white rot on *Peltophorum ferrugineum.* **Location :** Mulshi, Dist. Pune (08/10/2005), VH-PO-148. **Culture characters :** Growth characters: Growth rate moderate; advancing zone appressed to raised; Mat white, cottony to cottony woolly becoming low felty to velvety., colony whitish to cream; outline of colony even; hyphal rather distinct; odour none, reverse bleached to locally darker. Microscopic characters: Advancing hyphae hyaline thin, branched, 1-3 µm, wide, clamps lacking, with simple septa, aerial hyphae brownish yellow, simple septa, 3-5 µm, wide. **Tests :** α-napthal, p-cresol and pyrogallol, + H_2O_2 positive. **Species Code :** 1, 2, 3, 7, 12, 13, 14, 15, 21, 22, 25, 26, 30, 31, 35, 37, 38, 52, 53, 54. **Culture No :** PUCC. B: 57. **Remarks :** The species is recognized by its resupinate fruitbodies and the fairly large globose ultimately becoming rusty brown. The setae are quite wide, 8-11µm. wide, giving them a sub-ventricose shape. One may wonder whether it represents a resupinates stage of some pileate species. There is no doubt that it microscopically comes very close e. g. to *P. linteus.* This species has been earlier reported from India (The Western Ghats of Maharashtra) by Rabba (1994).

(14-2). ***Phellinus cesatii*** (Bres.) Ryv. *Norw. J. Bot.* **19:** 234, 1972.

=*Phellinus gilvoides* (Petch) Ryvarden, *Norw. Jl Bot.* **19:** 234, 1972.

=*Poria cesatii* Bres., *Stud. Trent.,* ser. **2 7:** 7, 1926.

=*Poria gilvoides* Petch, *Ann. R. bot. Gdns Peradeniya* **6:** 138, 1916. (Plate: 5, 10, Fig. 94).

Morphological Characters : Basidiomata perennial, resupinate flat to concave, wood hard, 17-19 × 12-15 cms., friable, yellow to yellowish brown; margin narrow, concolor. Pore surface dark yellowish brown, pores circular to angular in the margin, 6-7 per mm., dissepiments entire, thick:uni- to bistritified, 3-6 mm deep, concolorus, context very thin, 5 mm deep, fibrous, between the context and the substrate a black line may be observed. Hyphal system dimitic, generative hyphae hyaline to pale yellow, with thin to thickened walls, 2-4 µm wide; skeletal hyphae brownish, straight to slightly sinuous, thickwalled, 3-4 µm wide. Setae scant, ventricose to subulate, yellowish, 18-20 × 3-4 µm. Basidia clavate, hyaline, thin-walled, 9-10 × 4-5 µm. Basidiospore broadly ellipsoide to subglobose, with a slightly thickend yellow wall, 3-5 × 2.5-3.5 µm, non-amyloid. **Habitat :** On live standing tree. **Distribution :** Brazil, Sri Lanka. **Specimen examined :** Benzidine test positive, white rot on:*Cassia sayami.* **Location :** Pune university; Dist: Pune (24/10/2005), VH-PO-425. **Specimen examined :** Benzidine test positive, white rot on: *Atrocarpus integrifolia* L. **Location :** Baneshwar, Dist. Pune (13/01/2007), VH-590. **Culture characters :** Growth characters: Growth rate moderate; advancing zone appressed to raised; Mat white, low cottony to loose cottony, colony white; outline of colony even to uneven; odour none, reverse unchanged to darkened. Microscopic characters: Advancing hyphae hyaline thin, branched, 3-4.5 µm, wide, clamps lacking, with simple septa, aerial hyphae hyaline to yellowish, thin to thick-walled, branched, simple septate 4. 5-6 µm, wide. **Tests :** α-napthal, p-cresol and pyrogallol + H_2O_2 positive. **Species Code :** 1, 2, 3, 7, 12, 13, 14, 15, 21, 30, 31, 35, 37, 53, 54. **Culture No :** PUCC. B: 58. **Remarks :** The fairly small and hyaline to pale coloured spores separate this species from the complex around *P. umbrinella* where the spores are larger and rusty brown in maturity. The species is a new record from India.

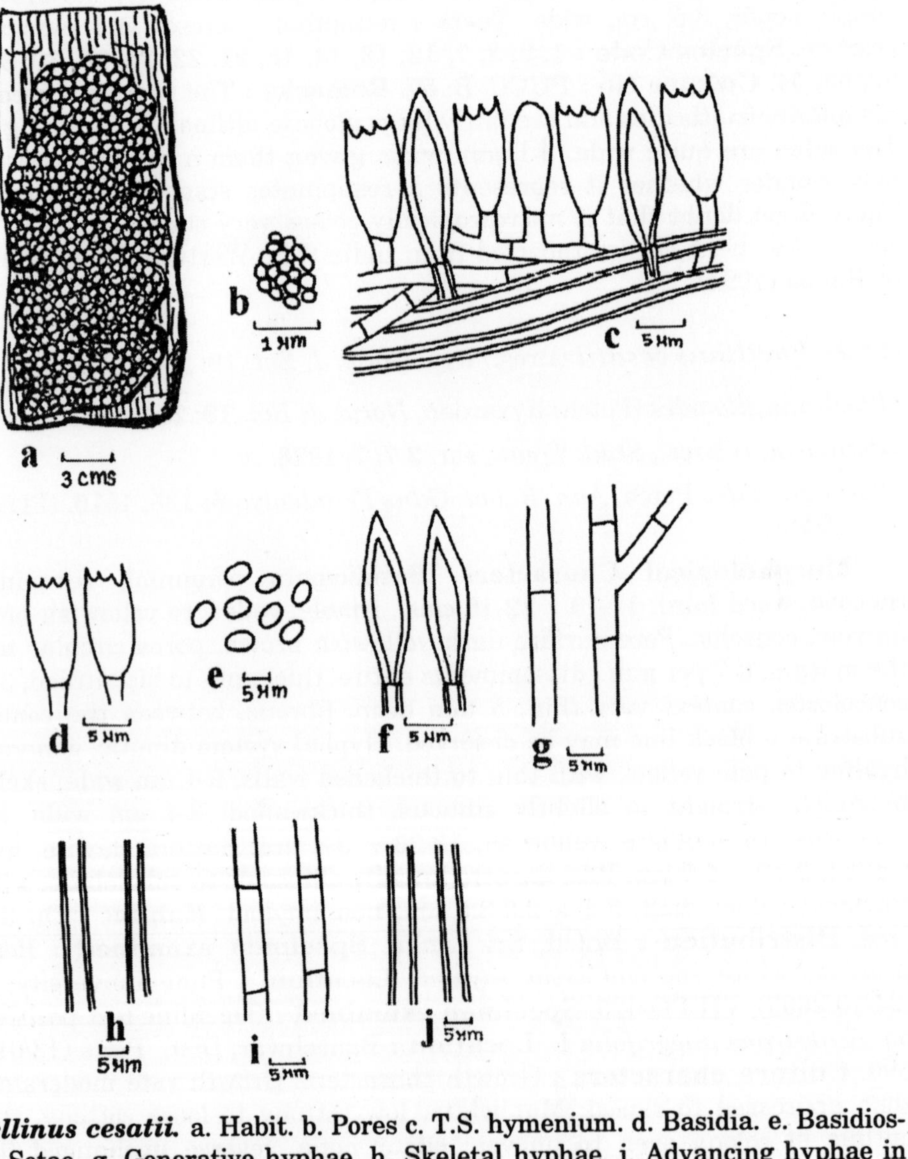

Fig. 94. ***Phellinus cesatii.*** a. Habit. b. Pores c. T.S. hymenium. d. Basidia. e. Basidiospores. f. Setae. g. Generative hyphae. h. Skeletal hyphae. i. Advancing hyphae in culture. j. Aerial hyphae in culture

(14-3). ***Phellinus chryseus*** **(Lév.) Ryvarden. A preliminary polypore flora of East Africa: 151, 1980.**

=*Polyporus chryseus* Lév., *Annls Sci. Nat.*, Bot., sér. **3 5:** 301, 1846.

=*Poria chrysea* (Lév.) Sacc., *Syll. fung.* (Abellini) **6:** 317, 1888. (Plate 5, Fig. 95).

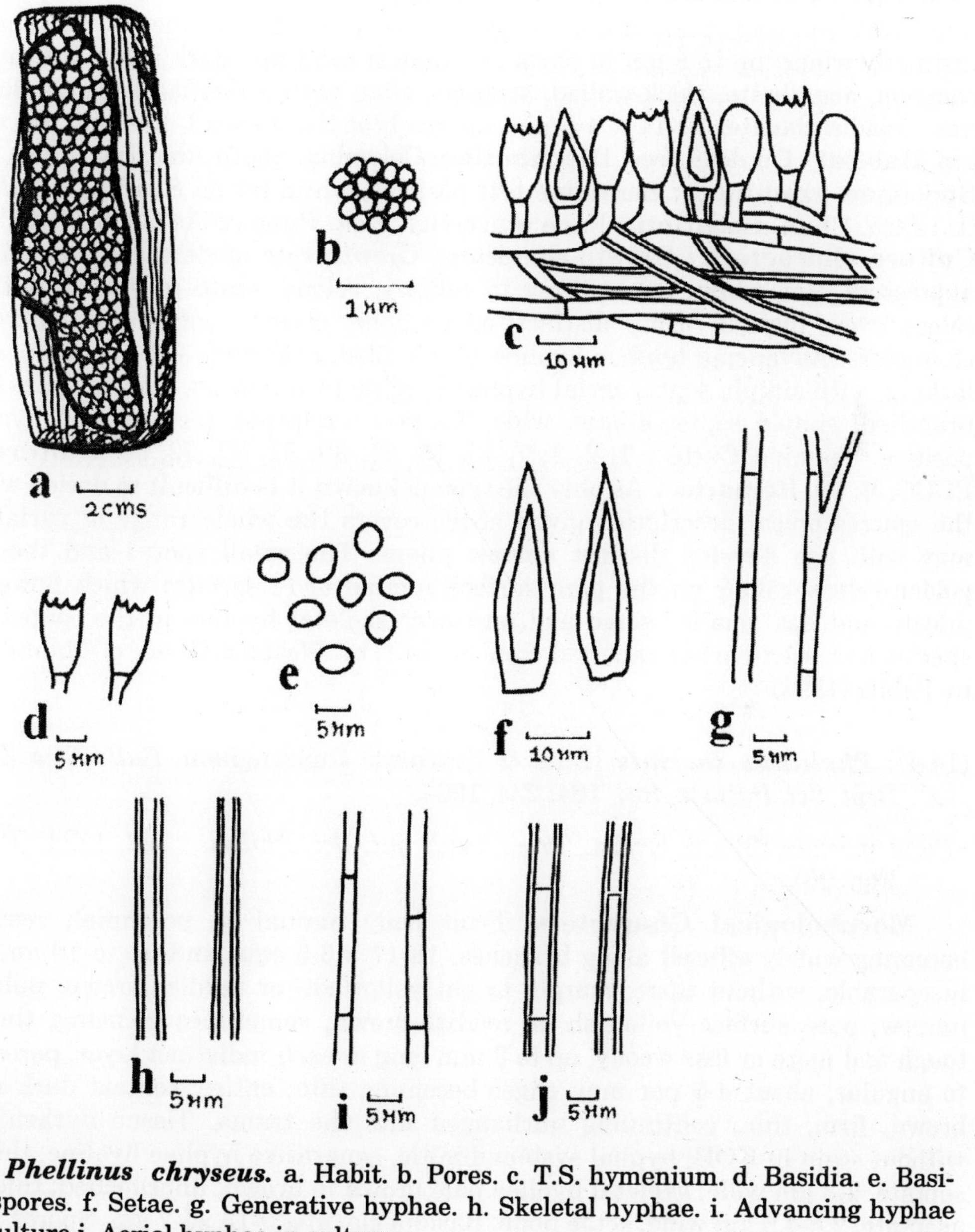

Fig. 95. *Phellinus chryseus.* a. Habit. b. Pores. c. T.S. hymenium. d. Basidia. e. Basidiospores. f. Setae. g. Generative hyphae. h. Skeletal hyphae. i. Advancing hyphae in culture.j. Aerial hyphae in culture

Morphological Characters : Fruitbody resupinate, effused, upper margin slightly reflexed, 10-12 × 4-6 cms. and up to 4 mm thick, dark brown and with some radial striae, finely, woody. Pore surface yellowish-brown, pores more or less round, about 6 per/ mm, tubes up to 1.5 mm deep, tubes ochraceous, trama cinnamon brown. Subiculum less than 1 mm, dark cinnamon brown. Hyphal system dimitic, generative hyphae delicately thin-walled, simple septate and 2-4 μm wide, skeletal hyphae dominating in the trama, rusty brown, up to 4.5 μm wide, in the subiculum distinctly wider, up to 8 μm in parts and almost solid and dark rusty brown. Setae common, acuminate, thick-walled, straight, often with a swollen base, 25-35 × 5-8 μm. Basidia clavate, 10-14 × 4-5 μm. Spores broadly ellipsoid, hyaline, 3-4 × 2-2.5 μm. **Habitat :** On dead tree. **Distribution:** Colombia, South America, Indies, India. **Specimen examined :** Benzidine test positive, white rot on *Swietenia mahogany* (L.) Jaca. Chun. **Location :** Pune university, Dist. Pune (26/08/2006), VH-PO-527. **Culture characters :** Growth characters: Growth rate moderate; advancing zone appressed; Mat white, low cottony to cottony, colony white to yellow; outline of colony even; hyphal rather distinct; odour none, reverse unchanged. Microscopic characters: Advancing hyphae hyaline thin-walled, branched, 2-3 μm, wide, clamps lacking, with simple septa, aerial hyphae hyaline to yellowish, thin to thick-walled, branched, simple septa, 4-5μm, wide. **Tests :** α-napthal, p-cresol and pyrogallol positive. **Species Code :** 1, 2, 3, 7, 13, 15, 21, 30, 31, 37, 52, 53. **Culture No :** PUCC. B: 59. **Remarks :** As only the type is known it is difficult to decide whether the macroscopical description given above covers the whole range of variation, it may with age develop distinct narrow pileus. The small spores and the bright golden-yellow colour on the pore surface remind of *P. sarcites* which however, is pileate and has smaller setae and narrower skeletal hyphae in the context. This species has been earlier reported from India (The Western Ghats of Maharashtra) by Rabba (1994).

(14-4). *Phellinus inermis* (Ellis & Everhart) Cunningham. *Bull. New Zealand Dept. Sci. Industr. Res.* **164:** 234, 1965.

=*Poria inermis* Ellis & Everh., *Proc. Acad. nat. Sci. Philad.* : 322, 1894. (Plate: 5, Fig. 96).

Morphological Characters : Fruit body annual to perennial, resupinate becoming widely effused along branches, 13-17 × 3-5 cms. and up to 10 mm thick, inseparable, without taste; margin bright yellowish- or reddish-brown, pubescent, narrow; pore surface yellowish or reddish-brown, sometimes glancing, the tubes tough and more or less woody, up to 3 mm long in each indistinct layer, pores round to angular, about 4-5 per mm, edges becoming thin, entire; context dark reddish brown, firm, thin, continuing unchanged into the trama. Tissue darkening but without stain in KOH; hyphal system dimitic, generative hyphae hyaline, thin-wall, septate, 2-3 μm wide; skeletal hyphae pale brown to brown, unbranched, thick-wall, aseptate, 2.5-4.5 μm wide, setae none. Basidia clavate, 9-13 ×5-7 μm. Spores reddish brown, smooth, non-amyloid, broadly oval or oblong, 4-6 ×3-5 μm. **Habitat :** On live standing tree.

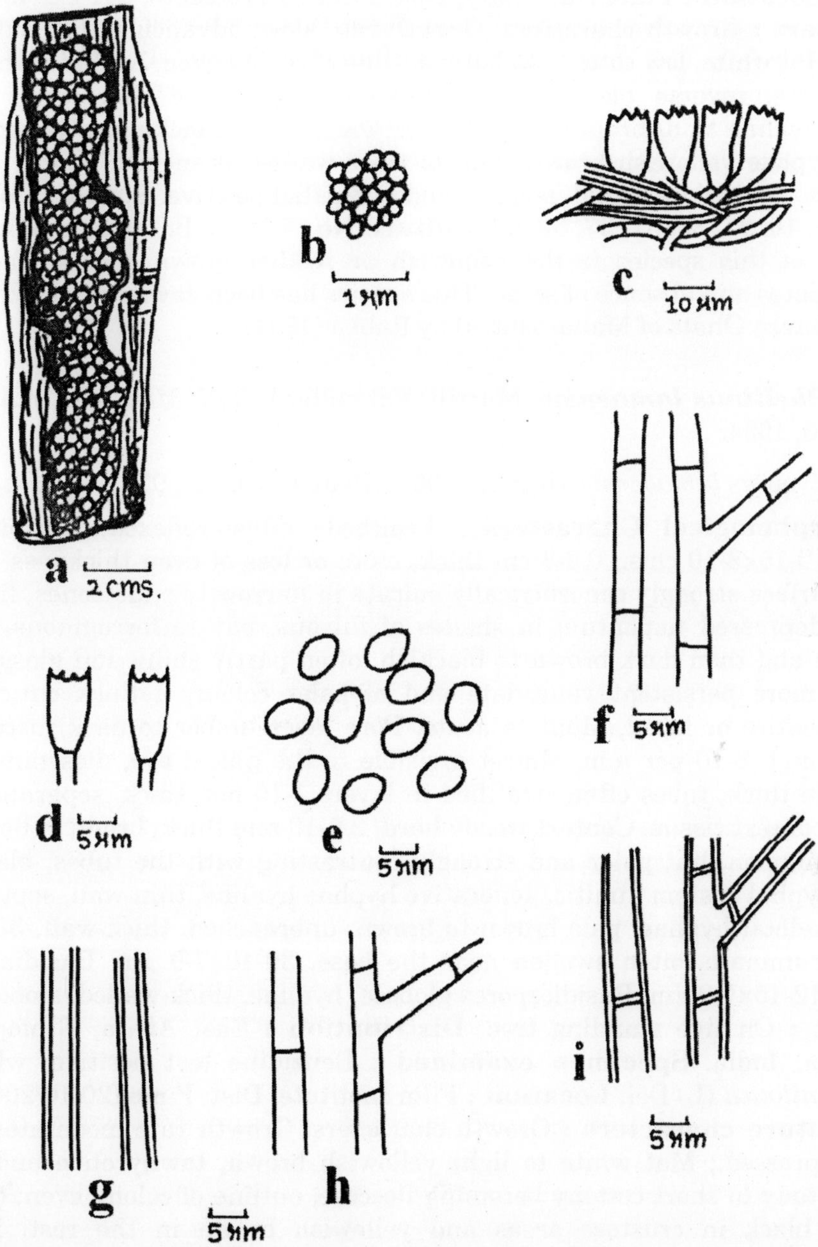

Fig. 96. *Phellinus inermis.* a. Habit. b. Pores. c. T.S. hymenium. d. Basidia. e. Basidiospores. f. Generative hyphae. g. Skeletal hyphae h. Advancing hyphae in culture. i. Aerial hyphae in culture

Distribution : North America, New Zealand, India. **Specimen examined :** Benzidine test positive, white rot on *Peltophorum pterocarpum* (DC.) Baker ex

Heyne. **Location :** Pune University; Dist: Pune (08/10/2005), VH-PO-497. **Culture characters :** Growth characters: Growth rate slow; advancing zone submerged to raised; Mat white, low cottony to felty; outline of colony even; hyphal rather distinct; odour none, reverse bleached to darkened. Microscopic characters: Advancing hyphae hyaline thin, branched, 3-3.5 μm, wide, clamps lacking, with simple septa, aerial hyphae yellowish brown, thin to thick-walled, branched, simple septate 3-6 μm, wide. **Tests :** α-napthal, p-cresol and pyrogallol positive. **Species Code :** 1, 2, 3, 8, 13, 15, 16, 21, 30, 31, 37, 52, 53. **Culture No :** PUCC. B: 60. **Remarks :** Specific features of this species is the yellowish or reddish-brown pore surface, reddish brown spores and absence of setae. This species has been earlier reported from India (The Western Ghats of Maharashtra) by Rabba (1994).

(14-5). *Phellinus lamaensis* (Murrill) Patouillard. *Bull. Mus. Hist. nat., Paris* **29:** 336, 1934.

=*Pyropolyporus lamaoensis* Murrill,, 1907. (Plate 6, 10, Fig. 97).

Morphological Characters : Fruitbody effuso-reflexed, subspathulate to lobate, 12-15×8-10 cms., 0.2-3 cm thick, more or less of even thickness, applanate, upper surface strongly concentrically sulcate in narrow to wide zones, first covered with a depressed tomentum in shades of fulvous, bay to ferruginous, soon more glabrous and then dark brown to blackish, often partly shiny and glossy, near the margin more persistent velutinate and of paler colour, a thick cuticle present. Margin entire or lobed, blunt to acute. Pore layer umber to dark fuscous or bay, pores round, 8-10 per mm, almost invisible to the naked eye, dissepiments entire and quite thick, tubes often stratified in layers 1-10 mm thick, separated by white lines of context tissue. Context woody hard, 2.5-10 mm thick, bright yellow-brown to dark cinnamon, but paler and strongly contrasting with the tubes, blackening in KOH. Hyphal system dimitic, generative hyphae hyaline, thin-wall, septate, 2-3 μm wide. Skeletal hyphae, pale brown to brown, unbranched, thick-wall, 3-5 μm wide. Setae acuminate, often swollen near the base, 30-40×7-9 μm. Basidia clavate to oblong, 12-15×7-9 μm. Basidiospores globose, hyaline, thick-walled, smooth, 4-5 μm. **Habitat :** On live standing tree. **Distribution :** East Africa, Ethiopia, Kenya, Tanzania, India. **Specimen examined :** Benzidine test positive, white rot on *Accacia nilotica* (L) Del. **Location :** Film Institute; Dist: Pune (20/10/2005) VH-PO-256. **Culture characters :** Growth characters: Growth rate moderate; advancing zone appressed.; Mat white to light yellowish brown, tawny olive and cinnamon buff, cottony to short cottony becoming floccose; outline of colony even; odour none, reverse black in crustose areas and yellowish brown in the rest. Microscopic characters: Advancing hyphae hyaline thin, branched, 3-3.5 μm, wide, clamps lacking, with simple septa, aerial hyphae yellowish, thin to thick-walled, simple septa 3-5 μm, wide. **Tests :** α-napthal, p-cresol and pyrogallol + H_2O_2 positive. **Species Code :** 1, 2, 3, 7, 13, 15, 16, 19, 21, 30, 31, 34, 35, 38, 53. **Culture No :** PUCC. B: 61. **Remarks :** The species is recognized by its obtuse setal hyphae, which, however, in the dissepiments and sometimes also in the context are difficult to recognize as they are much narrower than usual for setal hyphae, mostly 5-7 μm wide. From *P. noxius*, which also have obtuse setal hyphae, it is separated by having numerous hymenial setae. These are either absent from *P. noxius* or very rare. Further, the fruitbodies of *P. lamaensis* are frequently applanate, of an even thickness, and with numerous sulcate zones, while those of *P. noxius* are far more irregular, often effused reflexed, semi-ungulate etc. and with few wider zones. This species has been earlier reported from India (The Western Ghats of Maharashtra) by Rabba (1994).

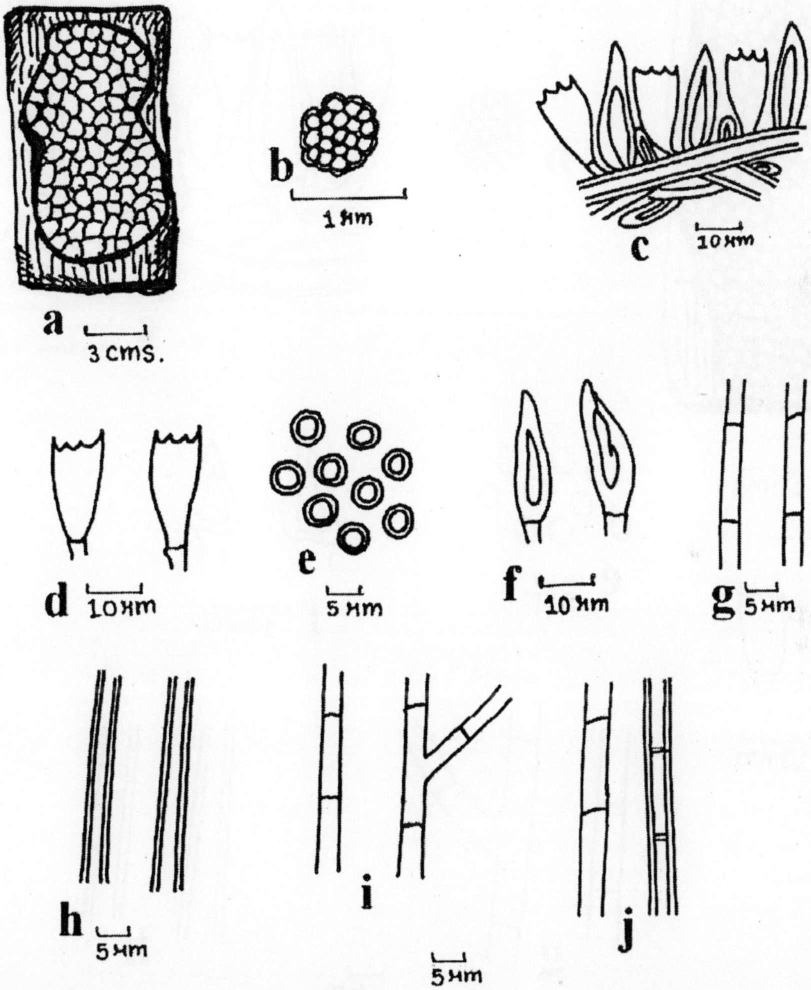

Fig. 97 ***Phellinus lamaensis.*** a. Habit. b. Pores c. T.S. hymenium. d. Basidia. e. Basidiospores. f. Setae. g. Generative hyphae. h. Skeletal hyphae. i. Advancing hyphae in culture. j. Aerial hyphae in culture

(14-6). ***Phellinus luctuosus*** (Cesati) Ryvarden. *Norweg. J. Bot.* **19:** 235, 1972.

=*Polyporus luctuosus* Ces., Atti dell'Accademia delle scienze fisiche e matematiche, Napoli 8 (8) : 7, 1879. (Plate: 6, Fig. 98).

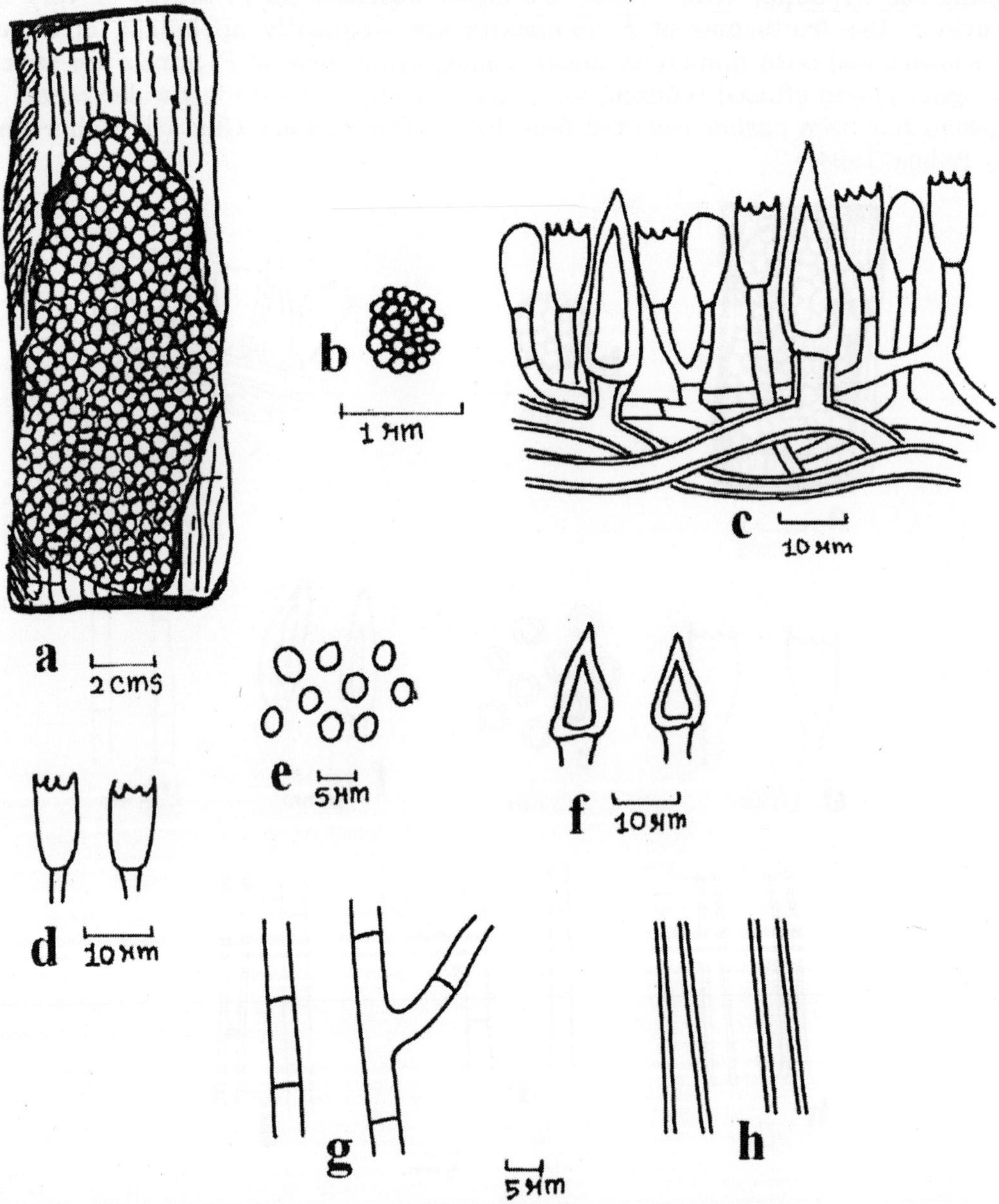

Fig. 98 ***Phellinus luctuosus.*** a. Habit. b. Pores c. T.S. hymenium. d. Basidia. e. Basidiospores. f. Setae. g. Generative hyphae. h. Skeletal hyphae

Morphological Characters : Fruitbody resupinate, effused, adnate, 9-12×5-8 cms. and up to 4 mm thick, woody hard, margin narrow and dark umber brown.

Pore surface deep umber brown in the type, pores tiny, 8-10 per mm, tubes concolorous, up to 3 mm deep. Subbiculum dark brown, 0.5 mm thick. Hyphal system dimitic, generative hyphae thin-walled and hyaline, only a few seen in the type, 2-3 μm wide, skeletal hyphae dominating, thick-walled and rusty brown, a few also golden-brown, 4-6 μm wide. Setae abundant, ventricose, straight, 10-15 ×5-8 μm. Basidia clavate, 15-18 × 5-8 μm. Basidiospores globose, hyaline, thick-walled, smooth, 4-5 μm. **Habitat :** On live standing tree. **Distribution :** South-East Asian countries. **Specimen examined :** Benzidine test positive, white rot on *Caesalpinia coriuria* (Jacq.) willd. **Location :** Baneshwar, Dist. Pune (13/01/2007), VH-PO-601. **Remarks :** The diagnostic characteristic is above all the small ventricose setae reminding strongly of those of *P. extensus*, which, however, is a distinctly pileate species. The culture was not obtained besides repeated trials. The species is a new record from India. Shape and size of the spore is reported for the first time.

(14-7). *Phellinus punctatus* (Karst.) Pil. Atlas Champignons Europe, 3, p. 530, 1942.

=*Polyporus punctatus* Fr., *Hymenomyc. eur.* (Upsaliae):572,1874 (Plate 6,10, Fig. 99)

Morphological Characters : Basidiomata perennial, resupinate to broadly effuso-reflexed, woody, difficult to separate from the substrate, 5-8×3-5 cms. Pore surface yellow to yellowish brown, dull; pores circular, 5-7 per/ mm, dissepiments thick, entire, slightly tomentose; margin yellow to brownish yellow, tomentose to rimose, up to 1 mm wide; tube concolorous with the hymenial surface, unistratified, up to 0.5 mm thick; context concolorous with the pore surface, up to 5 mm thick. Hyphal system dimitic, generative hyphae with thin to thick walls, hyaline, 2-3 μm wide; skeletal hyphae thick-walled, sometimes branching, 3-4 μm wide, both context and trama similar. Setae absent. Basidia clavate thin-walled, hyaline, 15-20 ×4-6 μm. Basidiospores ovoid to sub-globose, hyaline, smooth, thick-walled, 4- 6 ×4- 5 μm, dextrinoid. **Habitat :** On live standing tree. **Distribution :** Costa Rica, Panama, Cuba, Haiti, Jamaica, Puerto Rico, Brazil, Venezuela, Bolivia, Europe, Africa, Asia, Austrasia. **Specimen Examined:** Benzidine test positive, white rot on *Azadirachta indica* Juss. **Location:** Panchwati; Dist: Pune (24/09/2005), VH-PO-62. **Culture characters :** Growth characters: Growth rate slow; advancing zone raised.; Mat white, woolly; outline of colony uneven; odour none, reverse unchanged. Microscopic characters: Advancing hyphae hyaline, thin-walled, 1.5-4.5 μm, wide, clamp less, with simple septa, aerial hyphae yellowish, thin-walled, simple septa 1.5-2.0 μm, wide. **Tests :** α-napthal, p-cresol and pyrogallol+ H_2O_2 positive. **Species Code :** 1, 2, 3, 8, 12, 15, 22, 30, 37, 52, 53. **Culture No :** PUCC. B: 62. **Remarks :** The species is very similar to *P. hartigii* (Allesch. & Schnabl) Pat. which has effuse-reflexed basidiomata. Its resemblance to the *P. robustus* complex was noted by Lowe (1966) and Gilbertson (1979). The former author considers the present species to be a resupinate form of *P. robustus* (Karst.) Boud. & Galz., which is pileate. Larsen & Cobb-Poulle (1990) in their survey of the genus Phellinus considered *P. punctatus* (Fr. ex Karst.) Pilat as a valid name for the species. Ryvarden (1981) stated that the original spithet was published as *P. punctatus* Jungh. (=*Rigidoporus lineatus*) in 1838, whereby *P. punctatus* Fr. published in 1874 is a later homonym and illegitimate. However, according to Ryvarden (1991, p. 314) Karsten in 1882 erected the binomial *Poria puncta*, which should be the basionym for the present taxon, namely *P. punctatus* (Karst.) Pilát. We have adopted Ryvarden's criterion of using the latter authorship for the binomial. The species is a new record from India.

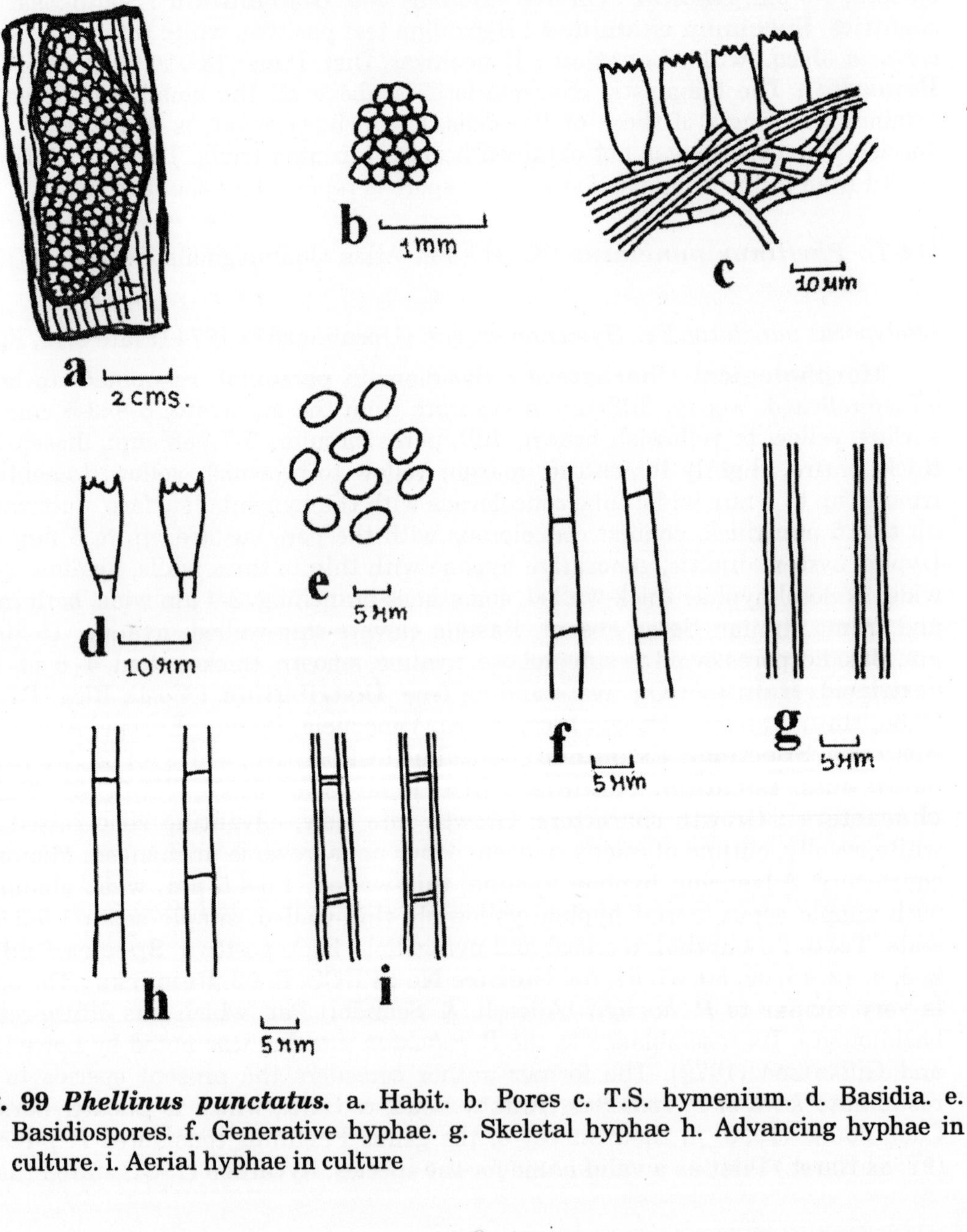

Fig. 99 ***Phellinus punctatus.*** a. Habit. b. Pores c. T.S. hymenium. d. Basidia. e. Basidiospores. f. Generative hyphae. g. Skeletal hyphae h. Advancing hyphae in culture. i. Aerial hyphae in culture

(14-8). *Phellinus purpureogilvus* (Petch) Ryvarden. *Norweg. J. Bot.* **19:** 235, 1972.

=*Poria purpureogilva* Petch, *Ann. R. bot. Gdns Peradeniya* **6:** 138, 1916. (Plate 6, 10, Fig. 100).

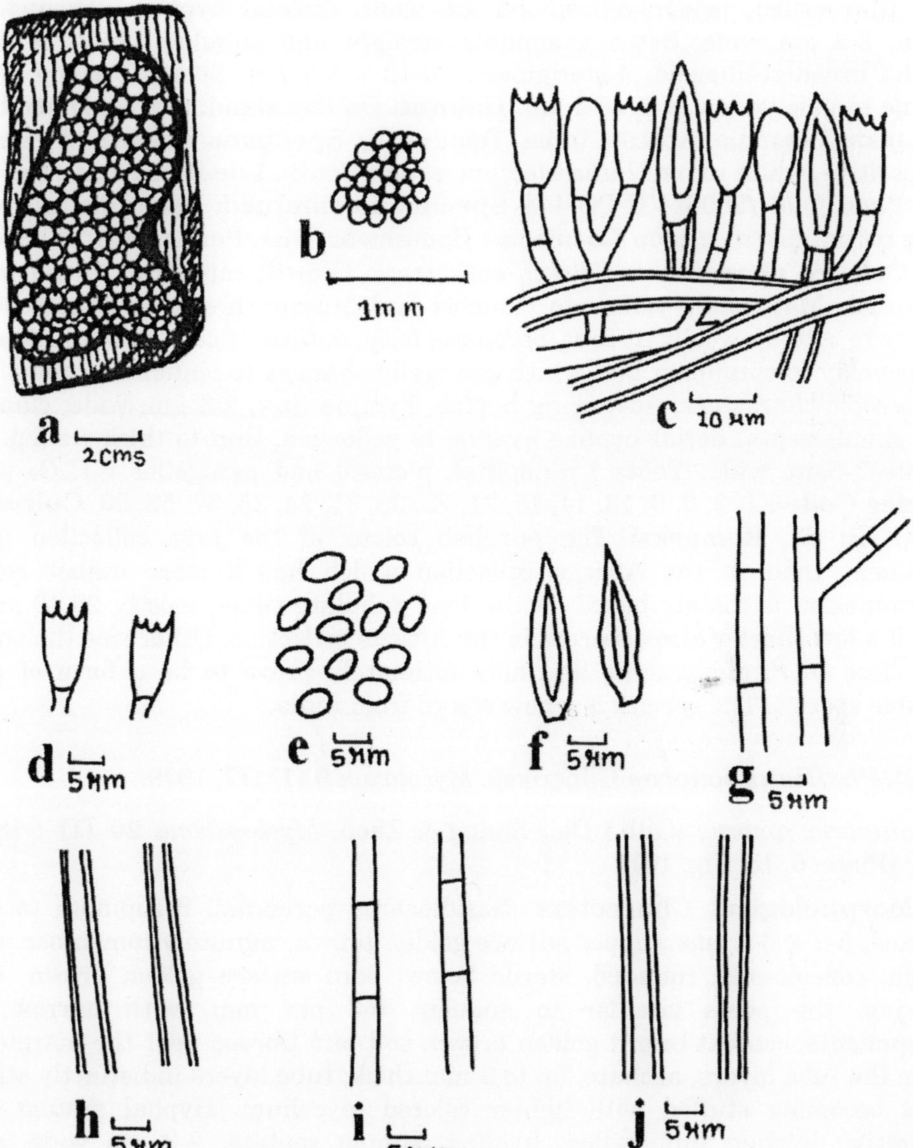

Fig. 100 *Phellinus purpureogilvus* a. Habit. b. Pores. c. T.S. hymenium. d. Basidia. e. Basidiospores. f. Setae. g. Generative hyphae. h. Skeletal hyphae. i. Advancing hyphae in culture. j. Aerial hyphae in culture

Morphological Characters : Fruitbody resupinate, perennial, adnate, 6-8 × 4-5 cms. and up to 2 mm thick, woody hard. Margin very thin, reddish-brown. Pore surface purplish-brown to umber, pores tiny, round, 7-8 per mm, tubes non-stratified in the type, up to 2 mm thick. Context reddish-brown, almost lacking in the type, 1-2 mm thick. Hyphal system dimitic, generative hyphae scanty in the type, thin-walled, golden-yellow, 2-3 μm wide, skeletal hyphae, yellow to rusty brown, 3-4 μm wide. Setae acuminate straight and subulate, 25-30 × 5-8 μm. Basidia broadly ellipsoid, 4-sterigmate, 10-12 × 5-7 μm. Spores broadly ellipsoid, hyaline to pale yellow, 4-5 ×3-4 μm. **Habitat:** On live standing tree. **Distribution**: Sri Lanka, Tanzania, African, India (Tamilnadu). **Specimen examined :** Benzidine test positive, white rot on: *Pithecolobium saman* Benth. **Location:** Pune University; Dist: Pune (14/01/2006) VH-PO-443. **Specimen examined:** Benzidine test positive, white rot on: *Areca catechu* **Location :** Baneshwar, Dist. Pune (10/01/2007) VH-PO-604. **Culture characters:** Growth characters: Growth rate slow; advancing zone appressed; Mat honey yellow to chamois and antique brown to ochraceous, low cottony to cottony-woolly at first, becoming felty; outline of colony even; odour none, reverse clay to cinnamon buff which gradually changes to cinnamon brown finally. Microscopic characters: Advancing hyphae hyaline thin, 2-3 μm, wide, clamp less, with simple septa, aerial hyphae hyaline to yellowish, thin to thick-walled, simple septate 3-5μm, wide. **Tests :** α-napthal, p-cresol and pyrogallol + H_2O_2 positive. **Species Code :** 1, 2, 3, 9, 13, 14, 15, 21, 22, 25, 31, 34, 35, 37, 52, 53. **Culture No :** PUCC. B: 63. **Remarks:** The purplish colour of the type collection is more prominent than in the African collection which has a more umber tint. The determination is mainly based on the long subulate setae, mostly 25-40 μm long, even if a few shorter also occurred in the African collection. Otherwise the species is very close to *P. glaucescens* and may ultimately prove to be a form of a fairly variable species. The species is a new record from India.

(14-9). *Phellinus sonorae* Gilbertson. *Mycotaxon* **9 (1)** :77, 1979.

=*Fomitiporia sonorae* (Gilb.) Dai, Zhang & Zhou, *Mycosystema* **20 (1)** : 18, 2001. (Plate 6, 10, Fig. 101).

Morphological Characters : Basidiocarps perennial, resupinate to slightly reflexed, 5-7 × 3-4 cms., upper surface golden brown, minutely tomentose, smooth; margin concolorous, rounded, sterile below; pore surface golden brown, smooth, glancing, the pores circular to angular, 5-7 per mm, with narrow, entire dissepiments; context bright golden brown, soft and fibrous near the margin, corky above the tube layers, azonate, up to 3 mm thick; tube layers indistinctly stratified, tubes becoming stuffed with lighter colored mycelium. Hyphal system dimitic, generative hyphae thin-walled, hyaline, simple septate, 2-3 μm wide, skeletal hyphae pale yellowish brown, thick-walled, 4-5 μm wide. Setae few, developing from cystidia, ventricose with long slender apical portion, 35-50 ×5-8 μm. Cystidia abundant, ventricose with elongated apical portion, thin-walled, hyaline, 40-60 ×5-7 μm. Basidia broadly ellipsoid, 4-sterigmate, 10-11 ×7. 5-8.5 μm. Basidiospores subglobose, hyaline, thick-walled, smooth, dextrinoid in Melzer's reagent, 5-5.5 ×4.5-5 μm.

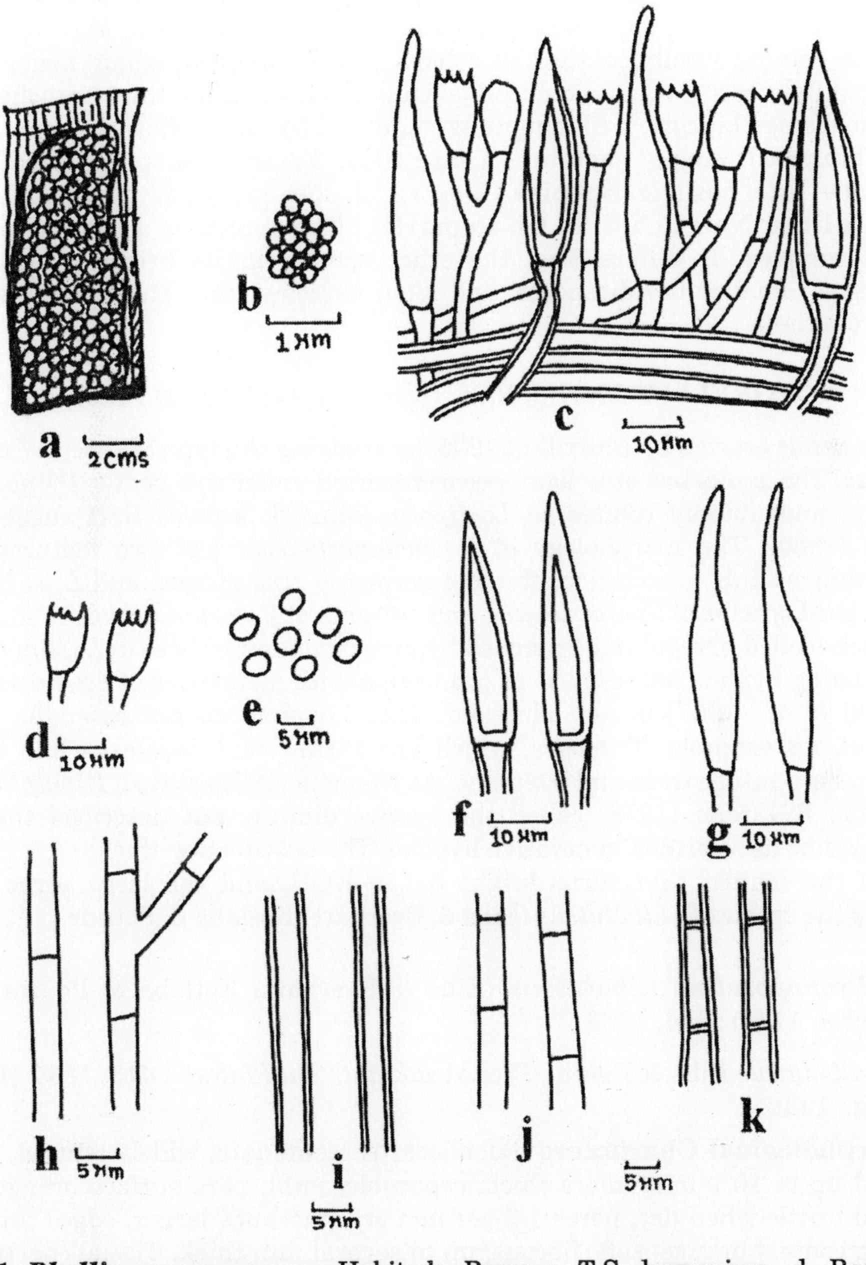

Fig. 101 ***Phellinus sonorae.*** a. Habit. b. Pores c. T.S. hymenium. d. Basidia. e. Basidiospores. f. Setae g. Thin-walled cystidia. h. Generative hyphae. i. Skeletal hyphae. j. Advancing hyphae in culture. k. Aerial hyphae in culture

Habitat: On live standing tree. **Distribution :** USA, Arizona. **Specimen examined:** Benzidine test positive, white rot on *Azadirachta indica* Juss. **Location :** Panchwati; Dist: Pune (15/07/2006), VH-PO-502. **Culture characters :** Growth characters: Growth rate slow; advancing zone appressed to raised; Mat white, loose cottony to cottony woolly; outline of colony even to uneven; odour none, reverse unchanged. Microscopic characters: Advancing hyphae hyaline thin, branched, 1.5-3 μm, wide, clamps lacking, with simple septa, aerial hyphae yellowish, thin to thick-walled, branched, simple septate 2-4 μm, wide. **Tests :** α-napthal, p-cresol and pyrogallol + H_2O_2 positive. **Species Code :** 1, 2, 3, 8, 12, 13, 15, 21, 22, 30, 35, 37, 52, 53. **Culture No :** PUCC. B: 64. **Remarks :** The species is a member of the *P. robustus* complex. It differs from the other species in its bright golden brown basidiocarps, smaller basidiospores, and long, narrow setae. The species is a new record from India.

(15). *PYCNOPORELLUS* Murrill. *Bull. Torrey Bot. Club* **32 (9) :** 489, 1905.

This genus erected by Murrill in 1905, by studying the type of species *Polyporus fibrillosus.* The genus has only four species reported under this genus. *Pycnoporellus* Murrill is undoubtedly related to *Laetiporus* Murrill, as was first suggested by Niemela (1980). The morphology of *P. metamorphosus* has two features which further support this association; the anamorphs of this species and *L. sulphureus* are both good species of *Sporotrichum,* and cultures of *P. metamorphosus* sometimes show thick-walled, irregularly branched hyphae resembling those of *Laetiporus*. The terms binding hyphae and dimitic in connection with structures of *Laetiporus* (e. g. Domanski *et al.,* 1967) is not adequate. These hyphae do not resemble binding hyphae of, for example, *Trametes*, which are narrow and regular, while those of *Laetiporus* are rather wide and irregular. As Niemela (1980) stated, *L. sulphureus* is monomitic. Ryvarden (1978) called the species dimitic, but described the thick-walled hyphae as sclerified generative hyphae. The specimen with characteristic the tissue of the context that turns bright red in KOH, and the large pores readily identifies the species *P. alboluteus* (Ellis & Everhart) Kotlaba & Pouzar.

(15-1). *Pycnoporellus alboluteus* (Ellis & Everhart) Kotlaba & Pouzar. *Ceská Mykol.* 17 (4) : 174, 1963.

=*Fomes alboluteus* Ellis & Everh., *Proc. Acad. nat. Sci. Philad.* : 413, 1895. (Plate: 6, Fig. 102).

Morphological Characters : Basidiocarps resupinate, widely effused, 5-7×2-3 cms. and up to 10 mm or more thick, separable, mild; pore surface orange, tubes rigid and brittle when dry, pores 1-3 per mm or sometimes larger, edges thin, often splitting; context orange, soft, fibrous, up to several mm thick. Tissue cherry red in KOH; hyphal system monomitic, without clamp, 4-6 μm wide. Cystidia immersed to projecting prominently, thin-walled, 30-35 × 5-7 μm. Basidia clavate, hyaline, with 4 sterigmata, 20-25×5-7 μm. Spores hyaline, smooth, non-amyloid, cylindrical, 7-10 ×

2-4 μm. **Habitat:** On dead standing tree. **Distribution**: Canada, U.S.A. **Specimen examined:** Benzidine test negative, brown rot on *Dimorphocalyx lawianus* Hook. **Location:** Pune University; Dist: Pune (27/08/2006), VH-PO-528. **REMARKS**: The tissue of the context that turns bright red in KOH and the large pores readily identifies the species. The species is a new record from India. The culture was not obtained besides repeated trials.

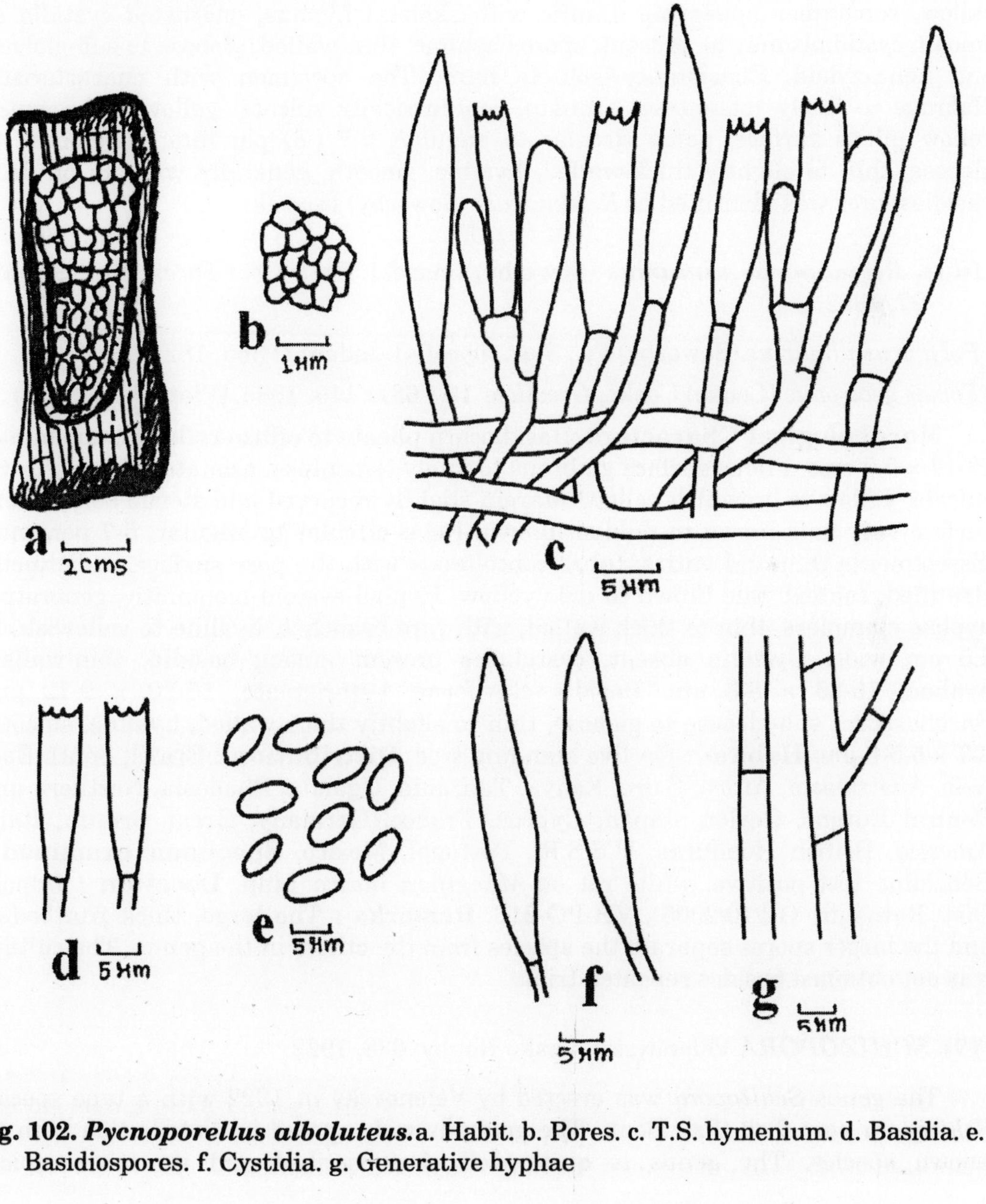

Fig. 102. ***Pycnoporellus alboluteus.*** a. Habit. b. Pores. c. T.S. hymenium. d. Basidia. e. Basidiospores. f. Cystidia. g. Generative hyphae

(16). *RIGIDOPORUS* Murrill. *Bull. Torrey Bot. Club* **32 (9) :** 478, 1905.

The genus *Rigidoporus* was erected by Murrill in 1905 with a type species *Polyporus micromegas* Mont (= *Rigidoporus microporus* (Fr.) Overeem.). The genus is wide spread in distribution with 74 known species. The genus is characterized by annual but more often perennial, pileate to resupinate fruitbodies, orange, buff to white, hyphal system monomitic with simple septa, thin-walled to solid, hyaline to yellow, sometimes appearing dimitic with skeletal hyphae, encrusted cystidia or smooth cystidiols may be present, spores hyaline, thin-walled, globose to sub-globose and nonamyloid. Consistency soft to hard. The specimen with characteristic glabrous to finely tomentose, azonate, concentrically sulcate; yellow to brownish yellow pileus surface; pores circular to angular, 5-7 (-8) per mm; subglobose to globose, thin to slightly thick-walled, hyaline, smooth, generally with an oil drop basidiospores was identified as *R. ulmarius* (Sowerby) Imazeki.

(16-1). *Rigidoporus ulmarius* (Sowerby) Imazeki. *Bull. Govt Forest Exp. Stn* **57:** 97, 1952.

=*Polyporus ulmarius* (Sowerby) Fr., *Syst. mycol.* (Lundae) **1:** 365, 1821.

=*Fomes geotropus* (Cooke) Cooke, *Grevillea* **13 (68) :** 119, 1885. (Plate 7, Fig. 103)

Morphological Characters : Basidiocarp pileate to effuso-reflexed, applanate, 13-17 × 5-7 cms. Pileus surface glabrous to finely tomentose, azonate, concentrically sulcate, yellow to brownish yellow. Margin slightly recurved and sterile below. Pore surface very pale brown to reddish brown. Pores circular to angular, 5-7 per/ mm, dissepiments thin and entire; tubes concolorous with the pore surface, indistinctly stratified; context pale brown to pale yellow. Hyphal system monomitic; generative hyphae clampless, thin to thick-walled, with rare branches, hyaline to yellowish, 3-5.5 μm wide. Cystidia absent, cystidioles present among basidia, thin-walled, hyaline, 15-25 × 5-8 μm. Basidia claviform, 4-sterigmate, 15-20 × 9-12 μm. Basidiospores sub-globose to globose, thin to slightly thick-walled, hyaline, smooth, 6-7 × 5.5-6 μm. **Habitat :** On live standing tree. **Distribution :** Brazil, South East Asia, Australasia, Africa, Zaire, Kenya, Tanzania, Uganda, Rhodesia, Northern and Central Europe, Ceylon, Japan, Cyprus, France, Germany, Great Britain, Italy, America, British Honduras, U.S.S.R., Portugal, Mexico. **Specimen examined :** Benzidine test positive, white rot on *Mangifera indica* Linn. **Location :** Dapoli; Dist: Ratnagiri (12/10/2005), VH-PO-216. **Remarks :** The large, thick fruitbodies and the larger spores separate the species from the others in the genus. The culture was not obtained besides repeated trials.

(17). *SCHIZOPORA* Velenovský. Ceske Houby: 638, 1922.

The genus *Schizopora* was erected by Velenovský in 1922 with a type species *Schizopora paradoxa* (Fr.) Donk. The genus is wide spread in distribution with 17 known species. The genus is quite well circumscribed and its microscopical characters fairly distinctive when learned. The hyphal system is almost identical with that of many *Hyphodontia* species, sharing with them the slightly thickened generative hyphae with numerous clamps, branching often occurring from them at an acute angle. Further, the variable cystidiols, often bottleshaped, ventricose, clavate to weakly vermicular (with a few constrictions) to hyphoid with a globose head are common to all three species known in the genus. Such organs are also common in *Hyphodontia* and no doubt *Schizopora* is one of the many links that connect to the two large families, Corticiaceae and Polyporaceae, both taken in a wide sense. From the investigation area four specimens were collected.

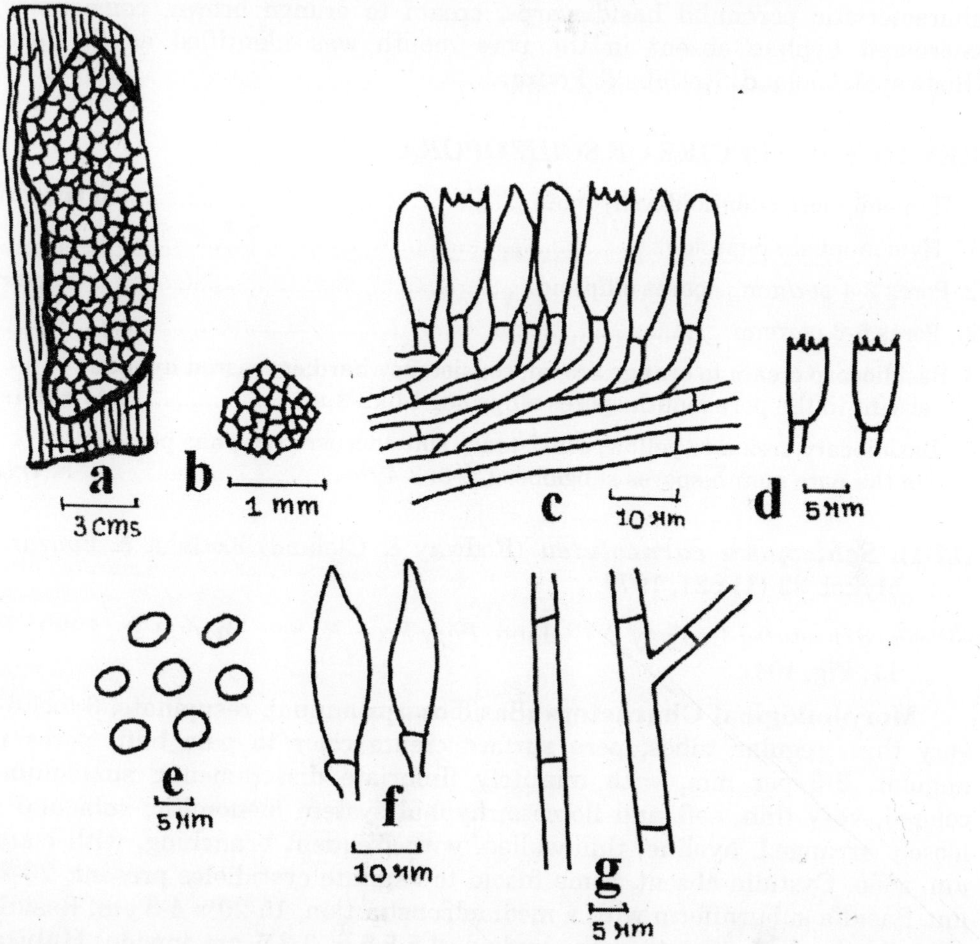

Fig. 103 ***Rigidoporus ulmarius*** a. Habit. b. Pores. c. T.S. hymenium. d. Basidia. e. Basidiospores. f. Cystidiole. g. Generative hyphae

The genus is characterized by annual to perennial, resupinate to pileate with narrow, imbricate pilei over a decurrent pore layer fruitbodies; pore surface and context white to cream; pores angular, round to split or as irregular teeth, often of variable length and partly flattened. Hyphal system dimitic, generative hyphae narrow to wide, thin to moderately thick-walled with distinct clamps, skeletal hyphae sparsely represented and mostly as very long cystidial organs, and it may be a matter of personal opinion and taste whether they should be classified as cystidia or relatively short skeletal hyphae, cystidia none, but subulate or bottle-shaped cystidiols, often present in the hymenium, enclosed or partly projecting, spores broadly ellipsoid, hyaline, thin-walled, smooth and non-amyloid. The specimen with characteristic odontioid to hydnoid hymenophore was identified as *S. roseotingens* Hjortstam & Ryvarden. *S. paradoxa* is easily distinguished from *S. trichiliae* by the much larger pores and microscopically by the larger spores. The specimen with characteristic perennial basidiocarps, cream to orange brown, consistency hard; encrusred hyphae absent in the pore mouth was identified as *S. carneolutea* (Rodway & Cleland) Kotlaba & Pouzar.

KEY TO THE SPECIES OF *SCHIZOPORA*

1. Hymenophore odontioid to hydnoid .... ***S. roseotingens***

1´. Hymenophore poproid 2

2. Pores 2-4 per/mm., spores ellipsoid. .... ***S. paradoxa***

2´. Pores 5-6 per/mm. 3

3. Basidiocarp cream to orange brown, consistency hard;encrusred hyphae absent in the pore mouth;spores ellipsoid, 4-5×2-3μm .... ***S. carneolutea***

3´. Basidiocarp cream to yellow, consistency soft;encrusred hyphae present in the pore mouth;spores subglobose, 3-5×3-4μm .... ***S. trichiliae***

(17-1). *Schizopora carneolutea* (Rodway & Cleland) Kotlaba & Pouzar. Ceská Mykol. **33 (1) :** 21, 1979.

=*Poria carneolutea* Rodway & Cleland, *Pap. Proc. R. Soc. Tasm.* : 18, 1929. (Plate: 7, 11, Fig. 104).

Morphological Characters : Basidiocarps annual, resupinate, 3-5×1.5-2 cms., very thin, regular tubes; pore surface cream color to pale buff, pores regular, angular, 3-5 per mm, with minutely fimbriate dissepiments; subiculum cream colored, very thin, soft and floccose; hyphal system monomitic; subicular hyphae loosely arranged, hyaline, thin-walled, with frequent branching, with clamps, 2-3 μm. wide. Cystidia absent, some fusoid to capitate cystidioles present, 20-25 × 4-6 μm. Basidia suburniform with a median constriction, 15-20 × 4-5 μm. Basidiospores ellipsoid, hyaline, smooth, non-amyloid, 4.5-5.5 × 3-3.5 μm inwide. **Habitat :** On fallen. branches. **Distribution:** Australia, India-Tamilnadu. **Specimen examined:** Benzidine test positive, white rot on *Mangifera indica* Linn. **Location :** Baneshwar; Dist: Pune (13/01/2007), VH-PO-605. **Culture characters :** Growth characters: Growth rate medium fast; advancing zone raised; Mat thin, sometimes first downy to cottony, locally farinaceous, outline of colony even; Hyphal rather distant; Reverse unchanged or bleached. Microscopic characters: Some aerial hyphae distinctly thick-walled, 2-3.5 µm wide. Marginal hyphae 1. 5-4 µm. wide, thin-walled, clamped, submerged hyphae 2-4 µm. wide, branched. **Tests :** α-napthal and pyrogallol + H_2O_2 positive; p-cresol negative. **Species Code :** 1, 2, 3, 8, 12, 14, 15, 17, 18, 21, 30, 31, 37, 39, 48, 52, 53. **Culture No :** PUCC. B: 65. **Remarks :** The species is close to *P. versipora* in many particulars, especially in the structure of the context hyphae. It differs mainly in the presence of abundant pyriform vesicles among the context hyphae, and surface colour. Vesicles are also present in some collections of *P. versipora* but are scanty, smaller and usually confined to apices of the dissepiments. This species has been earlier reported from India (The Western Ghats of Maharashtra) by Vaidya (1994).

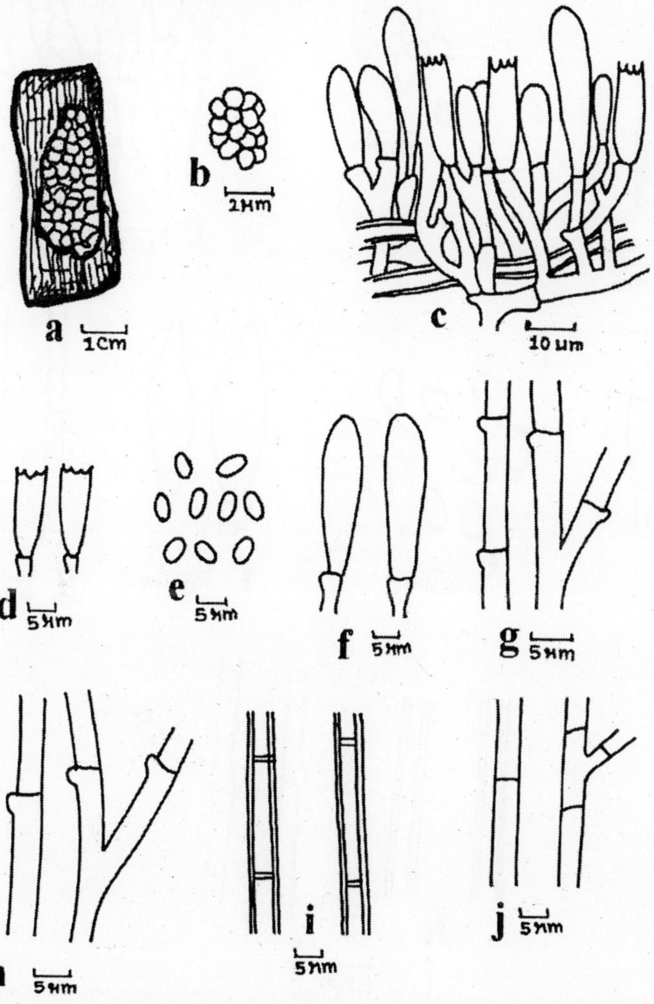

Fig. 104. ***Schizopora carneolutea.*** a. Habit. b. Pores c. T.S. hymenium. d. Basidia. e. Basidiospores. f. Thin-walled cystidia. g. Generative hyphae. h. Marginal hyphae in culture. i. Aerial hyphae in culture. j. Submerg hyphae in culture

(17-2). *Schizopora paradoxa* (Schrader) Donk. *Persoonia* **5 (1) :** 104, 1967.

=*Hydnum paradoxum* Schrad., *Spicil. Fl. Germ.* **1:** 179, 1794.

=*Polyporus versiporus* Pers., *Mycol. eur.* (Erlanga) **2:** 105, 1825. (Plate 7, Fig. 105).

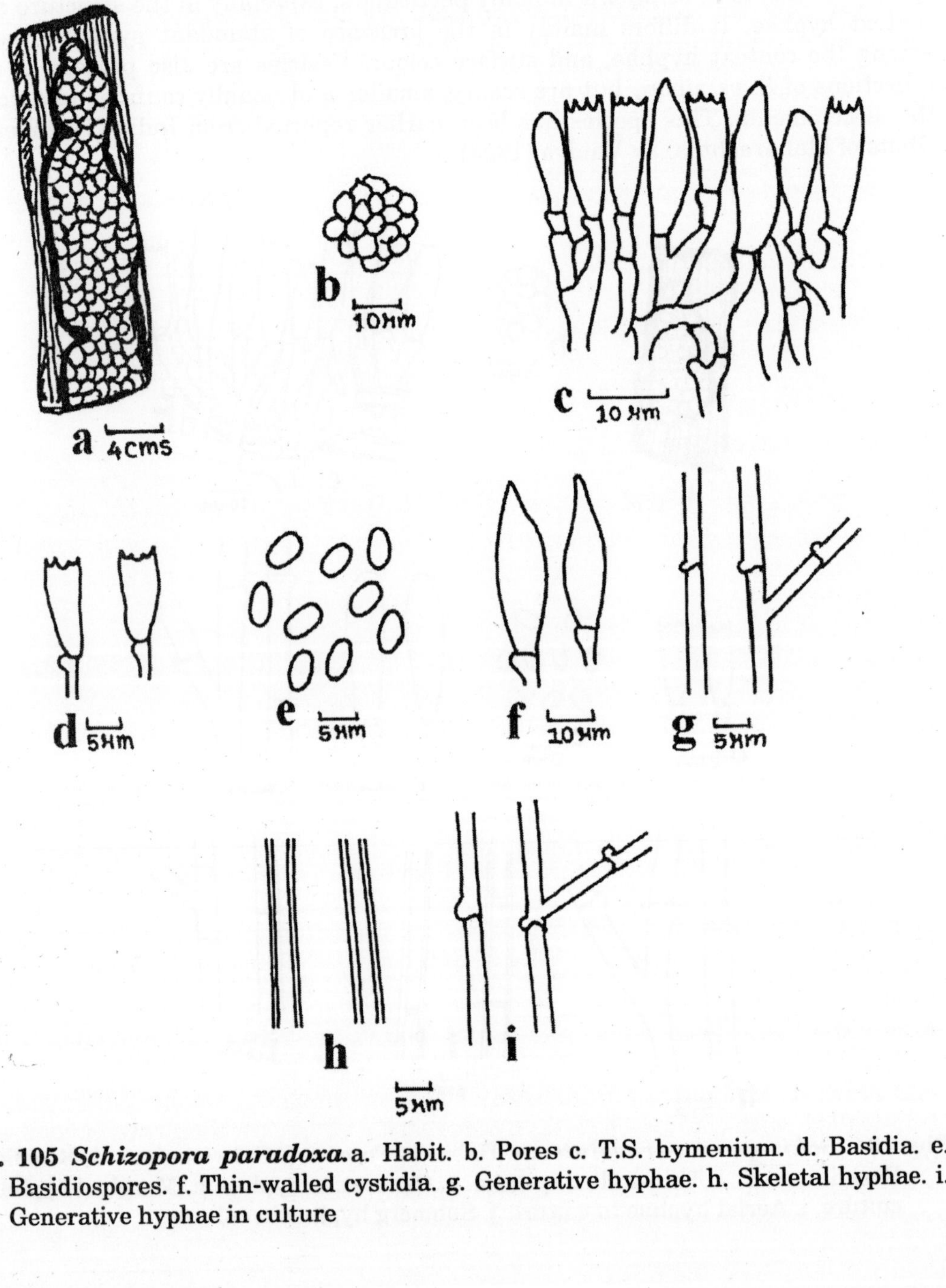

Fig. 105 *Schizopora paradoxa*. a. Habit. b. Pores c. T.S. hymenium. d. Basidia. e. Basidiospores. f. Thin-walled cystidia. g. Generative hyphae. h. Skeletal hyphae. i. Generative hyphae in culture

Morphological Characters : Fruitbody annual, resupinate, widely effused, 18-25 × 5-9 cms. and up to 5 mm thick, margin white to cream, finely floccose, narrow to wide, soft when fresh, brittle when dry. Hymenophore very variable, partly poroid with angular pores, 1-3 per/ mm, but soon lacerate and dentate as parts of the pore-walls grow faster than other parts, finally irpicoid to hydnoid with flattened crowded teeth, the pores or teeth are distinctly longer in central parts of the fruitbody than along the margin where they only may appear as a reticulate pattern or with a grandinoid surface, whitish cream to pinkish when fresh, cream to ochraceous when dry. Context white to cream, thin and dense, up to 1 mm thick. Hyphal system dimitic, generative hyphae 2-3.5 μm wide, with weakly thickened walls, with clamps, skeletal hyphae thick-walled except for the outer part, straight or flexuous, 3. 5-5 μm wide, dominating in the central part of the trama and projecting in the dissepiments. Cystidiol scattered in the hymenium, partly as hyphal ends, 30-35 × 8 μm, partly ventricose with a tapering rather acute tip or with a small tube-like extension, usually embedded in the hymenium. Basidia clavate, 20 × 6 μm., with four sterigmata and with basal clamp; Spores broadly ellipsoid, smooth, hyaline, thin-walled and non-amyloid, 5-6.5 × 3-4 μm. **Habitat :** On fallen branches. **Distribution :** Africa, Kenya, Tanzania. **Specimen examined:** Benzidine test positive, white rot on *Swietenia mahogany* (L.) Jacq. **Location :** Baneshwar; Dist: Pune (13/01/2007), VH-PO-803. **Culture characters :** Growth characters: Growth rate moderate; advancing zone appressed to slightly raised; Mats white, thin, subfelty to silky or raised and downy to cottony, outline of colony even appressed, fimbriate; no odor; Distance between hyphal tips distant; Aerial mycelium absent; Reverse bleached. Microscopic characters: Generative hyphae branched, 2-3.5 μm, wide, thin walled, with clamps, sparsely branched, usually branches arising opposite clamps. **Tests :** α-napthal and pyrogallol+ H_2O_2 positive; p-cresol negative. **Species Code :** 1, 2, 3, 7, 13, 14, 15, 16, 30, 36, 38, 39, 52, (53). **Culture No :** PUCC. B: 66. **Remarks:** Macroscopically the species is easily distinguished from *S. trichiliae* by the much larger pores and microscopically by the larger spores.

(17-3). *Schizopora roseotingens* Hjortstam & Ryvarden. *Mycotaxon* **20 (1) :** 142, 1984. (Plate 7, 11, Fig. 106).

Morphological Characters : Fruitbody resupinate, adnate, 7-8 × 2-3 cms., creamish to light brown or with rosy tint, distinctly odontioid to hydnoid with smooth and firm, hymenium between the aculei, individual aculei more or less conical or subcylindrical, rarely flattened, smooth or with a slightly fimbriate apex, mostly 0.5-1 mm long. Subiculum 0.3-0.5 mm thick, light rosy brown, fairly dense and tough. Margin not especially differentiated but loosening and tend to be more or less reflexed. Hyphal system dimitic. Generative hyphae branched, 3.5-4 μm wide, thin-walled, with clamps. Skeletal hyphae straight, thick-walled with fairly narrow, 3-5 μm wide. Cystidia lacking. Basidia subclavate, more or less suburniform, 20-25 × 4.5-5 μm, with 4 sterigmata and basal clamp. Spores ellipsoid, thin-walled, 4-4.5 ×

2-2.5 μm, non-amyloid, acyanophilous, non-dextrinoid. **Habitat :** On fallen branches. **Distribution :** Cosmopolitan species. Africa, Sierra Leone, Kenya, Tanzania. **Specimen examined :** Benzidine test positive, white rot on *Samanea saman* (Jacq.) Merr. **Location :** Koyna, Dist. Satara (11/02/2007), VH-PO-738. **Culture characters :** Growth characters: Growth rate medium fast; advancing zone raised; Mat thin, sometimes first downy to cottony, white to cream; outline of colony even; Hyphal rather distant; Aerial mycelium absent; Reverse unchanged or bleached. Microscopic characters: Generative hyphae branched, thin-walled, 3-5 μm wide, clamped. **Tests :** α-napthal and pyrogallol + H_2O_2 positive; p-cresol negative. **Species Code :** 1, 2, 3, 8, 12, 14, 15, 17, 18, 21, 30, 31, 37, 39, 48, 52, 53. **Culture No :** PUCC. B: 67. **Remarks :** Easily separated from S. *paradox* and *S. trichiliae* by the perennial fruitbodies. Pileus if present, without context, ochraceous to pink or rose buff when fresh.

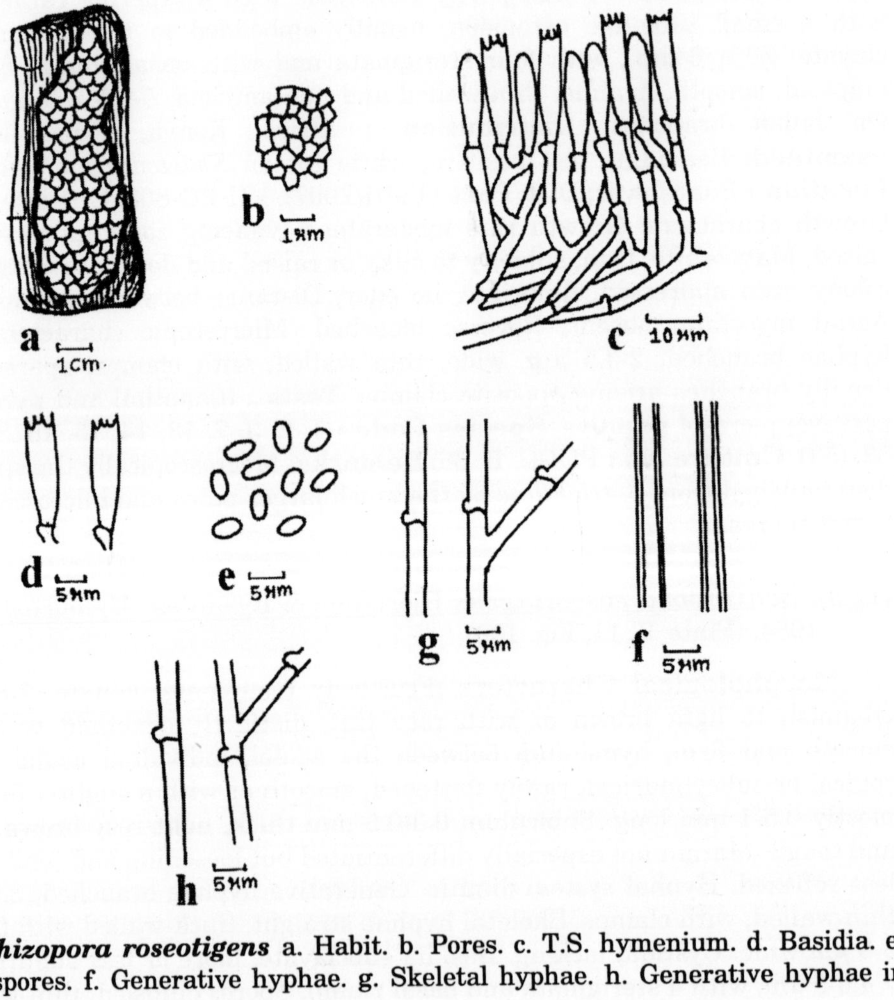

Fig. 106 ***Schizopora roseotigens*** a. Habit. b. Pores. c. T.S. hymenium. d. Basidia. e. Basidiospores. f. Generative hyphae. g. Skeletal hyphae. h. Generative hyphae in culture

(17-4). *Schizopora trichiliae* (Van der Byl) Ryvarden. A preliminary polypore flora of East Africa: 553, 1980.

=*Polyporus trichiliae* Van der Byl, *S. Afr. J. Sci.* **18(3-4)**:262,1922.(Plate 7, Fig. 107).

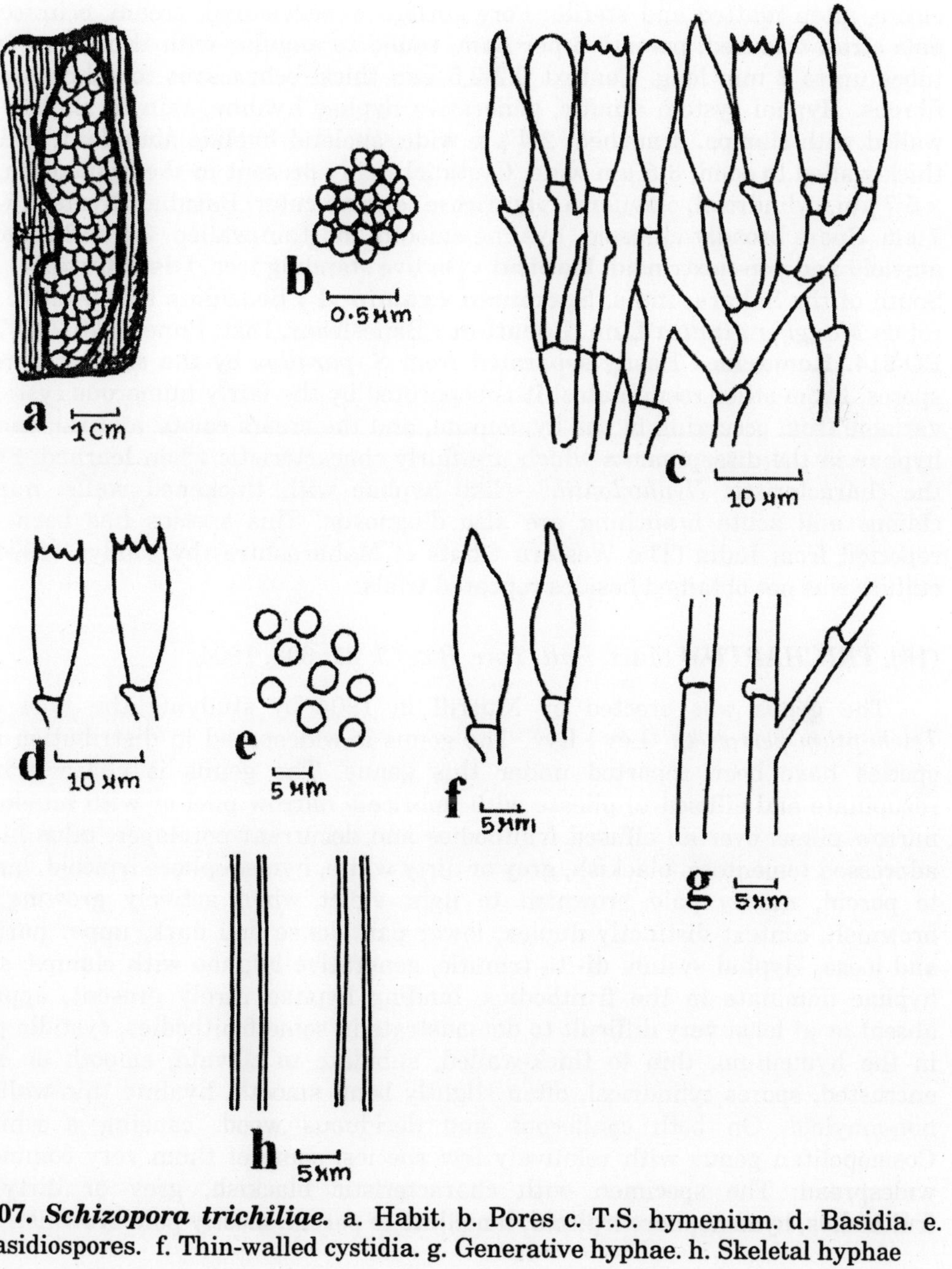

Fig. 107. *Schizopora trichiliae.* a. Habit. b. Pores c. T.S. hymenium. d. Basidia. e. Basidiospores. f. Thin-walled cystidia. g. Generative hyphae. h. Skeletal hyphae

Morphological Characters : Fruitbody annual, resupinate to effuseo-reflexed with narrow pilei, 5-6 × 2-3 cms. and up to 5 mm. thick. Consistency coriaceous to hard when dry. Pileus semicircular broadly attached, upper surface dull, finely tomentose to velvety, cream to ochraceous-buff, non-zoned. Margin round and entire, often matted and sterile. Pore surface woodcoloured, cream, ochraceous or pale straw-coloured, pores 5-6 per /mm, round to angular with thin dissepiments, tubes up to 2 mm long. Context 0.2-2.5 mm thick, ochraceous to ochraceous-buff, fibrous. Hyphal system dimitic, generative hyphae hyaline, thin to slightly thick-walled with clamps, branched, 2-4 μm wide, skeletal hyphae abundant, yellowish, thick-walled to solid, 3-5 μm wide. Cystidiol richly present in the hymenium, 15-25 × 5-7 μm, cylindrical, claviform, ventricose to vermicular. Basidia clavate, 15-18× 5-7 μm. Spore broadly ellipsoid, hyaline smooth and thin-walled, 3-5 × 2-3 μm, non-amyloid and non-dextrinoid. **Habitat :** On live standing tree. **Distribution :** Africa, South of the Sahara, India. **Specimen examined :** Benzidine test positive, white rot on *Mangifera indica* Linn. **Location :** Baneshwar, Dist. Pune (13/01/2007), VH-PO-614. **Remarks :** Easily separated from *S. paradox* by the smaller pores and spores. From other resupinates. It is separated by the fairly numerous cystidiols of variable from occurring in the hymenium, and the cream colour and the encrusted hyphae in the dissepiments which are fairly characteristic when learned. Further, the characteristic *Hyphodontia* – like hyphae with thickened walls, numerous clamps and acute branching are also diagnostic. This species has been earlier reported from India (The Western Ghats of Maharashtra) by Vaidya (1994). The culture was not obtained besides repeated trials.

(18). *TRICHAPTUM* Murr. *Bull. Torr. Bot. Cl.* **31:** 608, 1904.

The genus was erected by Murrill in 1904 by studying the type species *Trichaptum perrottetii* (Lév.) Ryv. The genus is widespread in distribution and 44 species have been reported under this genus. The genus is characterized by resupinate and effused to pileate with imbricate narrow pilei or with an elongated narrow pileus over an effused fruitbodies and decurrent porelayer, pilus hispid to adpressed tomentose, blackish, grey or dirty white, hymenophore irpicoid, lamellate to poroid, mostly pale brownish to light violet when actively growing, tubes brownish, context distinctly duplex, lower part dense and dark, upper part white and loose. Hyphal system di- to trimitic, generative hyphae with clamps, skeletal hyphae dominate in the fruitbodies, binding hyphae rarely present, apparently absent or at least very difficult to demonstrate in some fruitbodies, cystidia present in the hymenium, thin to thick-walled, subulate to clavate, smooth or apically encrusted, spores cylindrical, often slightly bent, smooth, hyaline thin-walled and non-amyloid. On both coniferous and deciduous wood, causing a white rot. Cosmopolitan genus with relatively few species, some of them very common and widespread. The specimen with characteristic blackish, grey or dirty white fruitbodies, hymenophore irpicoid, lamellate to poroid, mostly pale brownish to light violet; broadly ellipsoid spores was identified as *T. versatile* (Berkeley) Cunningham.

(18-1). *Trichaptum versatile* (Berkeley) Cunningham. *Bull. New Zealand Dept. Sci. Industr. Res.* **164:** 99, 1965.

=*Trametes versatilis* Berk., *J. Bot.,* London **1 (3) :** 150, 1842. (Plate 7, 11, Fig. 108).

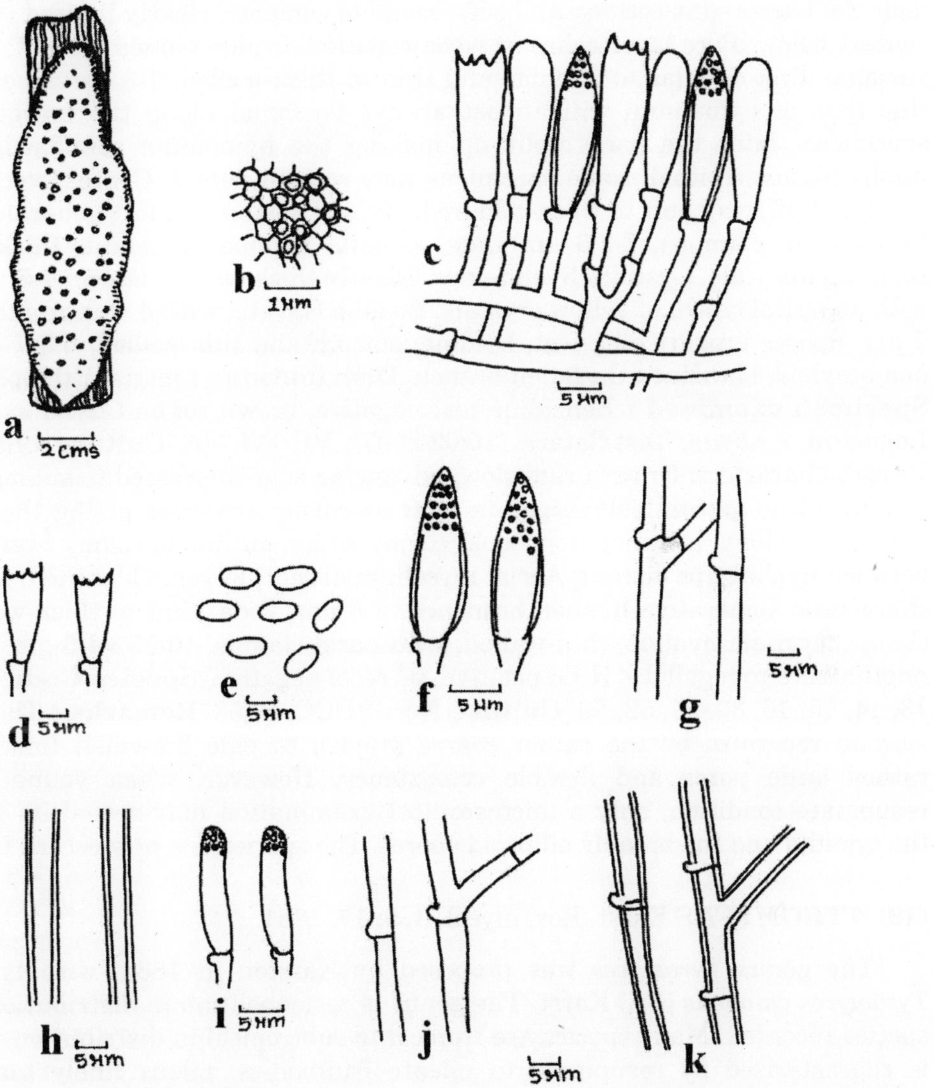

Fig. 108. *Trichaptum versatile.* a.Habit. b. Pores. c. T.S. hymenium. d. Basidia. e. Basidiospores. f. Encrusted cystidia. g. Generative hyphae. h. Skeletal hyphae. i. Cystidia in culture. j. Thin-walled generative hyphae in culture. k. Thick-walled generative hyphae in culture

Morphological Characters : Fruitbody annual to perennial, pileate-sessile, to reflexed and resupinate, 15-17×7-9 cms. and up to 10 mm. thick, solitary to imbricate or several specimens broadly attached to fused laterally in rows, consistency flexuous to tough. Pileus semicircular, flat and finely concentrically sulcate and radiately striate, hispid to tomentose, especially in the inner parts, more appressed along the margin, colour grey to ochraceous or milky coffee, margin thin, depressed to enrolled, even to crenate. The tomentum can be up to 7 mm thick near the base and is cottony and soft, easily to compact, clearly limited towards the context below. Pore layer ochre or wood-coloured or pale vinaceous buff, pores very variable, first circular to angular and thin to thick-walled, 1-3 pores per mm, and this type of hymenium will almost always be found along the margin of older specimens, later the pores split up making the hymenium semidaedaleoid and finally almost irpicoid, some specimens may remain poroid. Context very thin 0.1-0.3 mm, buff, clay buff to wood-coloured. Hyphal system dimitic, generative hyphae thin-walled, clamped, 2-3.5 μm wide, skeletal hyphae abundant, thick-walled to solid 4-6 μm wide. Cystidia hyaline and slightly thick-walled 12-15 ×4-5. 5 μm often with an apical crown of yellow crystals. Basidia clavate, with 4 sterigmata, 15-18×5-7 μm. Spores broadly elliptical, hyaline, smooth and thin-walled, 5.5-7 ×3.5-4 μm, non-amyloid. **Habitat :** On fallen branch. **Distribution :** Liberia, Ethiopia, Malawi. **Specimen examined :** Benzidine test negative, brown rot on *Cassia siamea* Lam. **Location :** Koyna; Dist:Satara (10/02/2007), VH-PO-745. **Culture characters :** Growth characters: Growth rate slow; advancing zone appressed to submerged; Mat cottony white getting submerged in agar as colony advances giving the plate and hence the colony, a transparent look; colony white; outline of colony even; Distance between hyphal tips distant; Aerial mycelium absent; Reverse bleached. Microscopic characters: Generative hyphae branched, 2-6 μm, wide, thin to thick-walled, with clamps, cystidia hyaline, thin-walled, with basal clamps, 10-15 ×4-5 μm. **Tests :** α-napthal and pyrogallol + H_2O_2 positive; p-cresol negative. **Species Code :** 1, 2, 3, 6, 13, 14, 15, 16, 30, 37, 52, 53. **Culture No :** PUCC. B: 68. **Remarks :** The species is easy to recognize by the rather coarse greyish to pale brownish tomentum, the rather large pores and flexible consistency. However, when young and in a resupinate condition, only a microscopical examination may reveal its identity by the cystidia and the broadly ellipsoid spores. The species is a new record from India.

(19). *TYROMYCES* Karst. Rev. mycol. **3, 9:** 17, 1881.

The genus *Tyromyces* was proposed by Karsten in 1881, with type species *Tyromyces chioneus* (Fr.) Karst. The genus is cosmopolitian in distribution with 301 species recorded. Many species are tropical to subtropical in distribution. The genus is characterized by resupinate to pileate fruitbodies, pileus finely tomentose to smooth, usually white, more rarely bluish, reddish to brownish, pore surface white to cream, rarely greyish brown or reddish by drying, pores regular, small to medium, round to angular, context or subiculum white to light brown when dry, consistency soft when fresh, mostly fragile and light weighted when dry. Hyphal system usually monomitic, more rarely dimitic, generative hypahe hyaline, normally with clamps, in one species with simple septa, cystidia absent or present, when dimitic either with rather few binding hyphae or hyaline skeletal hyphae, spores allantoid, cylindrical to ellipsoid, hyaline, smooth, thin-walled, non-amyloid to weakly amyloid. The specimen with characteristic pore surface white becoming cream to pale straw-coloured; spores allantoid, smooth, thin-walled, hyaline and non-amyloid was identified as *Tyromyces undosus* (Peck) Murrill.

(19-1). *Tyromyces undosus* (Peck) Murrill. *N. Amer. Flora* **9 (1)** : 34, 1907.

= *Polyporus undosus* Peck, *Ann. Rep. N.Y. state Mus.* **34:** 42, 1881. (Plate 7, Fig. 109).

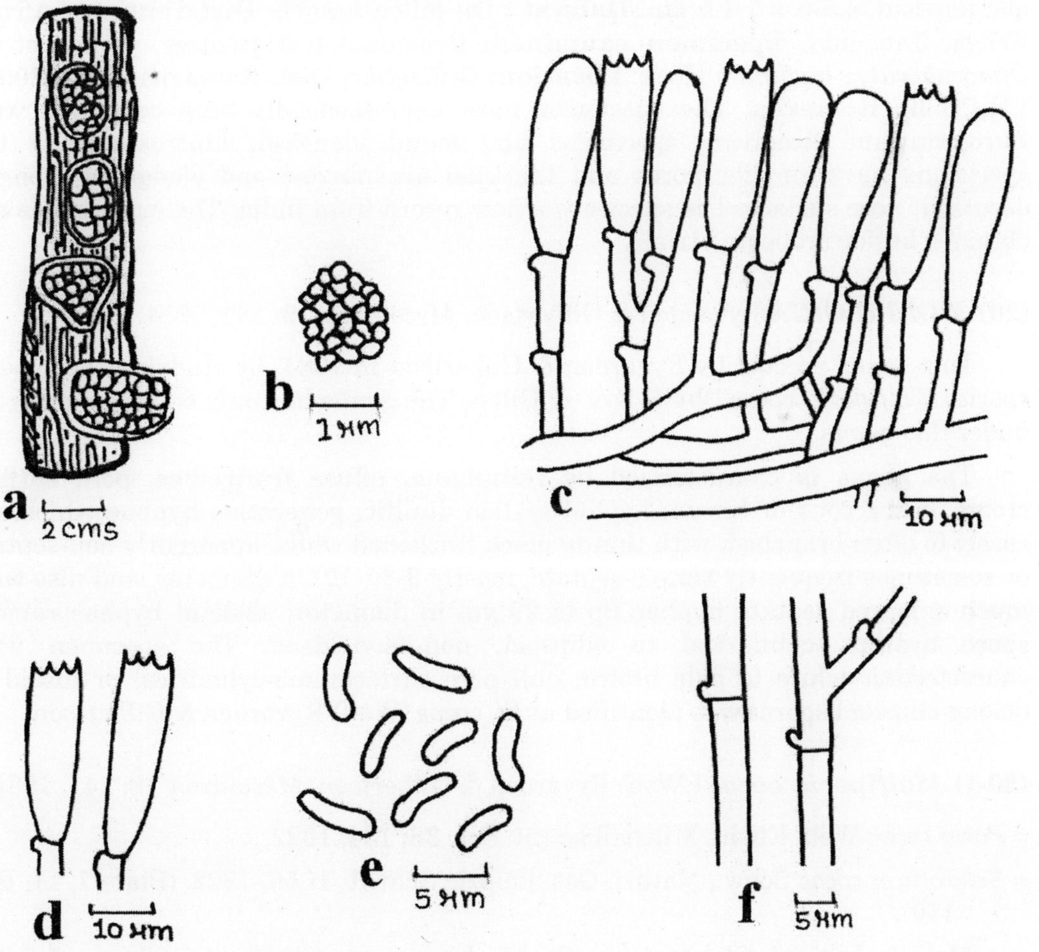

Fig. 109 *Tyromyces undosus* a. Habit. b. Pores. c. T.S. hymenium. d. Basidia. e. Basidiospores. f. Generative hyphae

Morphological Characters: Fruitbody annual, separable, resupinate to pileate with a narrow, elongated pileus along a decurrent pore surface, single or imbricate with numerous small pilei on a decurrent pore surface, 1-4 × 0.5-2 cms. and 2-10 mm thick measured vertically, soft when fresh, stiff and brittle when dry. Pileus white to light cream, finely appressed velutinate becoming glabrous and smooth with age or with very fine radial lines or streaks, azonate or with a few faint and narrow zones. Pore surface white becoming cream to pale straw-coloured, in reflexed-pileate specimens decurrent on the substrate and then pores often partly split, pores angular to round, 3-4 per/mm. Tubes concolorous and up to 10 mm deep in pileate specimens, rarely above 3 mm. Context white, up to 1 mm thick. Hyphal system monomitic, generative hyphae thin-walled, 3-5 μm wide, with clamps. Basidia clavate, 18-25 × 5-8 μm. Spores allantoid, smooth, thin-walled, hyaline and non-amyloid, 4.5-6 × 1-1.5 μm. **Habitat :** On fallen branch. **Distribution :** Africa, Kenya, Tanzania. **Specimen examined:** Benzidine test positive, white rot on *Dimorphocalyx lawianus* Hook. **Location:** Guhaghar; Dist: Ratnagiri (25/09/2005), VH-PO-39. **Remarks:** The specimens have microscopically been compared with European and American specimens and found identical. Macroscopically the specimens have smaller pores and the pilei are narrow and elongated along a decurrent pore surface. The species is a new record from India. The culture was not obtained besides repeated trials.

(20). *WOLFIPORIA* Ryvarden & Gilbertson. *Mycotaxon* **19:** 141, 1984.

This genus erected by Ryvarden & Gilbertson in 1984, by studying the type of species *Wolfiporia cocos* (Wolf) Ryv. & Gilbn. The genus has only 6 species reported under this genus.

The genus is characterized by resupinate, effuse fruitbodies, pore surface cream, ochraceous or brown; hyphal system dimitic, generative hyphae which are rarely to often branched, with thin to much thickened walls, apparently non-septate or sometimes frequently simple-septate, mostly 3-8 (-12) in diameter, and also with much enlarged septate hyphae up to 29 μm in diameter; skeletal hyphae ramose spore hyaline, cylindrical to ellipsoid, non-amyloideae. The specimen with characteristic white to pale brown, dull pore surface; sub-cylindrical or fusoid to oblong-ellipsoid spores was identified as *W. cocos* (Wolf) Ryvarden & Gilbertson.

(20-1). *Wolfiporia cocos* (Wolf) Ryvarden & Gilbertson. *Mycotaxon* **19:** 141, 1984.

= *Poria cocos* Wolf, Elisha Mitchell scient. Soc. **38:** 134, 1922.

= *Sclerotium cocos* Schw., Naturf. Ges. Leipzig Schrift. **1:** 56, 1822. (Plate 7, 11, Fig. 110).

Morphological Characters : Fruitbody annual, resupinate, adnate, 5-7×3-4 cms. and up to 7 mm thick, margin white to pale brown, membranous or velvety, at extreme edge fimbriate; pore surface white to pale brown, dull, the tubes tough and cartilaginous when fresh, corky or resinously impregated and fragile when dry, the pores angular, variable, 1-3 per/mm or some larger, edges becoming thin, usually more or less fimbriate; context whitish to cream next to the tubes, below darker, sometimes distinctly brown, firm, fibrous to corky, up to 2 mm thick, continuing without change into the trama. Sections not changing color in KOH. Hyphal system dimitic, generative hyphae, 3-5 wide, hyphae often gelatinizing more or less in KOH, skeletal hyphae 4-6 μm wide;thick-wall. Cystidia none. Basidia slender-clavate, 15-20 ×6-9 μm. Spores hyaline, smooth, non-amyloid, subcylindric or fusoid to oblong-ellipsoid, 6-11 × 2.5-4 μm. **Habitat :** On dead standing tree.

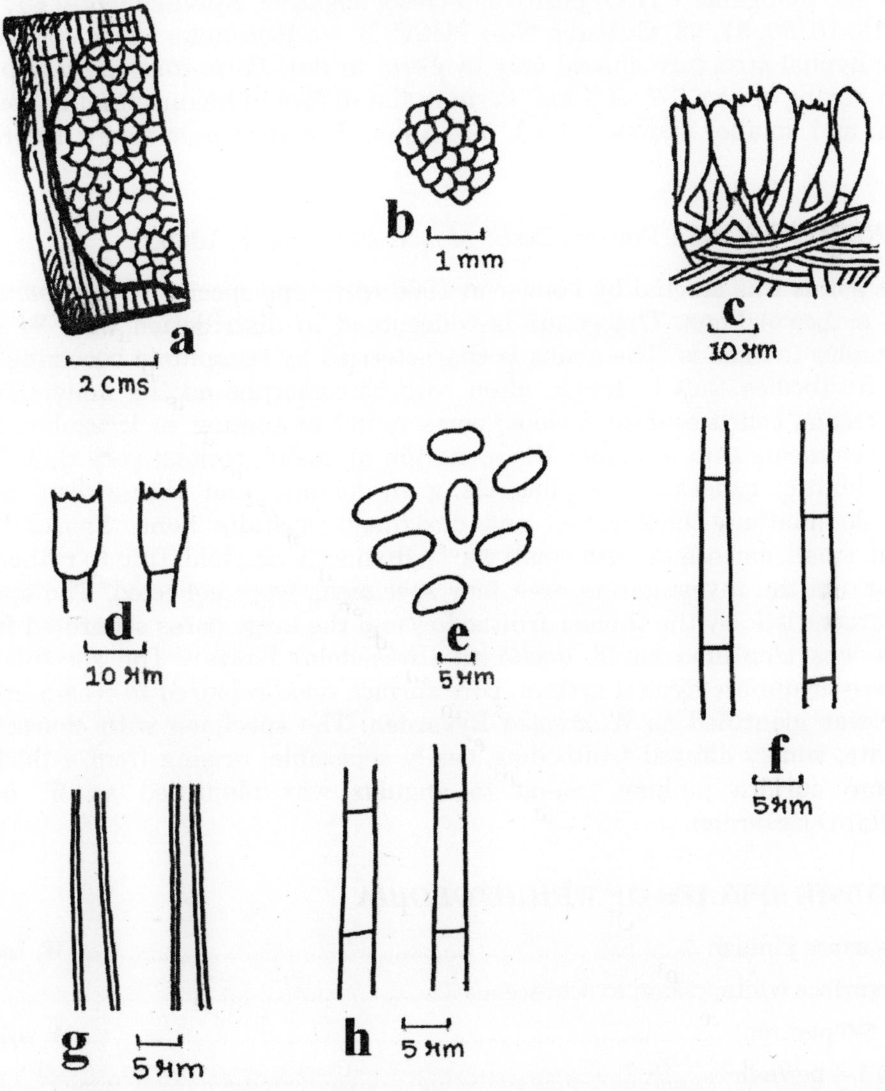

Fig. 110 ***Wolfiporia cocos*** a. Habit. b. Pores. c. T.S. hymenium. d. Basidia. e. Basidiospores. f. Generative hyphae. g. Skeletal hyphae. h.Generative hyphae in culture

Distribution: Japan, Australia. **Specimen examined :** Benzidine test negative, brown rot on *Ficus religiosa* L. **Location :** Khed, Dist. Ratnagiri (12/10/2005), VH-PO-223. **Culture characters:** Growth characters: Growth rate slow; advancing zone appressed to submerged; Mat cottony white getting submerged in agar as colony advances giving the plate and hence the colony, a transparent look; colony white; outline of colony even; Distance between hyphal tips distant; Aerial mycelium absent; Reverse bleached. Microscopic characters: Generative hyphae branched, 3-5 μm, wide, without clamps, with simple septa. **Tests:** α-napthal and pyrogallol + H_2O_2 positive; p-cresol negative. **Species Code :** 1, 2, 3, 6, 13, 14, 15, 16, 30, 37, 53. **Culture No :** PUCC. B: 69. **Remarks :** This species has a peculiar hyphal structure shared only by *Poria* in flat. *Poria dictyopora* appears to differ in having spores 6-7 ×4-5 μm. *Poria* in flat differs in having smaller pores, 3-5 per mm, and smaller spores 4-5 ×2. 5-3. 5 μm. The species is a new record from India.

(21). *WRIGHTOPORIA* Pouzar. *Ceská Mykol.* **20 (3)** : 173, 1966.

The genus was erected by Pouzar in 1966 with type species *Wrightoporia lenta* (Overh. & Lowe) Pouz. The genus is widespread in distribution with 34 species known under the genus. The genus is characterized by resupinate becoming widely effused fruitbodies, soft to tough, often with rhizomorphs on the underside; pore surface cream, ochraceous to fulvous; pores round to angular or irregular, 2-6 per mm, dissepiments thin to rather thick, margin myceloid, context very thin. Hyphal system dimitic, generative hyphae clamped, hyaline and thin-walled, skeletal hyphae dominating, thick-walled and dextrinoid, cystidia none. Spores broadly elliptical, small, smooth or with small warts, distinctly amyloid, thin to rather thick-walled. From the investigation area four specimens were collected. The specimen with characteristic by the thicker fruitbodies and the large pores separated from *W. africana* was identified as *W. avellanea* (Bresadola) Pouzar. The specimen with characteristic dimitic hyphal system, pore surface wood-coloured to cream, round to angular was identified as *W. cremea* Ryvarden. The specimen with characteristic resupinate, widely effused fruitbodies, easily separable, arising from a thick, pale lilac; pore surface pinkish, round to angular was identified as *W. iobapha* (Patouillard) Ryvarden.

KEY TO THE SPECIES OF *WRIGHTOPORIA*

1. Pore surface pinkish .... ***W. iobapha***

1´. Pore surface white, cream to ochraceous 2

2. Pores 5-6 per/mm. .... ***W. africana***

2´. Pores 1-4 per/mm 3

3. Pores 1-3 per/mm hyphal system trimitic .... ***W. avellanea***

3´. Pores 3-4 per/mm Hyphal system dimitic .... ***W. cremea***

(21-1). *Wrightoporia africana* Johansen & Ryvarden. *Trans. Brit. Mycol. Soc.* **72:** 196, 1979. (Plate 8, Fig. 111).

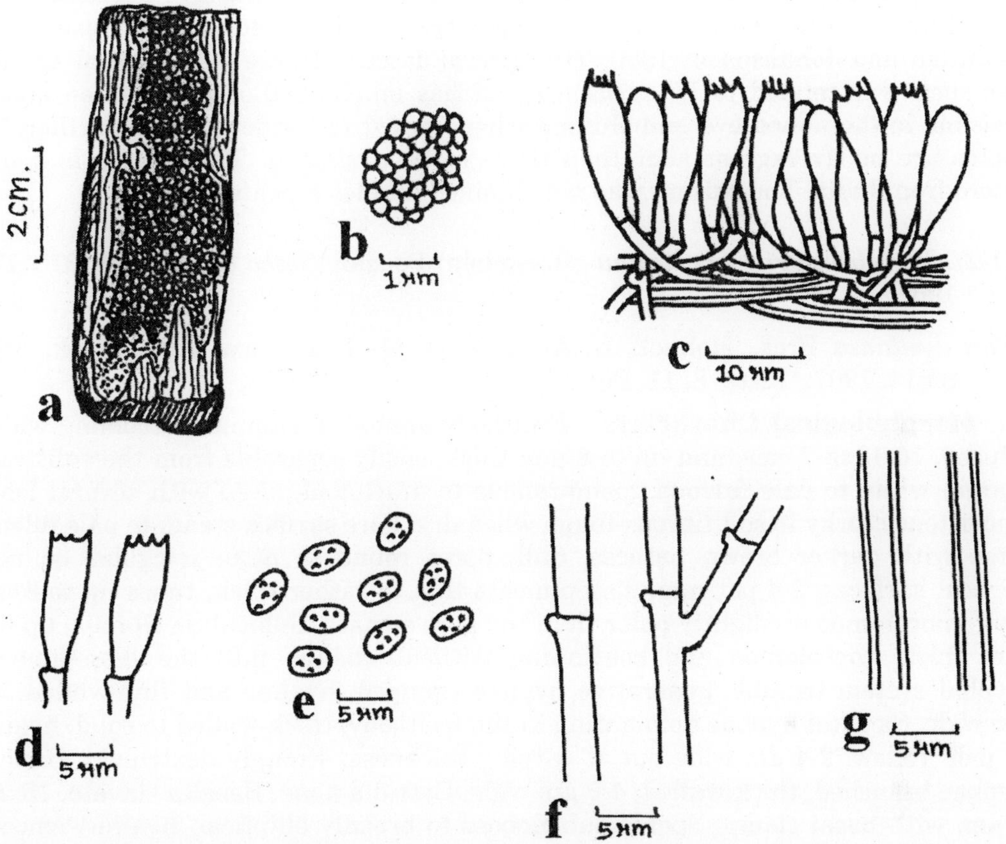

Fig. 111 ***Wrightoporia africana*** a. Habit. b. Pores. c. T.S. hymenium. d. Basidia. e. Basidiospores. f. Generative hyphae. g. Skeletal hyphae

Morphological Characters : Fruitbody annual, resupinate, becoming widely effused, 4-6 ×2-3 cms. and 2 mm thick, easily separable from the substratum, consistency soft, fibrous to tough when dry. Margin cream to white, cottony, fimbriate to slightly rhizomorphic, narrow to wide. Pore surface whitish cream to ochraceous, dull to slightly shiny when turned in incident light, pores first circular and regular, 3-4 per/mm with thinner and more fimbriate edges especially on sloping substrata. Tubes non-stratified, continuous with the context. Context very thin, fibrous, cream to white, with a few rhizomorphs penetrating into the substratum. Hyphal system dimitic, generative hyphae with clamps, hyaline, thin-walled, 2-3 μm wide, skeletal hyphae dominant throughout, dextrinoid, thick-walled, flexuous, unbranched, narrow and 2-4 μm, wall thickness often irregular. Cystidia none. Basidia clavate, 12-15× 4-5 μm, with basal clamp. Spores sub-globose to broadly ellipsoid, amyloid, thin to rather thick-walled, smooth or with small ridges and warts, 3-4 × 2-3 μm. **Habitat :** On live standing tree. **Distribution :** East. Africa, Kenya, Tanzania, Malawi. **Specimen examined :** Benzidine test positive, white rot on *Delonix regia* (Bojer ex. Hook) Raf. **Location :** Film Institute; Dist: Pune (19/09/2005), VH-PO-163. **Remarks :** The species was reported by Ryvarden and Johansen in 1980. The species described here is a tropical species. The species is related to *W. avellanea*, but has much smaller pores, often almost invisible to the naked eye and shining when turned in incident light. Further, the spores are on average smaller then those of *W. avellanea.* The species is a new record from India. The culture was not obtained besides repeated trials.

(21-2). *Wrightoporia avellanea* (Bresadola) Pouzar. *Ceská Mykol.* **20 (3) :** 173, 1966.

=*Pori avellanea* Bres., Hohnel., K. Akad. Wiss. Math. Naturw. Klas. Denk. Schr. **83:**14, 1907. (Plate 8, 11, Fig. 112).

Morphological Characters : Fruitbody annual, resupinate becoming widely effused, 10-13×5-7 cms. and up to 8 mm thick, easily separable from the substrate, margin white to pale fulvous, membranous to arachnoid, often with several lobes. Consistency corky to soft fibrous-tough when dry. Pore surface cream to pale fulvous often with darker brown pathces, dull, pores round to more irregular on near vertical surfaces, 2-4 per mm, dissepiments thin to rather thick, tubes up to 8 mm long, concolorous or slightly paler than the pore surface. Subiculum fibrous, up to 1 mm thick, concolorous and continuing without change into the dissepiments. Hyphal system trimitic, generative hyphae clamped, hyaline and thin-walled, 2-3 μm wide. Skeletal hyphae dominating in the fruitbody, thick-walled to solid, hyaline to pale yellow, 2-4 μm wide but of varying thickness, strongly dextrinoid. Binding hyphae branched, thick-walled, 4-6 μm wide. Cystidia none. Basidia clavate, 15-20× 7 μm, with basal clamp. Spores sub-globose to broadly elliptical, hyaline, smooth, thin to slightly thick-walled, 3. 5-5 × 3-4 μm, amyloid. **Habitat :** On live standing tree. **Distribution** : Cuba, Jamaica, Puerto Rico, Brazil, America, Africa, Tanzania. **Specimen examined :** Benzidine test positive, white rot on *Pongamia pinnata* (Linn.) Pierre. **Location:** Pune University; Dist: Pune (20/07/2006), VH-PO-506. **Culture characters :** Growth characters: Growth rate moderate; advancing zone raised; Mat cottony white to livid vinaceous, colony white; outline of colony even; distant between hyphal tips distant, Reverse unchanged. Microscopic characters: Generative hyphae branched, 3-5 μm, wide, clamps. Skeletal hyphae thick-walled, pale yellow to brown, 2-4 μm wide, basidia clavate, 12-17×5-7μm. **Tests :** α-napthal and pyrogallol+ H_2O_2 positive; p-cresol negative. **Species Code :** 1, 2, 3, 7, 12, 15, 14, 15, 21, 30, 33, 37, 53, 88. **Culture No :** PUCC. B: 70. **Remarks :** The species can easily be separated from *Wrightoporia africana* by the thicker fruitbody and the large pores. The species is a new record from India. The culture was not obtained besides repeated trials.

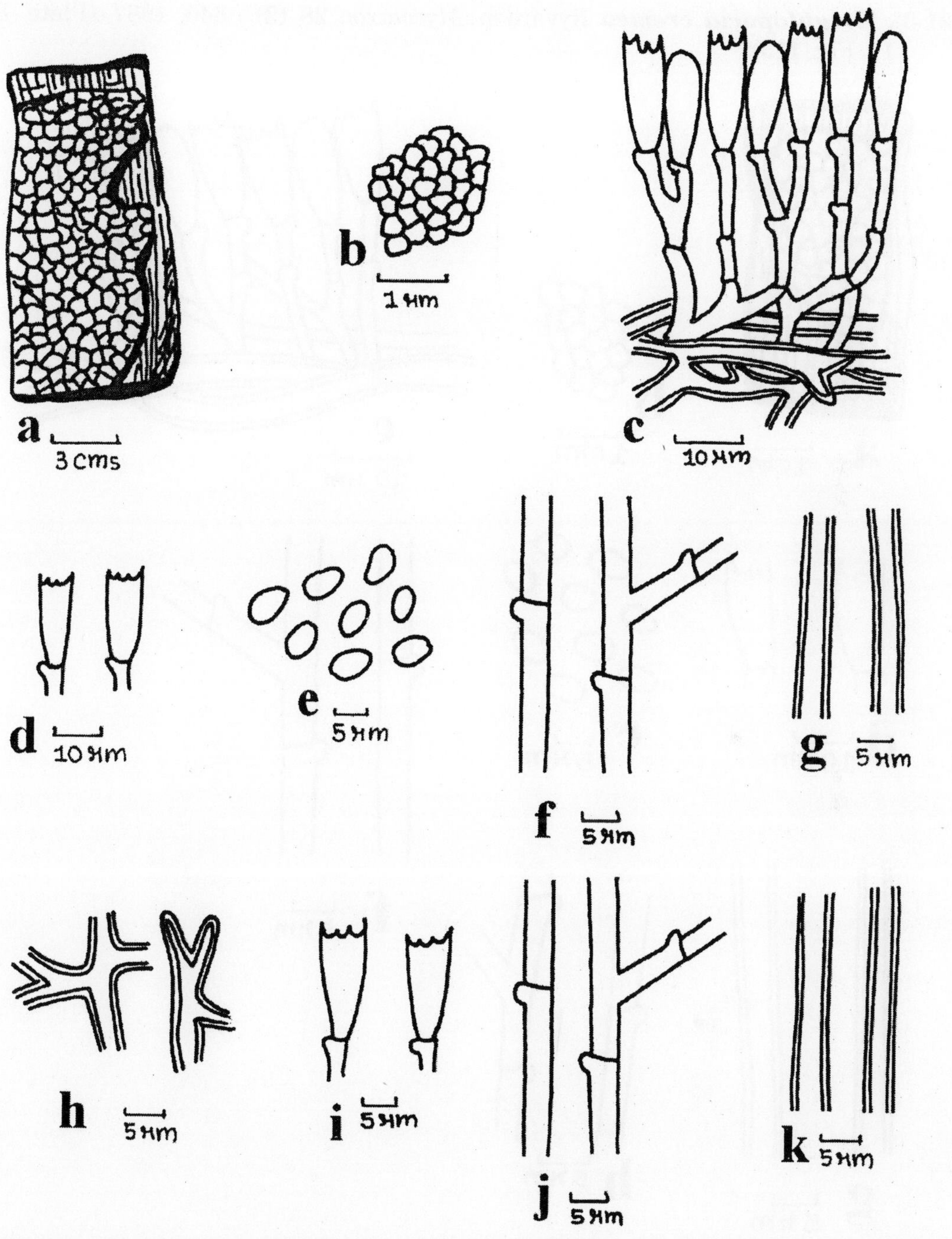

Fig. 112. ***Wrightoporia avellanea.*** a. Habit. b. Pores c. T.S. hymenium. d. Basidia. e. Basidiospores. f. Generative hyphae. g. Skeletal hyphae. h. Binding hyphae. i. Basidia. j. Generative hyphae in culture. k. Skeletal hyphae in culture

(21-3). *Wrightoporia cremea* Ryvarden. *Mycotaxon* **28 (2)** : 540, 1987. (Plate 8, 11, Fig. 113).

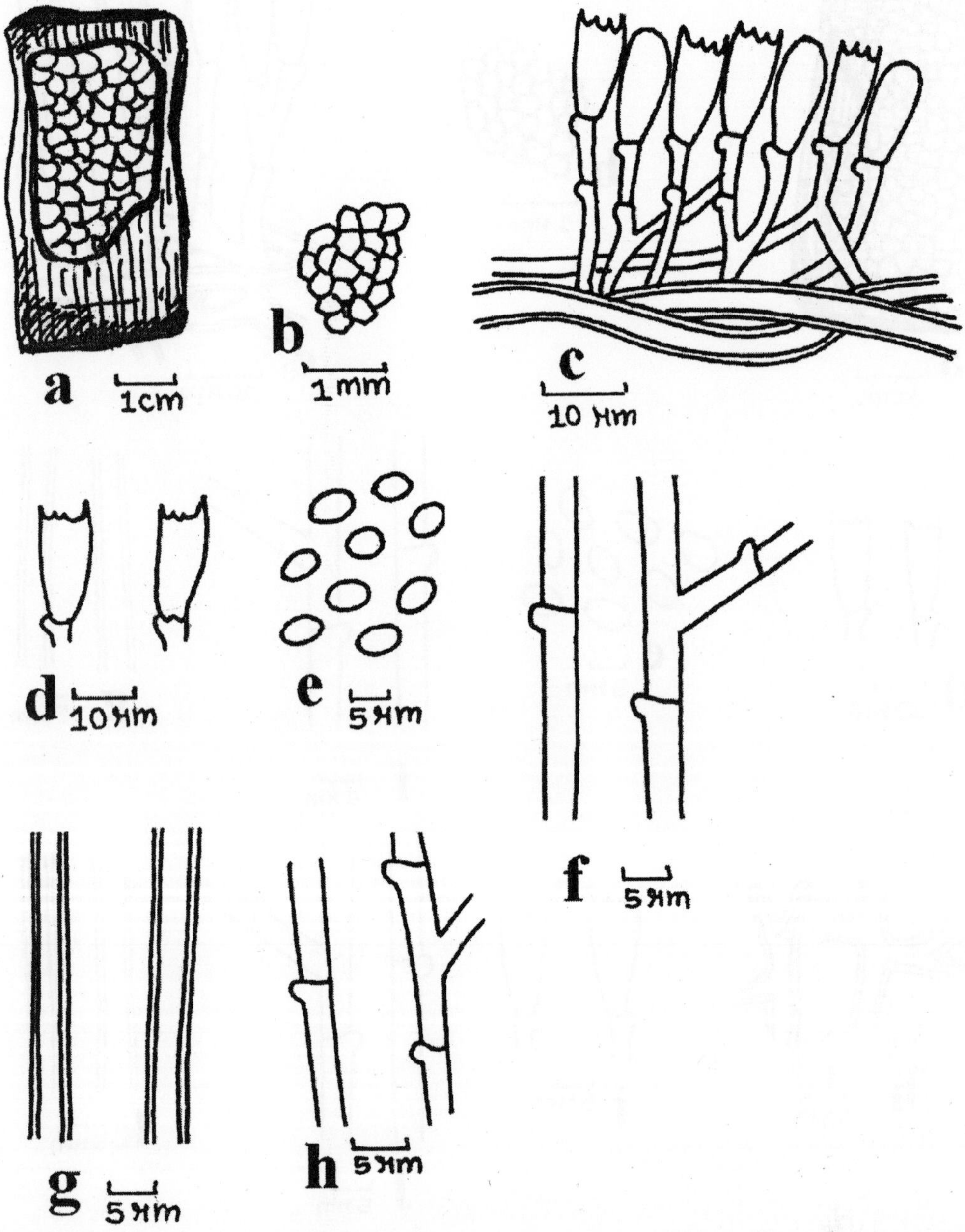

Fig. 113 *Wrightoporia cremea*. a. Habit. b. Pores. c. T.S. hymenium. d. Basidia. e. Basidiospores. f. Generative hyphae. g. Skeletal hyphae. h. Generative hyphae in culture

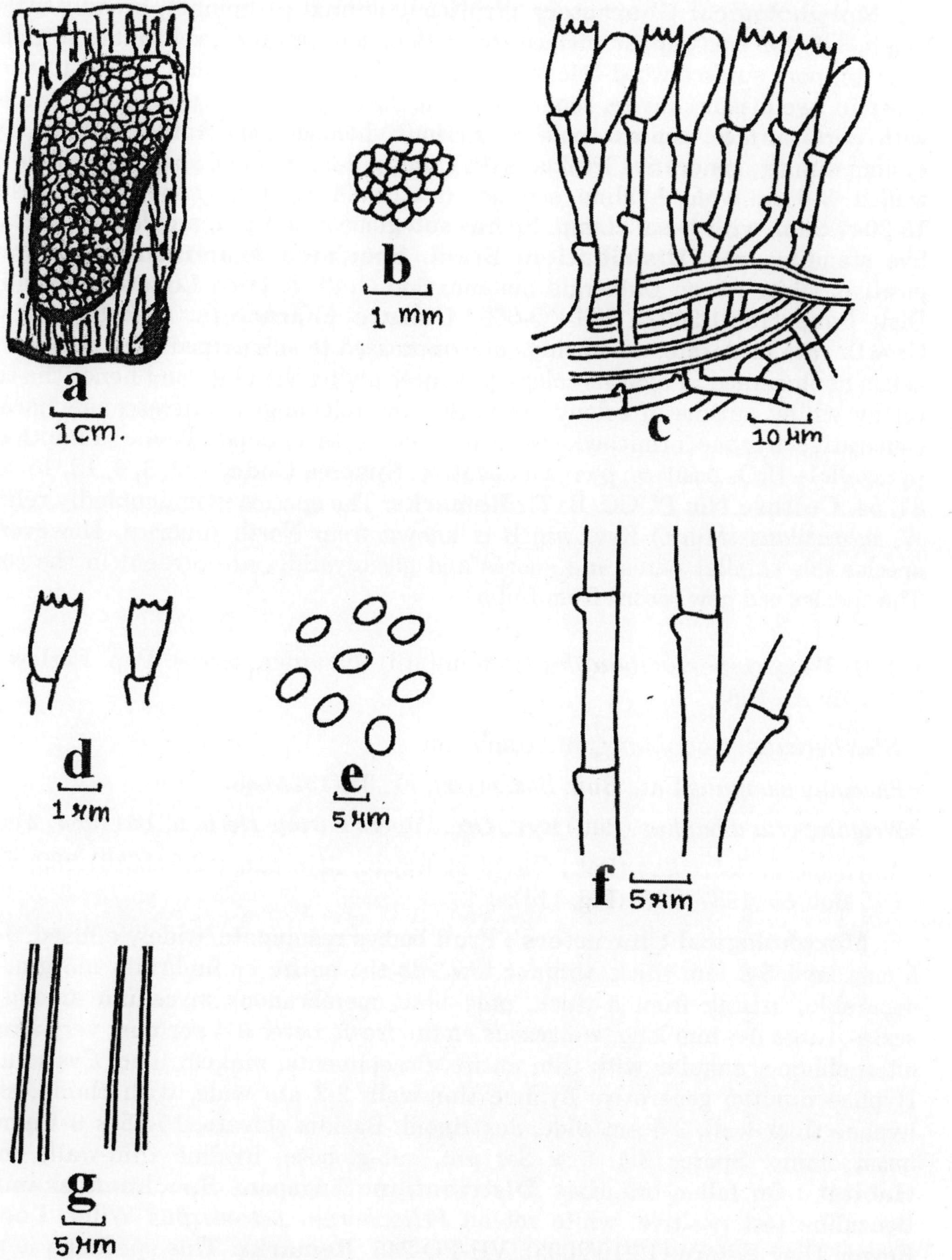

Fig. 114. ***Wrightoporia iobapha*** a. Habit. b. Pores. c. T.S. hymenium. d. Basidia. e. Basidiospores. f. Generative hyphae. g. Skeletal hyphae

Morphological Characters : Fruitbody annual resupinate, broadly attached, tough, 3-4 ×2-3 cms., upper surface cream to pale ochraceous, velutinate to glabrous, azonate, pore surface wood-coloured to cream, pores 3-4 per/mm, round to angular, tubes in two distinct layers separated by a thin layer of context, tubes concolorous with pore surface, context pale ochraceous, homogenous, rather dense. Hyphal system dimitic, generative hyphae with clamps, 2-3 μm wide, skeletal hyphae thick-walled, 3-5 μm wide, hyaline, strongly dextrinoid. Cystidia none. Basidia clavate, 15-20×7-9μm, with basal clamp. Spores sub-globose, 3-4 μm, amyloid. **Habitat:** On live standing tree. **Distribution:** Brazil. **Specimen examined:** Benzidine test positive, white rot on Dalbergia melanoxylon Guill. & Perr. **Location:** Kondhona; Dist: Pune (13/01/2007), VH-PO-565. **Culture characters:** Growth characters: Growth rate moderate; advancing zone appressed to submerged; Mat cottony white getting submerged in agar as colony advances giving the plate and hence the colony, colony white; outline of colony even; Reverse unchanged. Microscopic characters: Generative hyphae branched, 5-6 μm, wide, with clamps. **Tests:** α-napthal and pyrogallol+ H_2O_2 positive; p-cresol negative. **Species Code:** 1, 2, 3, 6, 13, 15, 21, 30, 37, 54. **Culture No:** PUCC. B: 71. **Remarks:** The species is undoubtedly related to *W. subrutilans* (Murr.) Ryv. which is known from North America. However, this species has smaller pores and spores and gloeocystidia are present in the context. The species is a new record from India.

(21-4). *Wrightoporia iobapha* (Patouillard) Ryvarden. Occas. Pap. Farlow Herb. **18:** 21, 1983.

= *Steccherícium iobaphum* (Pat.) comb. nov.

=*Phaeolus iobaphus* Pat., *Bull. Soc. mycol. Fr.* 38 (1923) 85.

=*Wrightoporia iobaphus* (Pat.) Ryv., *Occ. Papers Farlow Herb*. n. 18 (1983) 21.

=*Amylosporus iobaphus* (Pat.) David et Rajch., *Mycotaxon* 22 (1985) 289; Can. J. Bot. 65 (1987) 208. (Fig. 114).

Morphological Characters : Fruit bodies resupinate, widely effused, 5-7 × 3-5 cms. and 5-8 mm thick, thinner towards the entire or fimbriate margin, easily separable, arising from a thick, pale lilac, membranous mycelium among bark-scales, tubes 6-8 mm long, violaceous as the fresh; pores 3-4 per/mm, very irregular, often oblique, angular with thin entire dissepiments, pinkish lilac. Cystidia none. Hyphae dimitic; generative hyphae thin-wall, 2-3 μm wide, with clamp; skeletal hyphae thick-wall, 3-5 μm wide, dextrinoid. Basidia clavate, 15-25 × 9-11μm, with basal clamp. Spores 3-4. 5 × 3-4 μm, sub-globose, hyaline thin-wall, amyloid. **Habitat :** On fallen branches. **Distribution:** Singapore. **Specimen examined :** Benzidine test positive, white rot on *Peltophorum petrocarpus* Willd. **Location:** Koyna; Dist: Satara (12/10/2005), VH-PO-245. **Remarks:** This specimen is readily recognized by pale lilac fruitbody, pinkish lilac pore surface, cystidia absent. The species is a new record from India. The culture was not obtained besides repeated trials.

FLORISTIC AFFINITIES

The flora of India is diverse and the central India is rich from surrounding areas like Indo-Gengetic Planes, Malabar, East Assam to contribute to the composition of the floras in Maharashtra. In addition to the original elements already existing. Flora of Maharashtra, on the whole, is well represented by dry deciduous species in vegetation with climatic climax. With regard to various elements it has been noticed that they are mostly of south subtropical distribution (Mahabale, 1987), with few temperate elements but they are of secondary succession. Majority of plants at present are of post-Miocene period, while large number of plants found in Quarternary are a mixture of dry and a few temperate species. Slowly the temperate genera like *Podocarpus* and *Salix* disappeared finally getting succeeded by present day dry deciduous species.

The tropical plants seems to be lingered in semi-evergreen pockets while others have migrated towards South India. The temperate species have migrated to higher altitudes. Many species disappear from the Western Ghats due to uncertainty of rainfall and rise in temperature (Mahabale, 1987). While Meherhomji (1976-1978) considers that the deciduous vegetation is of degenerate type. On the other hand when there is more rain and low temperature, this deciduous vegetation develops into moist deciduous or semi-evergreen vegetation.

The lichen flora of South West India shows affinity with that of African continent (Awasthi, 1965). Similarly various theories proposed for drifting of continent Wegener and Davidson (1954) and Bharucha (1983). Tectonic plate theory Krishnan (1968). Wadia (1975) revealed that in past there was an existence of the continuity of continent in Gondwanaland which slowly separated into Lemuria followed by separation of four Southern continent namely, S. America, Africa, South East Asia and Australia. It is therefore, confirmed that there must be some connection regarding the distribution of the investigation areas in the parts of the Western Ghats with that of above tropical Gondwanaland countries.

In the present study a total of 110 species of resupinate Aphyllophorales (27 families and 54 genera) from the 46 respective hosts were identified out of 355 colected specimens throughout the Western Ghats.

From these 110 fungal species 2 different species of *Aleurodiscus*, 2 different species of *Amylocorticium*, 2 species of *Amylocystis*, 1 species of *Amylosporus*, 1 species of *Amylostereum*, 1 species of *Anomoporia*, 1 species of *Auriporian*, 1 species of *Botryohypochnus*, 1 species of *Candelabrochaete*, 1 species of *Cejpomyces*, 1 species of *Cerrena*, 1 species of *Cystostereum*, 1 species of *Diplomitoporus*, 1 species of *Earliella*, 1 species of *Fibriciellum*, 1 species of *Flavodon*, 16 species of *Gloeocystidiellum*, 1 species of *Grammothelopsis*, 4 species of *Hymenochaete*, 1 species of *Hymenogramme*, 1 species of *Hyphoderma*, 1 species of *Hyphodontia*, 4 species of *Hypochnicicum*, 1 species of *Junghuhnia*, 1 species of *Kavinia*, 1 species of *Laeticorticium*, 1 species of *Laxitextum*, 1 species of *Lepidomyces*, 1 species of *Leucogyrophana*, 1 species of *Lopharia Metulodontia*, 2 species of *Oxyporus*, 7 species of *Peniophora*, 2 species of *Perenniporia*, 7 species of *Phanerochaete*, 9 species of *Phellinus*, 2 species of *Phlebia*, 4 species of *Phlebiopsis*, 1 species of *Pseduoxnasma*, 1 species of *Pycnoporellus*, 1 species of *Ramaricium*, 1 species of *Rigidoporus*, 4 species of *Schizopora*, 2 species of *Scopuloides*, 1 species of *Trichaptum*, 1 species of *Tubulicrinis*, 1 species of *Tyromyces*, 1 species of *Vararia*, 1 species of *Vuilleminia*, 1 species of *Wolfiporia*, 4 species of *Wrightoporia*, 1 species of *Xylobolus* have been identified (Table 1).

Out of 110 species including; *Aleurodiscus aberrans*, *A. cremeus*, *Amylocorticium cebennense*, *A. olivaceoalbum*, *Amylocystis sericeomollis*, *Amylosporus bracei*, *Amylostereum laevigatum*, *Anomoporia dumontii*, *Auriporian aurulenta*, *Botryohypochnus anomalus*, *Candelabrochaete verruculosa*, *Cerrena unicolor*, *Cystostereum murraii*, *Diplomitoporus hondurensis*, *Earliella scabrosa*, *Flavodon flavous*, *Gloeocystidiellum fistulatum*, *G. flammeum*, *G. furfuraceum*, *G. insidiosum*, *G. irpiscescens*, *G. kenyense*, *G. lacticolor*, *G. luridum*, *G. luteocystidiatum*, *G. percuriosum*, *G. turpe*, *Grammothelopsis puiggarii*, *Hymenogramme javensis*, *Hyphodontia alienata*, *Hypochnicicum cymosum*, *H. eichlerie*, *H. globosum*, *Kavinia himantia*, *Laeticorticium simplicibasidium*, *Laxitextum lutescens*, *Lepidomyces subcalceus*, *Leucogyrophana mollis*, *Metulodontia flavidoalba*, *Oxyporus latamarginatus*, *Peniophora aurantiaca*, *P. gladiola*, *P. laurentii*, *P. pithya*, *P. gomezii*, *Phanerochaete calotricha*, *P. joseferroirae*, *P. tuberculata*, *Phlebiopsis galochroa*, *P. peniophoroides*, *Phlebia rufa*, *P. subcretacea*, *Phlebiopsis galochroa*, *P. peniophoroides*, *Phellinus cesatii*, *P. luctuosus*, *P. purpureogilva*, *P. punctatus*, *P. sonorae*, *Pseduoxnasma verrucisporum*, *Pycnoporellus alboluteus*, *Schizopora carneolutea*, *S. trichiliae*, *Scopuloides hydnoides*, *Scytinostromella cerina*, *Steccherinum setulosum*, *Steccherinum setulosum*, *Tubulicrinis ellipsoideus*, *Tyromyces undosus*, *Vuilleminia acerina*, *Wolfiporia cocos*, *Wrightoporia Africana*, *W. avellanea*, *W. cremea*, *W. iobaphus*, *Xylobolus apricans* and *X. frustulatus* are reported for the first time in India and to Maharashtra as well, in this investigation (Table 1).

Table 1. Table denoting-Poroidness, Habitat, Rot type, Amyloidness, Fruit body type, Season, New records to India and Maharashtra.

Name of the Fungus	Poroido		Host			Rot type		Amyloid test			Resupinate type			Season		NTI	NTM
	P	NP	LST	FBT	DST	WR	BR	AM	NAM	DEX	RSU	EF	EFR	RS	WS		
Aleurodiscus aberrans	–	●	–	●	–	●	–	●	–	–	●	–	–	–	●	●	–
Aleurodiscus cremeus	–	●	●	–	–	●	–	●	–	–	●	–	–	–	●	●	–
Amylocorticium cebennense	–	●	●	–	–	–	●	●	–	–	●	–	–	●	–	●	–
Amylocorticium olivaceoalbum	–	●	●	–	–	–	●	●	–	–	–	●	–	●	–	●	–
Amylocystis sericeomollis	●	–	●	–	–	–	●	●	–	–	●	–	–	●	–	●	–
Amylosporus bracei	●	–	–	–	●	–	●	●	–	–	–	●	–	–	●	●	–
Amylostereum laevigatum	–	●	–	–	●	–	●	●	–	–	●	–	–	–	●	●	–
Anomoporia dumontii	●	–	●	–	–	–	●	●	–	–	●	–	–	●	–	●	–
Auriporia aurulenta	●	–	●	–	–	–	●	–	●	–	●	–	–	–	●	●	–
Botryohypochnus anomalus	–	●	●	–	–	●	–	–	●	–	●	–	–	–	●	●	–
Candelabrochaete verruculosa	–	●	–	●	–	●	–	–	●	–	●	–	–	–	●	●	–
Cejpomyces terrigenus	–	●	●	–	–	●	–	–	●	–	●	–	–	●	–	–	–
Cerrena unicolor	●	–	–	●	–	●	–	–	●	–	●	–	–	●	–	●	–
Cystostereum murrayi	–	●	●	–	–	●	–	–	●	–	–	–	●	●	–	●	–

Diplomit-oporus hondurensis	●	–	–	●	–	●	–	–	●	–	–	●	–	–	●	●	–
Earliella scabrosa	●	–	–	●	–	●	–	–	●	–	–	–	●	–	●	●	–
Fibriciellum silvae-ryae	–	●	–	●	–	●	–	–	●	–	●	–	–	●	–	●	–
Flavodon flavus	●	–	–	●	●	●	–	–	●	–	–	–	●	●	–	–	●
Gloeocystidiel lum citrinum	–	●	●	–	–	●	–	●	–	–	●	–	–	●	–	–	–
Gloeocys-tidiellum convolvens	–	●	●	–	–	●	–	●	–	–	●	–	–	●	–	–	–
Gloeocy-stidiellum fistulatum	–	●	●	–	–	●	–	●	–	–	●	–	–	●	–	●	–
Gloeocyst-idiellum flammeum	–	●	●	–	–	●	–	●	–	–	●	–	–	●	–	●	–
Gloeocyst-idiellum furfuraceum	–	●	●	–	–	●	–	●	–	–	●	–	–	●	–	●	–
Gloeocyst-idiellum insidiosum	–	●	–	●	–	●	–	●	–	–	●	–	–	●	–	●	–
Gloeocysti-diellum irpiscescens	–	●	●	–	–	●	–	●	–	–	●	–	–	●	–	●	–
Gloeocyst-idiellum kenyense	–	●	●	●	–	●	–	●	–	–	●	–	–	●	–	●	–
Gloeocysti-diellum lactescens	–	●	●	–	–	●	–	●	–	–	●	–	–	●	–	–	–
Gloeocysti-diellum lacticolor	–	●	●	–	●	●	–	●	–	–	●	–	–	●	–	●	–

Gloeocystidiellum leucoxanthum	–	●	●	–	–	●	–	●	–	–	●	–	–	●	–	–	–
Gloeocystidiellum luridum	–	●	●	–	–	●	–	●	–	–	●	–	–	●	–	●	–
Gloeocystidiellum luteocystidiatum	–	●	●	–	–	●	–	●	–	–	●	–	–	●	–	●	–
Gloeocystidiellum percuriosum	–	●	●	–	–	●	–	●	–	–	●	–	–	●	–	●	–
Gloeocystidiellum sulcatum	–	●	●	–	–	●	–	●	–	–	●	–	–	●	–	–	–
Gloeocystidiellum turpe	–	●	●	–	–	●	–	●	–	–	●	–	–	●	–	●	–
Grammothelopsis puiggarii	●	–	–	●	–	●	–	–	–	●	●	–	–	●	–	●	–
Hymenochaete attenuata	–	●	–	●	–	●	–	–	●	–	–	–	●	●	–	–	–
Hymenochaete cinnamomea	–	●	–	●	–	●	–	–	●	–	●	–	–	●	–	–	–
Hymenochaete rubiginosa	–	●	–	–	●	●	–	–	●	–	–	–	●	●	–	–	–
Hymenochaete tabacina	–	●	–	●	–	●	–	–	●	–	–	–	●	●	–	–	–
Hymenogramme javensis	●	–	–	●	–	●	–	–	●	–	●	–	–	●	–	●	–
Hyphoderma radula	–	●	–	●	–	●	–	–	●	–	●	–	–	●	–	–	–
Hyphodontia alienata	–	●	●	–	–	●	–	–	●	–	●	–	–	–	●	●	–
Hypochnicium cymosum	–	●	●	–	–	●	–	–	●	–	●	–	–	–	●	●	–

Hypochnicium eichleri	–	●	–	●	–	●	–	–	●	–	●	–	–	●	–	●	–
Hypochnicum globosum	–	●	–	–	●	●	–	–	●	–	●	–	–	–	●	●	–
Junghuhnia crustacea	●	–	●	–	–	●	–	–	●	–	●	–	–	●	–	–	–
Kavinia himantia	–	●	●	–	–	–	●	–	●	–	–	●	–	●	–	●	–
Laeticorticium simplicibasidium	–	●	–	●	–	●	–	–	●	–	–	●	–	●	–	●	–
Laxitextum lutescens	–	●	●	–	–	●	–	–	●	–	●	–	–	●	–	●	–
Lepidomyces subcalceus	–	●	–	●	–	●	–	–	●	–	–	●	–	–	●	●	–
Leucogyrophana mollis	–	●	●	–	–	–	●	●	–	–	●	–	–	●	–	●	–
Lopharia fulva	–	●	●	●	●	●	–	–	●	–	–	–	●	–	●	–	–
Metulodontia flavidoalba	–	●	●	–	–	●	–	–	●	–	●	–	–	–	●	●	–
Oxyporus cervinogilvus	●	–	●	–	–	–	●	–	●	–	–	–	●	–	●	–	–
Oxyporus latemarginatus	●	–	–	●	–	–	●	–	●	–	●	–	–	–	●	●	–
Peniophora aurantiaca	–	●	–	●	–	●	–	–	●	–	●	–	–	●	–	●	–
Peniophora farinosa	–	●	●	–	–	●	–	–	●	–	●	–	–	●	–	●	–
Peniophora gladiola	–	●	●	–	–	●	–	–	●	–	●	–	–	●	–	●	–
Peniophora laurentii	–	●	–	●	●	●	–	–	●	–	●	–	–	●	–	●	–

Species																	
Peniophora nuda	–	●	●	–	–	●	–	–	●	–	●	–	–	●	–	–	–
Peniophora pithya	–	●	–	●	–	●	–	–	●	–	●	–	–	●	–	●	–
Peniophora violaceolivida	–	●	●	–	–	●	–	–	●	–	●	–	–	●	–	–	–
Perenniporia gomezii	●	–	●	–	–	●	–	–	–	●	●	–	–	–	●	●	–
Perenniporia medullapanis	●	–	●	–	–	●	–	–	–	●	–	●	–	–	●	–	–
Phanerochaete cacaina	–	●	●	–	–	●	–	–	●	–	–	●	–	●	–	–	–
Phanerochaete calotricha	–	●	–	–	●	●	–	–	●	–	●	–	–	●	–	●	–
Phanerochaete jose-ferreirae	–	●	–	●	–	●	–	–	●	–	●	–	–	●	–	●	–
Phanerochaete laevis	–	●	●	–	–	●	–	–	●	–	–	●	–	●	–	–	–
Phanerochaete sanguinea	–	●	●	–	–	●	–	–	●	–	●	–	–	●	–	●	–
Phanerochaete sordida	–	●	–	–	●	●	–	–	●	–	●	–	–	●	–	–	–
Phanerochaete tuberculata	–	●	●	–	–	●	–	–	●	–	–	●	–	●	–	●	–
Phellinus chryseus	●	–	–	–	●	●	–	–	●	–	●	–	–	–	●	–	–
Phellinus luctuosus	●	–	●	–	–	●	–	–	●	–	●	–	–	–	●	●	–
Phellinus purpureogilvus	●	–	●	–	–	●	–	–	●	–	●	–	–	–	●	–	●
Phellinus carteri	●	–	●	–	–	●	–	–	●	–	●	–	–	–	●	–	–

Phellinus cesatii	●	–	●	–	–	●	–	–	●	–	●	–	–	–	●	●	–
Phellinus inermis	●	–	●	–	–	●	–	–	●	–	–	●	–	–	●	–	–
Phellinus lamaensis	●	–	●	–	–	●	–	–	●	–	–	–	●	–	●	–	–
Phellinus punctatus	●	–	●	–	–	●	–	–	●	–	–	●	–	–	●	●	–
Phellinus sonorae	●	–	●	–	–	●	–	–	●	–	●	–	–	–	●	●	–
Phlebia rufa	–	●	–	–	●	●	–	–	●	–	●	–	–	–	●	–	●
Phlebia subcretacea	–	●	–	–	●	●	–	–	●	–	●	–	–	–	●	●	–
Phlebiopsis galochroa	–	●	–	●	–	●	–	–	●	–	●	–	–	–	●	●	–
Phlebiopsis gigantea	–	●	–	●	–	●	–	–	●	–	–	●	–	●	–	–	–
Phlebiopsis penioph-oroides	–	●	–	●	–	●	–	–	●	–	●	–	–	–	●	●	–
Phlebiopsis roumeguerei	–	●	●	–	–	●	–	–	●	–	–	●	–	–	●	–	–
Pseduox-enasma verrucis-porum	–	●	●	–	–	●	–	●	–	–	●	–	–	●	–	●	–
Pycnoporellus alboluteus	●	–	–	–	●	–	●	–	●	–	●	–	–	●	–	●	–
Ramaricium polypor-oideum	–	●	●	–	–	●	–	–	●	–	●	–	–	●	–	–	–
Rigidoporus ulmarius	●	–	●	–	–	●	–	–	●	–	–	–	●	●	–	–	–
Schizopora carneolutea	●	–	–	●	–	●	–	–	●	–	–	●	–	●	–	–	●

Schizopora paradoxa	●	–	–	●	–	●	–	–	●	–	●	–	–	●	–	–	–
Schizopora roseotigens	●	–	–	–	●	●	–	–	●	–	●	–	–	●	–	–	–
Schizopora trichiliae	●	–	●	–	–	●	–	–	●	–	–	–	●	●	–	–	●
Scopuloides hydnoides	–	●	–	●	–	●	–	–	●	–	–	–	●	●	–	●	–
Scopuloides rimosa	–	●	–	●	–	●	–	–	●	–	–	●	–	●	–	–	–
Scytinostromella cerina	–	●	–	●	–	●	–	●	–	–	●	–	–	–	●	●	–
Steccherinum setulosum	–	●	●	–	–	●	–	–	●	–	●	–	–	●	–	●	–
Trichaptum versatile	●	–	–	●	–	–	●	–	●	–	–	–	●	–	●	●	–
Tubulicrinis ellipsoideus	–	●	●	–	–	●	–	–	●	–	●	–	–	–	●	●	–
Tyromyces undosus	●	–	–	●	–	●	–	–	●	–	●	–	–	–	●	●	–
Vararia sphaericospora	–	●	–	–	●	–	●	–	●	–	●	–	–	●	–	–	–
Vuilleminia acerina	–	●	–	●	–	●	–	–	●	–	●	–	–	–	●	●	–
Wolfiporia cocos	●	–	–	–	●	–	●	–	●	–	●	–	–	●	–	●	–
Wrightoporia africana	●	–	●	–	–	●	–	●	–	–	●	–	–	●	–	●	–
Wrightoporia avellanea	●	–	●	–	–	●	–	●	–	–	●	–	–	●	–	●	–
Wrightoporia cremea	●	–	●	–	–	●	–	●	–	–	●	–	–	●	–	●	–
Wrightoporia iobapha	●	–	●	●	–	●	–	●	–	–	●	–	–	●	–	●	–

Xylobolus apricans	–	●	–	–	●	●	–	●	–	–	●	–	–	●	–	●	–
Xylobolus frustulatus	–	●	●	–	–	●	–	●	–	–	●	–	–	●	–	●	–

PO : Poroid, **NPO**: Non – Poroid, **LST** : Live Standing Tree, **FBT** : Fallen Branches of Tree, **DST** : Dead Standing Tree, **WR** : White Rot, **BR** : Brown Rot, **AM** : Amyloid, **NAM** : Non Amyloid, **DEX**: Dextrinoid, **RSU**: Rsupinate, **EF** : Effused, **EFR**: Effuso Reflexed, **RS**: Rainy Season, **WS**: Winter Season, **NTI**: New To India, **NTM**: New to Maharashtra.

It was observed that out of 110 reported species, 73 species were non-poroid and 37 species were poroid resupinate Aphyllophorales (Table 1).

As comparison, *Mangifera indica* Linn have demonstrated to be the most susceptible host, which harbors the highest number fungi. Where as the least number of fungi were found on *Adhatoda vasica* Nees. and *Albizzia lebbek* (L.) Benth. The mixed vegetation of *Artocarpus heterophyllus* Lam, *Azadirachta indica* A. Juss, *Bauhinia purpurea* L., *Ficus benganina* L., *Mangifera indica* Linn. and *Memecylon umbellatum* Burm. shows 46 species of investigation areas in which non-poroid members are dominant (27 non-poroid and 11 poroid). Resupinate Aphyllophorales occur almost in equal proportion; non-poroid members are more common on the plants which show either well developed bark or hard wood or under second layer canopy trees.

It is also confirmative that most of the hosts which is decaying or wounded in lower side of a standing tree shows luxuriant growth of non-poroid resupinate Aphyllophorales, mainly the species of *Hyphoderma, Phanerochaete, Laeticorticium* and *Gloeocystidiellum,* which shows white to pinkish-white resupinate basidiocarps. The observations made also conclude that species of *Hymenochaete, Peniophora, Phanerochaete* and *Phlebiopsis* occur more frequently on the fallen branches of second layer trees of bushes or climbers. The habitat of fungi can be grouped in three major categories namely live standing tree, dead standing tree and fallen branches (Table 1). It was observed that species belonging to *Aleurodiscus, Hymenochaete, Peniophora, Phlebiopsis* and *Schizopora* occurred mainly on the dead decaying branches, either attached to the tree or fallen on the forest floor. Species of *Gloeocystidiellum, Hymenochaete, Perenniporia, Phellinus* and *Wrightoporia* occurred on standing tree.

Among the members of resupinate Aphyllophorales, 85% are white rot fungi and the rest are brown rot fungi (Table 1). Species of *Amylocorticium, Amylocystis, Anomoporia, Oxyporus, Trichaptum* and *Vararia* exhibit brown rot type decay. The proportion of white rot fungi to brown rot fungi is not in concurrence with that of North American polypore worked out by Gilbertson and his co-workers (1979), as the area under investigation calls under typical southern tropical vegetation. To confirm the rot, Benzidine test was carried out and it was revealed that 15 species belongs to brown rot and 95 to white rot. Amongst white rot, majority of that species belonging to *Gloeocystidiellum, Hymenochaete, Peniophora, Phanerochaete, Phellinus* and *Wrightoporia* (Table 1).

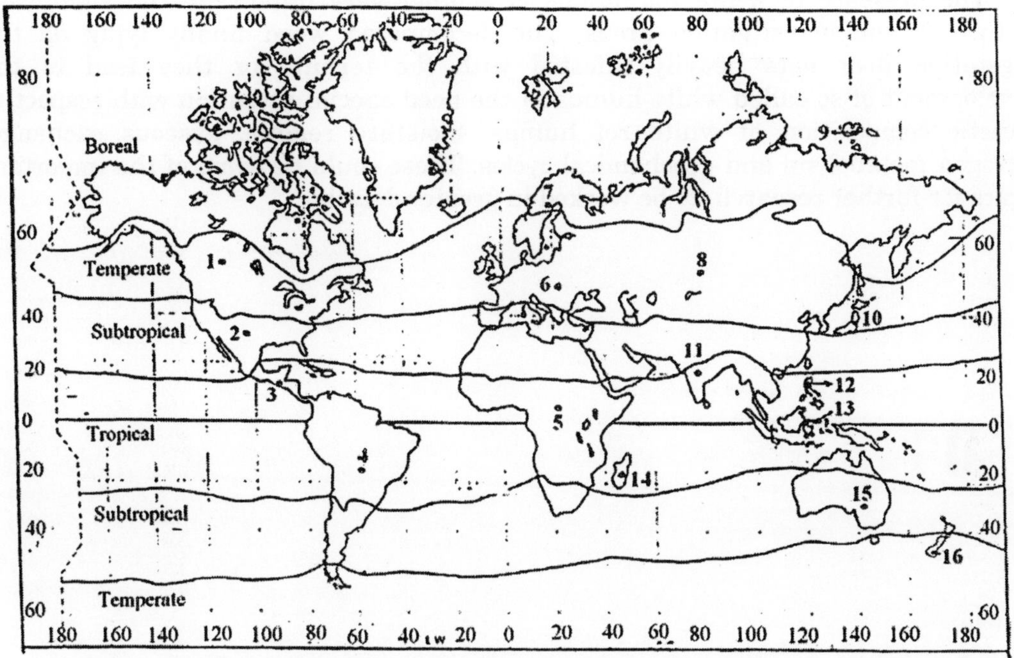

1) Canada 2) USA 3) Central America 4) South America 5) Africa 6) Europe 7) Finland 8) USSR 9) Siberia 10) Japan 11) India 12) The Philippines 13) East Indies 14) Madagascar 15) Australia 16) New Zealand

Fig. 115

Among the members of resupinate Aphyllophorales 74 species are non-amyloid (Species of *Hymenochaete, Hypochnicium, Oxyporus, Peniophora, Phanerochaete, Phell-inus, Rigidopus, Schizopora*, etc.), 33 species are amyloid (*Amylocorticium, Amylo-stereum, Gloeocystidiellum, Leucogyrophana, Wrightoporia, Xylobolus*, etc.) and 3 species are dextronoid (*Perenniporia* and *Grammothelopsis*) which is shown in Table 1.

It was observed that 81 identified species belong to resupinate basidiocarps (species of *Aleurodiscus, Amylocorticium, Amylostereum, Cejpomyces, Gloeocystid-iellum, Hymenochaeta, Peniophora, Phanerochaete, Phellinus, Vararia, Wrighto-poria*, etc.) 16 species belong to effused (species *of Diplomitoporus, Kavinia, Laeticorticium, Perenniporia, Perenniporia, Phanerochaeta, Phellinus, Phlebiopsis, Scopuloidesshow*) and 13 species belong to effuso-reflexed (Table 1). Another important observation shows species of *Gloeocystidiellum, Hymenochaeta, Peniophora, Phanerochaeta, Schizopora, Vararia, Wrightoporia* etc. occur in the rainy season. The members of *Aleurodiscus, Hypochnicicum, Oxyporus,*

Perenniporia, Phellinus, Phlebia, Phlebiopsis, Tyromyces, etc. appear in the winter season. (Table 1).

The presence of white rot humus is another very important and significant feature in the investigation areas. The decomposed wood finally lying on the vegetation floor gets heavily infested with the termite or they lead to the development of so called white humus in the need special attention with respect to floristic composition of white rot humus, moisture regime, gaseous exchange, nitrogen metabolism and geochemical cycles. These could be some of the important topics for further research, to be worked in greater details.

LITERATURE CITED

Alfredsen G., Solheim, H. and Jenssen K.M. (2005). Evaluation of decay fungi in Norwegian buildings. The international research group on wood protection. Bangalore, India. Pp.1-12.

Ames, A. (1913). A consideration of structure in relation to genera of the polyporaceae. *Annual of Mycology* 11: 211 -253.

Awasthi, D.D. (1965). Catalogue of the Lichens from India, Nepal, Pakistan and Ceylon. Weinheim, 1-137.

Bagchee, K.D. and Bakshi, B.K. (1950). Some fungi as wound parasite on Indian trees. *Indian Forester* 76: 244- 253.

Bagchee, K.D. and Bakshi, B.K. (1954). Studies on Indian Thelephoraceae 1. Some species of *Stereum*, *Peniophora* and *Corticium*. *Ind. for Bull,* 166: 1-2.

Bagchee, K.D., Puri, Y.N. and Bakshi, B.K. (1954). Principal diseases and decay of Oaks and other hardwood in India- II. *Indian Phytopathology* 7: 18- 42.

Bagchee, K.D., Singh, U. (1954). List of common names of fungi attacking Indian forest trees, timber and the herbaceous and shrubby undergrowths and list of cultures of forest fungi. *Indian Forest Records* 1: 199- 348.

Bakshi, B.K. (1958). New records of Hymenomycetes in India. *Indian Phytopathology* 11: 88.

Bakshi, B.K. (1963). Dangerous forest diseases in India, pp. 55-61. In Internationally dangerous forest tree diseases. U.S. Dep. Agric. Misc. Plbl. 939

Bakshi, B.K. (1971). Indian Polyporaceae (on trees and lumber). Indian Council of Agricultural Research: New Dehli. 246 pp.

Bakshi, B.K. and Singh, S. (1970). Heart rot in trees. Idt. Rev. *Forest Research* 3:197-251.

Bakshi, B.K., Sen, M. and Singh, B. (1970). Cultural diagnosis Indian Polyporaceae- II. Genera *Fomes* and *Trametes*. *Indian Forest Records* 2: 245- 276.

Bakshi, B.K., Singh, S. and Sing H.B. 1963. A re-examination of *Fomes lignosus* and *Polyporus zonalis*. *Trans Brit Mycol. Soc.* 46: 426-430.

Basham, J.A. (1975). Biotechnology and Bioengineering Symposium pp 9 -19 ed. By C.R. Wilkes.

Beneke, E.S. (1958). Laboratory manual of medical mycology V + 186 p., Pl.I. XIII.

Berkeley, M.J. (1839).Description of exotic fungi in the collection Sir W.J. Hooker – from memoirs and notes of J.F. Klotzsch, with additions and corrections. *Annuals of natural History* 3: 375- 401.

Bharucha, F.R. (1983). A text book of the plant Geography of India: Oxford University Press, India, 179 pp.

Bhosle, S.R., Ranadive, K.R., Bapat, G., Garad, S., Deshpande, G. and Vaidya, J. (2010). Taxonomy and Diversity of *Ganoderma* from the Western parts of Maharashtra (India). *Mycosphere* 1(3):249–262.

Boidin J., Lanquetin P. and Gilles G. (1997). Le genre *Gloeocystidiellum* sensu lato (Basidiomycotina). *Bull. Soc. Mycol. France* 113:1-80.

Boidin, J. (1951). Recherche de la tyrosinase et de la lacease chezles Basidiomycetes in culture pure. *Revue. Mycol.* 16: 173- 197.

Boidin, J. (1958 a). Essai biotaxonomique sur les Hydnés résupinés et les Corticiés. Rev. *Mycol. Mém. Hors-Sér.* 6: 1-388.

Boidin, J. (1958 b). Hetero basidiomycetes saprophytes et Homobasidiomycetes resupines-V. Essai surle genre Stereum Pers. ex. S.F. Gray, Revue. *Mycol.* 23; 318- 346.

Boidin, J. (1959 a). Ibid VI, Essai surle genre *Stereum*. Pers. ex. S.F. Gray. Revue. *Mycol.* 24: 197-225.

Boidin, J. (1959 b). Ibid VII, Essai surle genre *Stereum*. Pers. ex. S.F. Gray. Bull. Mens. Soc. Linn. Lyon. 28: 205- 222.

Boidin, J. (1966). Basidiomycètes Corticiaceae de la République Centrafricaine. 1–Le Genre *Gloeocystidiellum* Donk. *Cah. Maboké* 4: 5-17.

Boidin, J. and Gilles, G. (1986). Basidiomycetes Aphyllophorales de l'ile de la Réunion. I Introduction. II Les genres *Tubulicrinis*, *Tubulicium* et *Litschauerella*. III Aleurodiscoideae a spores amyloides. IV Les genres *Epithele* (Pat.) Pat. et *Pteridomyces*. V Famille des Ceratobasidiaceae Martin et genre *Suillosporium* Pouzar. VI Le genre *Cerinomyces* Martin., *Bull Trimestriel Soc. Mycol. France* 102: 273-319.

Boidin, J. and Gilles, G. (2000). Basidiomycetes Aphyllophorales de l'Ile de La Reunion XXI-suite. *Mycotaxon* 75: 357-387.

Boidin, J. and Languetin, P. (1975). *Vararia* subgenus *Vararia* especies d Afrique intertropicale. *Bull. Soc. Mycol. France*. 91:457-513.

Bondarzew, A.S. and Singer, R. (1941). Zur Systematik der Polyporaceen. *Annals of Mycology* 39: 43-65.

Boquiren, D.T. (1971). The genus *Epithele*, *Mycologia* 63: 937-957.

Bose, S.R. (1919-1928). Description of fungi in Bengal. Proceedings of the Indian Association for the Cultivation of Science. 4: 109.

Bose, S.R. (1944). Journ. Indian. Bot. Soc. 23: 153- 157.

Bourdot, H. and Galzin, A. (1928). Contributions a la Flore Mycologique de France I. Hymenomycetes de France. Heterobasidies - Homobasidies - Gymnocarpes. M. Bry, Sceaux, 761 pp.

Bourdt, H. and Galzin, A. (1911). Hyménomycètes de France (III - Corticiés: Corticium, Epithele, Asterostromella) (Groupe Botryodea). *Bull. Soc. Myc. Fr.* 27: 22 266.

Bresadola, G. (1897). Hymenomycetes hungarici Kmetiani. Atti. Imp. *Regia Accad. Rovereto* 3(3): 66-120.

Buchanan, P.K. and Hood, I.A. (1992). New species and new records of Aphyllophorales (Basidiomycetes) from New Zealand, *New Zealand J. Bot.* 30: 95-112.

Buchanan, P.K. and Ryvarden, L. (1993). Type studies in the Polyporaceae 24. Species described by Cleland, Rodway and Cheel, *Austral. Syst. Bot.* 6: 215-235.

Buchanan, P.K. and Ryvarden, L. (1998). New Zealand polypore fungi, (Aphyllophorales): Three new species and a new record, *New Zealand J. Bot.* 36: 219-231.

Buchanan, P.K. and Ryvarden, L. (2000a). New Zealand polypore fungi: six new species and a redetermination, *New Zealand J. Bot.* 38: 251-263.

Buchanan, P.K. and Ryvarden, L. (2000b). An annotated checklist of polypore and polypore-like fungi recorded from New Zealand, *New Zealand J. Bot.* 38: 265-323.

Buller, (1922). Researches of Fungi, Vol. I. Longmans.

Buller, A.H.R. (1924). Researches of Fungi (3) Longmans, Green et co edit. , London, 611 pp.

Burdsall, H.H. (1985). A contribution to the taxonomy of the genus *Phanerochaete*, *Mycol. Mem.* 10: 1-165.

Burt, E.A. (1918) Corticiums causing *Pellicularia* disease of the coffee plant, hypochnose of pomaceous fruits and *Rhizoctonia* disease. *Annals of the Missouri Botanical Garden* 5: 199-132.

Burt, E.A. (1920). The Thelephoraceae of North America. XII. *Stereum*. *Annals of the Missouri Botanical Garden* 7 (2-3): 81-248.

Burt, E.A. (1924). The Thelephoraceae of North America. *Annals of the Missouri Botanical Garden* 11: 1-35.

Burt, E.A. (1926). The Thelephoraceae of North America. XV. *Corticium*. *Annals of the Missouri Botanical Garden* 13 (3): 173-354.

Butler, E.J. (1905 a). Some Indian Forest fungi. Indian Forester 31: 487-494.

Butler, E.J. (1905 b). Some Indian Forest fungi. Indian Forester 31: 548- 556.

Butler, E.J. (1905 c). Some Indian Forest fungi. Indian Forester 31: 611- 617.

Butler, E.J. (1905 d). Some Indian Forest fungi. Indian Forester 31: 670- 679.

Butler, E.J. (1918). Fungi and diseases in plants. Thacker Spink and Company, Calcutta VI 547 pp.

Butler, E.J. and Bisby, G.R. (1931). The fungi of India. Scientific monograph Number 1. The Imperial Council of Agricultural Research: Calcutta. 237 pp.

Cartwright, K. and Findlay .W.P.K. (1958) Decay of timber and its preservation. Her Majestry stationary office. London.

Champion, H.G. and Seth, S.K. (1968). Revised survey of the forest types of India. Government of India publication, New Dehli.

Chamuris, G.P. (1988).The nonstipitate stereoid fungi in the northeastern United States and adjacent Canada. Mycologia Memoir 14. Berlin, Stuttgart: J. Cramer. 247 pp.

Chang, T.T. and Chou, W.N. (2004). *Antroida cinnamomea* reconsidered and *A. salmonea* sp.nov. on *Cunninghamia konishii* in Taiwa. *Bot. Bull. Acad. Sin.* 45:347-352.

Christiansen, M.P. (1960). Danish resupinate fungi. Part II. Homobasidiomycetes. *Dansk Bot. Ark.* 19: 57-388.

Coner, E.J.H. (1932).The fruit body of *Polystictus xanthopus*. Fr. *Annals of Botany* 46: 71- 111.

Cooke, M.C. (1886). Praecursors of monographia polypororum. Grew 15: 50- 56.

Cooke, M.C. (1891 a). *Trametes* and its allies. Grev. 19: 92-93.

Cooke, W.A. and Rayner, A.D.M. (1984). Ecology of saprophytic fungi. London and New York: Longman.

Corner, E.J.H. (1932a). A *Fomes* with two systems of hyphae. Transactions of the British Mycological Society 17: 51- 81.

Corner, E.J.H. (1947b). Variation in the size and shape of spores, basidia and cystidia in Basidiomycetes. *New Phytologist* 46: 195 -228.

Corner, E.J.H. (1948). *Asterodon* a due to the morphology of the fungus fruit body, with notes on Asterostromella; *Trans. Br. Mycol. Soc.* 31: 234- 245.

Corner, E.J.H. (1950). A monograph of *Clavaria* and allied genera -Ann. Bot. Memoirs 1: 740.

Corner, E.J.H. (1953). The construction of polypores I. Introduction: *Polyporus sulphureus, P. squamosus, P. betulinus* and *Polystictus microcyclus*. *Phytomorphology* 3: 152-167.

Corner, E.J.H. (1987). Ad Polyporaceas IV, Nova Hedwigia, Beih. 86: 256Pp.

Corner, E.J.H. (1989). Ad Polyporaceas VI, Nova Hedwigia, Beih. 97: 197Pp.

Cowling, E.B. (1961). Comparative biochemistry of the decay of Sweetgum sapwood by white rot and brown rot fungi. United State Department of Agriculture Technical Bulletin 1258 79 p.

Cunningham, G.H. (1945). Hyphal system as aids in identification of species and genera of polyporaceae. Trans. Br. *Mycol. Soc.* 37: 47- 50.

Cunningham, G.H. (1946). Notes on classification of the Polyporaceae. *New Zealand Journal of Science and Technology* A 28: 238- 251.

Cunningham, G.H. (1947). New Zealand Polyporaceae.I. The genus *Poria* . DSIR Plant Diseases Division Bulletin 72.

Cunningham, G.H. (1947). Notes on classification of the polyporaceae. *New Zealand Journal of Science and Technology* A 28: 238- 251.

Cunningham, G.H. (1948). New Zealand Polyporaceae. The genus Fomes. *Plant Disease Division Bulletin* 79: 24.

Cunningham, G.H. (1949). The genus *Irpex*. Plant. Dis. Div. Bull. 82: 8.

Cunningham, G.H. (1955). Thelephoraceae of New Zealand Part IV. The genus *Peniophora* op. cit. Pp. 247 -293.

Cunningham, G.H. (1955). Thelephoraceae of New Zealand Part IV. The genus *Vararia*. Trans. *Roy. Soc. N.Z.* 82 (5): 973 -985.

Cunningham, G.H. (1956). Thelephoraceae of New Zealand. Parts IX, X and XI. Trans. *Roy. Soc. New Zealand* 84: 201-268.

Cunningham, G.H. (1959). The genus *Odontia*, Trans. Roy Soc. N.Z. 86:65-103.

Cunningham, G.H. (1963). Thelephoraceae of Australia and New Zealand. D.S.T.R. New Zealand Bull. 145- 389.

Cunningham, G.H. (1965). Polyporaceae of New Zealand. Bull. *New Zealand Dept. Sci. Industr. Res.* 164:1-304.

Dai, Y.C., Wei, Y.L. and Zhang, X.Q. (2004). An annotated checklist of non-poroid Aphyllophorales in China. *Ann. Bot. Fennici* 41: 233–246.

Dai, Y.C., Wu, S.H. and Chou W.N. (2002). Two new polypores (Basidiomycota) from Taiwan. *Mycotaxon* 83: 209-216.

Dai, Y.C., Yu, C.J. and Wang, H.C. (2007) Polypores from eastern Xizang (Tibet), Western China. *Ann. Bot. Fennici* 44: 135–145.

Dai, Yu-Cheng and Bao-Kai Cui (2006). Two new species of *Wrightoporia* (Basidiomycota, *Aphyllophorales*) from southern China. *Mycotaxon* 96: 199 - 206.

David, A. (1967). Caracteres myceliens de quelques *Trametes* (Polyporaceae). *Naturaliste Can.* 94:557-572.

Davison, R.W., Campbell, W.A. and Blaisdell, D.J. (1938). Differentiation of wood decaying fungi by their reactions on gallic or tannic acid medium. *Journal of Agricultural Research* 57: 683-695.

Davison, R.W., Campbell, W.A. and Vaughan, D.B. (1942). Fungi causing decay of living oaks in the eastern United States and their cultural identification. *Tech. Bull. U.S. Dep. Agric., Washington.* 785: 65.

Decock, C. (2001). Studies in *Perenniporia*. Some Southeast Asian Taxa Revisited. *Mycological Research*. 93: 774-795

Decock, C. and Herrera Figueroa, S. (2000). Studies in *Perennipora*. *Navisporus ortizii*, a synonym of *Perenniporia martius*, and a note on *Navisporus* and *Perenniporia*in Cuba, *Cryptog. Mycol*. 21: 153-162.

Decock, C. and Ryvarden, L. (1997). *Phellinus anchietanus* (Basidiomycetes, sp. nov. from Brazil. *Cryptogamie Mycologie*. 18(3): 221-225.

Decock, C. and Ryvarden, L. (1998). *Microporellus straminellus* comb. nov. and a note on *Perenniporia stipitata*. Cryptogamie Mycologie 19

Decock, C. and Ryvarden, L. (1999). Studies in neotropical polypores. 1. Some colored resupinate species of *Perenniporia*. *Mycological Research* 103: 1138-1144.

Decock, C. and Ryvarden, L. (1999). Studies in *Perenniporia*: *Perenniporia detrita* and its taxonomic synonyms, *Mycologia* 91: 386-395.

Decock, C. and Ryvarden, L. (2000). Studies in neotropical polypores. 6. New resupinate *Perenniporia* species with small pores and small basidiospores. *Mycologia* 92(2): 354-360.

Decock, C. and Ryvarden, L. (2003). *Perenniporiella* gen. nov. segregated from *Perenniporia*, including a key to neotropical Perenniporia species with pileate basidiomes. *Mycol. Res.* 107: 93-103.

Decock, C. and Stalpers, J.A. (2006). Studies in *Perenniporia*: *Polyporus unitus*, *Boletus medulla-panis*, the nomenclature of *Perenniporia*, *Poria* and *Physisporus*, and a note on European *Perenniporia* with a resupinate basidiome, Taxon, Volume 55, Number 3, August 2006 , pp. 759-778 (20).

Decock, C., Buchanan, P.K. and Ryvarden, L. (2000). Revision of some Australasian taxa of *Perenniporia* (Basidiomycetes, Aphyllophorales), *Austral. Syst. Bot.* 13: 823-844.

Domanski, S. (1965). Grazyby (Zagwiowate I, Szczeciniakowate I). Panstwowe Wydawnic two Naukowe, 278 pp. (1972 - Fungi: Polyporaceae resupinate), Mucronoporaceace I (resupinate) U.S. Dep. Commerce, Springfield. 235 pp..

Donk, M.A. (1930). Revisie de Niederlondischen. Homobasidiomycetes en Homobasidiomycetae - Aphyllophoraceae. I, Meded. Ned. Mycol. Ver. 18-20, 68-200.

Donk, M.A. (1933). Revisie de Niederlondischen. Homobasidiomycetes Aphyllophorales II. Meded. Bot. Mus. Herb. Rijksuniv. *Utrecht* 9: 1 - 278.

Donk, M.A. (1954). Notes on resupinate Hymenomycetes I. On *Pellicularia* Cooke. *Reinwardtia* 3: 363-379.

Donk, M.A. (1956b). Notes on resupinate Hymenomycetes III. *Fungus* 26: 3 -24.

Donk, M.A. (1957). Notes on resupinate Hymenomycetes IV. *Fungus* 27: 1 -29

Donk, M.A. (1958). Notes on resupinate Hymenomycetes V. *Fungus* 28: 16 -36.

Donk, M.A. (1960). The generic names proposed for Polyporaceae. *Persoonia* 1: 173-302.

Donk, M.A. (1964). A conspectus of the families of Aphyllophorales. *Persoonia* 3:199-324.

Donk, M.A. (1967). Notes on European Polypores - II. Notes on *Poria*, *Persoonia* 5: 47-130.

Donk, M.A. (1971). Multiple convergences in the Polyporaceous Fungi. Evolution in the higher Basidiomycetes. Knoxville Editted by Peterson, R.H. 393-422.

Donk, M.A. (1973). Check list of European polypores. North- Holland Publishing Company, Amsterdam - London, 469 pp.

Donk, M.A. (1973). Notes on European Polypores -XII. In Proc. K. Ned. Akad. Wet. (c).

Donk, M.A. (1974). Check list of European polypores Verh. Koninkl. Nederlandsche Akad. Wetensch. Afd. Natuurk, *Tweede Reeks* 62: 1-469.

Donk, M.A. (1974a). Check list of European hymenomycetous Heterobasidiae. Supplements and corrections, *Persoonia* 8: 33-50.

Edward Angus Burt (1918). "The Thelephoraceae of North America: X. Hymenochaete" in *Annals of the Missouri Botanical Garden* 5: 301 – 372.

Edward Angus Burt (1920). "The Thelephoraceae of North America: XII. Stereum" in *Annals of the Missouri Botanical Garden* 7: 81 – 248.

Edward Angus Burt (1924). "The Thelephoraceae of North America: XIII. Cladoderris, Hypolyssus, Cymatella, Skepperia, Cytidia, Solenia, Matruchotia, Microstroma, Protocoronospora, and Asterostroma" in Annals of the Missouri Botanical Garden 11: 1 – 36.

Edward Angus Burt (1926). "The Thelephoraceae of North America: XV. *Corticium* and supplement to the whole series" in *Annals of the Missouri Botanical Garden* 13: 173 – 354.

Erikson, J. (1950). *Peniophora* Cke Sect. Coloratae Bourd. and Galz. A taxonomical study with special reference to the Swedish species. *Symb. Bot.* Upsal. 10: 176.

Erikson, J. (1958). Studies in the Heterobasidiomycetes and Homobasidiomycetes Aphyllophorales and Muduus National park in North Sweden. *Symb. Bot. Upsaliensis* 16: 1- 172.

Erikson, J. and Ryvarden L. (1973). The Corticiaceae of North Europe 2: 59-286. *Aleurodiscus - Confertobasidium*. Fungiflora Oslo.

Erikson, J. and Ryvarden L. (1975). The Corticiaceae of North Europe 3:287-546. *Coronicium - Hyphoderma*. Fungiflora Oslo.

Erikson, J. and Ryvarden L. (1976). The Corticiaceae of North Europe 4: 547-886. *Hyphodermella - Mycoacia*. Fungiflora Oslo.

Erikson, J., Hjortstam, K. and Ryvarden L. (1978). The Corticiaceae of North Europe Oslo. pp. 889-1047.

Farr, M.L. (1973). An annotated list of Spegazzini's fungus taxa, *Biblioth. Mycol.* 35: 1-1661.

Flalck, R. (1926) Uber Koroslve and biologische Bedeutung. *Ber. dt. bot.Ges.* 44: 652-664.

Fries, E.M. (1821). Systema mycologium, 1: 326- 393.

Fries, E.M. (1828). Elenchus Fungorum 1: 1-238.

Fries, E.M. (1849). Sumwa Veg. scand 2: 319 -321.

Fritz, C.W. (1923). Cultural criteria for the distinction of wood -destroying fungi. *Tran. R. Can. Inst.* 17: 191-288.

Ganesh, P.N. and Leelavathy, K.M. (1986). New records of *Phellinus* from India. Current Science 55: 727- 728.

Gilbertson, R.L. (1977-78). Synopsis of wood rotting fungi on spruce in North America I and II. *Mycotaxon* 6: 43-77 and 7: 337- 356.

Gilbertson, R.L. (1979). The genus *Phellinus* (Aphyllophorales: Hymenochaetaceae) in Western North America. *Mycotaxon* 9: 51- 89.

Gilbertson, R.L. (1980). Wood -Rotting Fungi of North America. Mycologia 72: 1-49.

Gilbertson, R.L. and Adaskaveg, J.E. (1993). Studies on wood-rotting basidiomycetes of Hawaii. *Mycotaxon* 49: 369-397.

Gilbertson, R.L. and Ryvarden, L. (1986). North American Polyporaceae Vol. I. Fungiflora - Oslo - Norway. 433.

Gilbertson, R.L. and Ryvarden, L. (1987). North American Polyporaceae Vol. II. Fungiflora - Oslo - Norway. 435- 885.

Ginns, J. (1976). *Merulius:* s.s. and s.l., taxonomic disposition and identification of species. Canad. J. Bot. 54: 100-167.

Ginns, J. (1982). A monograph of the genus *Coniophora* (Aphyllophorales, Basidiomycetes) *Opera Botanica* 61: 1-615.

Ginns, J. and Freeman, G.W. (1994). The Gloeocystidiellaceae (Basidiomycota, Hericiales) of North America, *Biblioth. Mycol.* 157: 1-118.

Greslebin A.G. and Rajchenberg, M. (1997). Corticioid Aphyllophorales (Basidiomycota) from the Patagonian Andes forests of Argentina. 1. Lachnocladiaceae on *Nothofagus pumilio,* 65: 197-203.

Greslebin A.G. and Rajchenberg, M. (1997). Corticioid Aphyllophorales (Basidiomycota) from the Patagonian Andes forests of Argentina. 2. *Hyphodontia decorticans* sp. nov., 65: 205-209.

Greslebin A.G. and Rajchenberg, M. (1998). Corticioid Aphyllophorales (Basidiomycota) from the Patagonian Andes forests of Argentina 3. The genus *Dendrothele,* 67: 469-486.

Greslebin A.G. and Rajchenberg, M. (1999). Corticioid Aphyllophorales (Basidiomycota) from the Patagonian Andes forests of Argentina 4. *Nothocorticium patagonicum gen. et sp.nov.,* 70: 371-375.

Greslebin A.G. and Rajchenberg, M. (1999). Corticioid Aphyllophorales (Basidiomycota) from the Patagonian Andes forests of Argentina 5. *Some new taxa,* 73: 9-17.

Hallenberg N., (1980). New taxa of Corticiaceae from N. Iran (Basidiomycetes). *Mycotaxon* 11:447-475.

Hallenberg, N. (1981). *Phlebia centrifuga* Karst. (Corticiaceae, Basidiomycetes) – compatibility between specimens from Sweden and Canada. Göteborgs Svampklubb Arsskrift. Seite Pp33-37.

Hallenberg, N. (1983). On the *Schizopora paradoxa* complex (Basidiomycetes). *Mycotaxon.* 18 (2): 303-313.

Harkin, J.M. and Obst, J.R. (1973). Syringaldazine, an effective reagent for detecting laccase and peroxidase in fungi. *Experientia* 29: 381-387.

Harsh, N.S.K. (1982). Studies on wood-decaying fungi of Kumaun hills. Ph.D. Thesis, Department of Botany, Government Science College, Jabalpur.

Hawksworth, D.L., Sutton, B.C. and Ainsworth, G.C. (1983), *Ainsworth and Bisby's Dictionary of Fungi.* 7th ed. Commonwealth Mycological Institute, Kew.

Hennings, P. (1901). Fungi India Orientalis –II. CI. W. Gollan. 1900 Collecti. *Hedwigia* 40:323-342.

Hintikka, V. (1970). Studies on white rot humus formed by higher fungi in forest soils. *Metsant Julk* 67: 1-68.

Hjortstam, K. (1973). Studies in the Corticiaceae (Basidiomycetes) and related fungi of Vastergoland in South West Sweden I Svensk *Bot. Tidskr.* 67: 97-126.

Hjortstam, K. (1980). Notes on Corticiaceae VII. A synopsis of the genus *Amylocorticium* Pouz. *Mycotaxon* 11: 430- 434.

Hjortstam, K. (1983). Studies in tropical Corticiaceae V. Specimen from East Africa, collected by L. Ryvarden. *Mycotaxon* 9: 505- 519.

Hjortstam, K. and Larsson, K.H. (1977). Notes on Corticiaceae (Basidiomycetes). *Mycotaxon* 5: 475-480

Hjortstam, K. and Ryvarden, L. (1979). Notes on Corticiaceae 4., *Mycotaxon* 9(2): 505-519.

Hjortstam, K. and Ryvarden, L. (1980). *Ibid.* 12: 168-184.

Hjortstam, K. and Ryvarden, L. (1980). Studies in tropical Corticiaceae I. *Mycotaxon* 10: 269- 287.

Hjortstam, K. and Ryvarden, L. (1982). Aphyllophorales from Northern Thailand. *Nordic Journal of Botany* 2: 273- 281.

Hjortstam, K. and Ryvarden, L. (1984). Some new and noteworthy Basidiomycetes (Aphyllophorales) from Nepal, *Mycotaxon* 20: 133 - 151.

Hjortstam, K. and Ryvarden, L. (1985). Some new and noteworthy Basidiomycetes (Aphyllophorales) from Tierra Del Fuego, Argentine. *Mycotaxon* 22: 159 167.

Hjortstam, K. and Ryvarden, L. (1988). *Tomentellago* gen. nov. (Thelephoraceae, Basidiomycetes). *Mycotaxon* 31: 39- 43.

Hjortstam, K. and Ryvarden. L. (1987). A note on *Anomoporia* Pouz. (Polyporaceae, Aphyllophorales). *Mycotaxon* 28(2): 553-555.

Hjortstam, K. and Staplers, J.A. (1982). Notes on Corticiaceae XI. *Boidina*, a new genus segregated from *Gloeocystidiellum*. *Mycotaxon* 14: 75-181.

Hjortstam, K., Telleria, M.T., Ryvarden, L. and Calonge, F.D. (1981). Notes on Aphyllophorales of Spain II. *Nova Hedw.* 34: 525-538.

Hjortstam, K., Tellería, M.T., Ryvarden, L. and Calonge, F.D. (1981) Notes on the Aphyllophorales of Spain. II. *Nova Hedwigia* 34(3-4): 525-538.

Jackson, H.S. (1948). Studies of Canadian Thelephoraceae II. Some new. species of.*Corticium*. *Canad. J. Res,, C,* 26: 143-157.

Jagdale, R.P. (1994). Ecology of Bhimashankar Forest Western Ghats, Maharashtra state. Ph. D. thesis, department of Botany, University of Poona, Pune.

Job, D.J. (1990). Le genre *Hymenochaete* dans les zones tempérées de l'hémisphère sud, *Mycol. Helvetica* 4: 1-51.

Jülich, W. (1972). Monographie der Athelieae (Corticiaceae, Basidiomycetes). *Willdenowia Beih.* 7:1-283.

Julich, W. (1984). Basidiomycetes of South- East Asia. *Persoonia* 12: 107-117.

Jülich, W. and J.A. Stalpers (1980). The resupinate Non-poroid Aphyllophorales of the Temperate Northern Hemisphere. North-Holland Publishing Company, Amsterdam, Oxford, New York.

Kaarik, A. (1965). The identification of the mycelia of wood - decay fungi by theiroxidation reaction with phenolic compounds. *Studia Forestalia Suecia* 31:1 -38.

Karsten, P.A. (1881). Eumeratia Boletinearum et. *Polyporum fennicarm*, Systemate novo dispostorum. Rev. *Mycol.* 36: 15-23.

Karsten, P.A. (1889). Kritisk of oversight of Finlands Basisvompar (Basidiomycetes:Gastero and Hymenomycetes) Bider. *Kann. Finl. Nat. Och. Folk* 48: 1 -470.

Khara, S. (1978). Some stipitate Hydnums form North Western Himalayas. *Indian journal of mycology and Plant Pathology* 7 (2): 127 -134.

Khara, S. (1978). The Hydnaceae of North Western Himalayas V. *Indian Phytopath* 30: (1), 94-98.

Khundalker and Patil (1985). Proc. 72nd Indian Sci. Congr. Abstr; p. 81.

Khundalker (1984). Proc. 71st Indian Sci. Congr. Abstr; p. 20.

Kirk, P.M., Cannon P.F., David, J.C. and Stalpers J.A. (2001). Ainsworth and Bisbs Dictionary of the fungi,9th ed. CABI International , Wallingford, 655 pp.

Kotlaba, F. and Pouzar, Z. (1978). Notes on *Phellinus rimosus* complex (Hymenochaeta ceae). Acta. *Bot. Croat.* 37: 171-182.

Krishnamurthy, P. and Udas, G.R. (1981). Regional geochemical characters of Deccan trap lavas and their genetic implications in subbarao, K.V. and Snkheswala, R.N. (ed.) Deccan Volcanism G. Society India, Memoir No. 3, Banglore.

Kulkarni, B.G. (1979). Flora of South Ratnagiri. Unpublished Ph.D. thesis submitted to Univ. of Poona, pp. 248.

Larsen, M.J. and Cobb-Poulle, L.A. (1990). *Phellinus* (Hymenochaetaceae) A survey of the world taxa [Synopsis Fungorum 3]. Fungiflora, Oslo.

Léger, J.C. (1998). Le genre *Hymenochaete* Léveillé. Biblioth. *Mycol.* 171: 1-319.

Lemke, P.A. (1964a). The genus *Aleurodiscus* S. str. In North America *Can. J. Bot.* 42: 213- 285.

Lemke, P.A. (1964b). The genus *Aleurodiscus* Str. In North America *ibid* 42: 723- 768.

Lentz, P.L. (1955). *Stereum* and allies genera of fungi in the upper Mississippi valley. U.S. Dep., Agric. *Mon.* 24: 1-74.

Liberta, A.E. (1973). The genus *Trechispora* (Basidiomycetes, Corticiaceae) *Can. J. Bot.* 51: 1871-1892.

Lindsey, J.P. and Gilbertson, R.L. (1977). New species of corticioid fungi on quaking aspen, *Mycotaxon* 5(1): 312-319.

Linneaous, C. (1753). Species Plantarum II.

Lloyd, C.G. (1898-1925). Mycological Notes nos. 1-75:1-1346. Cincinnati, Ohio.

Loguercio-Leite, C. and Wright, J.E. (1995). The genus *Phellinus* (Hymenochaetaceae) on the island of Santa Catarina, Brazil. *Mycotaxon* 54: 361-388.

Lohwag, Kurt (1940). Zur Anatomie des Deckgeflechtes der Polyporaceen. *Ann. Mycol. Berl.* 38 (5, 6): 401- 452.

Lombard, F.F. and Gilbertson, R.L. (1965). Studies on some western Porias with negative or weak oxidase reactions. *Mycologia* 57: 43-76.

Lombard, F.F. and Gilbertson, R.L. (1966). *Poria luteo-alba* and some related species in North-America. *Mycologia* 58: 827- 845.

Long, W.H. and Harsch, R.M. (1918). Pure cultures of wood -rotting fungi on artificial media. *Journal of Agriculture Research* 12: 33-82.

Lowe, J.L. (1957). Polyporaceae of North America. The genus *Fomes*. Technical Publication No. 80, 97 P. New-York State College of Forestry.

Lowe, J.L. (1963a). The Polyporaceae of the world, *Mycologia* 55: 1-12.

Lowe, J.L. (1963b). A synopsis of *Poria* and similar fungi from the tropical regions of the world, *Mycologia* 55: 453-486.

Lowe, J.L. (1966). Polyporaceae of North America. The genus *Poria*. State University of New York Coll. For. *Tech. Publ.* 90:1-183.

Lowe, J.L. (1975). Polyporaceae of North America. The genus *Tyromyces. Mycotaxon*.2: 1-82.

Maas Geesteranus, R.A. (1971). Hydnaceous fungi of the eastern old world, *Verh. Kon. Ned. Akad. Wetensch., Afd. Naturk., Tweede Reeks* 60 (3): 1-175.

Maekawa, N. (1993). Taxonomic study of Japanese Corticiaceae. (Aphyllophorales) I.Rep Tottori Mycol Inst 31:1–149.

Maekawa, N. (1994). Taxonomic study of Japanese Corticiaceae. (Aphyllophorales) II.Rep Tottori Mycol Inst 32:1–125.

Maekawa, N., Arita, I. and Hayashi, Y. (1982). Corticiaceae in Japan. Rep Tottori Myc Inst. 20: 33-41

Mahabale, T.S. (1987). Botany and Flora of Maharashtra. Maharashtra State Gazetteers, Government of Maharashtra, Bombay, 873 pp.

Massee, G. (1901). Fungi Exotici III. *Kew. Bull.* 150-164

Massee, G. (1906). Fungi Exotici, IV, *Bull. Misc. Inform.*91-94.

Massee, G. (1908).Fungi exotici VIII . Kew Bulletin.216-219.

Massee, G. (1910). Fungi exotici X . Kew Bulletin.249-253.

Meherhomji, V.M. (1977). History of Deciduous Forests of Western India and Central India in "Ecological and Archeology of Western India". Concept Publishing Co., New Delhi.

Montagne, J.F.C. (1842). Cryptogamae Nilgherensis. Ann. Sci. Nat. Nat. Ser. II, 18: 12-23.

Murrill, W.A. (1902-1905). The Polyporaceae of North America IXIII. *Torrey Botany Club Vols.* pp 29-32.

Murrill, W.A. (1903). A historical review of the genera of Polyporaceae. Journal of *Mycology* 9: 87-102.

Murrill, W.A. (1907). (Agaricales) Polyporaceae- North American Flora 9: 1-72.

Murrill, W.A. (1915). Tropical Polypores. 113 pp. New-York. (Published by author).

Naik-Vaidya, C.D. (1990). Wood Rotting Fungi from Karnala and Kankeshwar, Ph.D. Thesis, Department of Botany. University of Pune. pp. 1-283.

Nakasone, K.K. (2003). Type studies of resupinatehydnaceous Hymenomycetes described by Patouillard, Cryptogamie. *Mycologie.* 24:131-145.

Nakasone, K.K. (1990). Cultural studies and identification of wood-inhabiting Corticiaceae and selected Hymenomycetes from North America. Mycologia Mem. 15: 412 pp.

Nakasone, K.K. and Burdsall, H.H. Jr .(1995). *Phlebia* species from eastern and southeastern United States. *Mycotaxon* 54: 335-359.

Nanda, M.K. (1996). Wood Rotting Fungi from Bhimashankar Ph.D. thesis, Department of Botany. University of Pune. pp 1-397.

Natarajan, K. and Kolandavelu, K. (1985). Resupinate Aphyllophorales from South India -I. Kavaka 12: 71-76.

Natarajan, K. and Kolandavelu, K. (1998). Resupinate Aphyllophorales of Tamilnadu, India, University of Madras. pp 1-133.

Natarajan, K. and Raman, N. (1980). South Indian Agaricales-IX. *Sydowia* 33: 225-235.

Nobles, M.K. (1948). Studies in forest pathology VI. Identification of cultures of wood-rotting fungi, *Can. J. Res.* 26: 281–431.

Nobles, M.K. (1958 a). A rapid test for extracellular oxidase in cultures of wood-inhabiting Hymenomycetes. *Canadian Journal of Botany* 36: 91-99.

Nobles, M.K. (1958b). Cultural characters as a guide to the taxonomy and phylogeny of the Polyporaceae. *Canadian Journal of Botany* 36: 883-926.

Nobles, M.K. (1965). Identification of cultures and wood-inhabiting Hymenomycetes. *Canadian Journal of Botany* 43: 1097-1139.

Núñez, M. and Ryvarden, L. (1997). The genus *Aleurodiscus* (Basidiomycotina_Syn. *Fung.* 12: 1-164.

Oberwinkler, F. (1965). Primitive Basidiomyceten. Revision einiger Formenkreise von Basidienpilzen mit plastischer Basidie. Sydowia, Ann Myc Ser II 19:1-72.

Oberwinkler, F. (1972). The relationship between the Tremellales and the Aphyllophorales. *Persoonia* 7: 1-16.

Omar, M.B., Bolland, L. and Heather. W.A. (1979). A permanent mounting medium for fungi. *Bulletin of the British Mycological Society,* 13: 31-32.

Overholts, L.O. (1929). Research methods in the taxonomy of Hymenomyceaes. Inst. Congr. *Plant. Sci. Proc.*2:1688-1712.

Parmasto, E.H. (1968). Conspectus Systematis Corticiacearum Tartu 262 Pp.

Patil and Thite (1980). J. Shivaji Unvi. 20: 61-67.

Patil and Thite, (1979). Proc. 66th Indian Sci. Congr. Abstr.p 28.

Patouillard, N. (1900). Essal taxonomique sur les families et les genres des Hymenomycetes. (These) Lonse Ie Saunier, 184pp.

Pegler, D.N. (1973). Aphyllophorales IV. Poroid families. In the fungi and advanced treatise 4(b) Academic Press. New York San Francisco and London.

Pegler, D.N. and Waterston, J.M. (1968). *Phellinus noxius*: C.M.I. Descriptions of pathogenic fungi and bacteria, No. 195. C.M.I., UK.

Persoon, C.H. (1801). Synopsis methodica fungorum Gottingan, 1-706.

Petch, T. and Bisby, G.R. (1950). The fungi of Ceylon. Colombo. Peradeniya Manual 6, Government Publications Bureau, 111 Pp.

Pilat, A. (1926). Monographic de mitteteuropaischen o disceen, Ann. *Mycol.* 24: 203-230.

Puri, Y.N. (1956). Studies on Indian Poria, *J. Indian Bot. Soc.* 35 (3): 277-283.

Rabba, A.S. (1994). Studies in the genus *Phellinus* Quel from Maharashtra Ph.D. thesis, Department of Botany. University of Pune. Pp. 1-221.

Rajchenberg, M. (1983). Cultural studies of resupinate polypores. *Mycotaxon* 17: 275-293.

Rajchenberg, M. (1983). New South American resupinate polypores. *Mycotaxon* 16(2): 500 -506.

Rajchenberg, M. (1986). On *Trametes aethalodes* and other species of *Deadalea* (Polyporaceae). *Can. J. Bot.* 64: 2130-2135.

Rajchenberg, M. (1987). Type studies of Polyporaceae (Aphyllophorales) described by J. Rick, *Nordic J. Bot.* 7: 553-568.

Rajchenberg, M. (1987a). New South American polypores. *Mycotaxon* 28: 111-118.

Rajchenberg, M. (1987b).Type studies of Polyporaceae (Aphyllophorales.) described by J. Rick. *Norwegian Journal of Botany*. 7: 553-568.

Rajchenberg, M. (1988). Xylophilous Aphyllophorales (Basidiomycetes) from the southern Andean forests. Additions and corrections. II., *Sydowia* 40: 235-249.

Rajchenberg, M. (1989). Polyporaceae (Aphyllophorales, Basidiomycetes) from Southern South America: a mycogeographical view. *Sydowia* 41:277-291.

Rajchenberg, M. (1994). A taxonomic study of the subantarctic *Piptoporus* (Polyporaceae, Basidiomycetes) I, *Nordic J. Bot.* 14: 435-447.

Rajchenberg, M. (1995a). Notes on New Zealand polypores (Basidiomycetes). 2. Cultural and morphological studies of selected species, *New Zealand J. Bot.* 33: 99-109.

Rajchenberg, M. (1995b). A taxonomic study of the subantarctic *Piptoporus* (Polyporaceae, Basidiomycetes) II, *Nordic J. Bot.* 15: 105-119.

Rajchenberg, M. (1995c). New Polypores from the Nothofagus forests of Argentina, *Mycotaxon* 54: 427-453.

Rajchenberg, M. (2002). Corticoid and polyporoid fungi (Basidiomycota) that decay *Austrocedrus chilensis* in Patagonia, Argentina, *Mycotaxon* 81: 215-227.

Rajchenberg, M. (2003). Taxonomic studies on selected Austral polypores, *Austral. Syst. Bot.* 16: 473-485.

Rajchenberg, M. and Meijer, A.A.R. 1990. New and noteworthy polypores from Paraná and São Paulo States, Brazil. *Mycotaxon* 38:173-185.

Rajchenberg, M. and Wright, J.E. (1987). Two new South American species of *Perenniporia* (Polyporaceae). *Mycotaxon* 15: 306-310.

Rajchenberg, M. and Wright, J.E. (1982). Two new South American species of *Perenniporia* (Polyporaceae). *Mycotaxon* 15: 306-310.

Ramakrishnan, T.S. (1959). Notes on some fungi from South India. Proc. Indian *Acad. Sci*. 49: 124-128.

Ranadive, K.R. *et al.* (2011). Checklist of Aphyllophorales from the Western Ghats of Maharashtra State, India. *Mycosphere* 2(2): 91–114.

Rattan, S.S. (1977) .The resupinate Aphyllophorales of the "North Western Himalayas". *Bibliothica Mycol.* 60:1-247.

Rayner, A.D.M. (1977a). Fungal colonization of wood stumps from natural sources. I non Basidiomycetes, *Trans. of the British Mycol. Soc.,* 69:291-302.

Rayner, A.D.M. (1977b). Fungal colonization of hard woods stumps from natural sources. II Basidiomycetes *Trans. Br. Mycol. Soc.*, 69:303.

Reaid, D.A., Thind, K.S. and Chatrath, M.S. (1959). The Polyporaceae of Mussoorie Hills: Indian, IV. *Trans of the British Mycological Society* 42: 40-44.

Rehill, P.S. and Bakshi, B.K. (1965). Studies on Thelephoraceae-II.Indian species of *Peniophora* and *Corticium*. Indian Forest Bull., Dehra Dun (242): 1-30.

Rehill, P.S. and Bakshi, B.K.(1966). Studies on Indian thelephoraceae III. The genus *Stereum*. Indian For. Bull. 250:20 pp.

Roberts, P. (1999). A taxonomic guide. The Herbarium, Royal Botanic Gardens, Kew.

Roy, Anjali (1972). Some microstructures in relation to polyporaceae. Mycopathol et. *Mycol appl.* 48: 111-119.

Roy, Anjali (1973). Record of *Poria xylostromatoides* from India. Sci. and Cult., 39:179-398.

Roy, Anjali (1975). Anatomy of Indian Polyporaceae, VI. *Hexagoan discopoda* and H. sulcate. *Bull Botan. Soc. Beng.* 29:57-64.

Roy, Anjali (1976). Structure of zones in fruiting bodies of polyporaceae. *Nova Hedwigia* 27: 801-804.

Roy, Anjali (1979). Taxonomy of *Fomes durissimus*. *Mycologia* 71:1005-1009.

Roy, Anjali (1981). Studies on Indian Polypores IV, Morphological and cultural characters of *Polyporus grammocephalus*. *Mycologia* 73(1): 150-156.

Roy, Anjali (1981). Studies on Indian Polypores VI, Morphological and cultural characters of *Irpex flavus*, Nova Hedwigia 34 (1-2):259-263.

Roy, Anjali (1982). Hyphal system in Aphyllophorales and their respective evolution. Science and Culture 48:372-376.

Roy, Anjali (1983). Wood-rotting fungi and their role in tree ecosystem. Science and Culture 48:246-268

Roy, Anjali (1987). Lignin Biodegradation- Present status and future. *Current Science* 56: 350-353.

Ryvarden, L. (1973). New genera in the Polyporaceae, *Norw. J. Bot.* 20:1-5.

Ryvarden, L. (1978). Polyporaceae of North Europe. Vol. 2: Fungi Flora-Oslo-Norway. 219-507.

Ryvarden, L. (1983).Type studies in the Polyporaceae 14 Species described by N. Patouillard, either alone or with other mycologist. Occas. Pap. *Farlow Herb. Crypt. Bot.* 18:1-39.

Ryvarden, L. (1987). New and noteworthy polypores from tropical America. *Mycotaxon* 28: 525-541.

Ryvarden, L. (1991). Genera of Polypores. Nomenclature and taxonomy. Synopsis Fungorum 5: Fungi flora Oslo Norway. pp. 363

Ryvarden, L. (1992). Type studies in the Polyporaceae 25-26. Species described by P. Henning's and E. Fries, respectively. Mycotaxon.

Ryvarden, L. (2000). Studies in Neotropical polypores 5. New and noteworthy species from Puerto Rico and Virgin Islands. *Mycotaxon* 74: 119-129.

Ryvarden, L. (2000). Studies in Neotropical polypores 7. *Wrightoporia* (Hericiaceae, Basidiomycetes) in tropical America. *Karstenia* 40: 153-158

Ryvarden, L. and Johansen, I. (1980). A preliminary polypore flora of East Africa. Fungi flora-Oslo-Norway. Pp.636.

Salcedo Larralde, I. (1994). Contribution to our knowledge of the Aphyllophorales (Basidiomycotina) of the Basque country. *Mycotaxon* 50: 1-7.

Sathe, A.V. (1969). A new rust record of *Pennisetum typhoides* Stapf. and Hub. *Mycologia* 61: 198-199.

Sen, M. (1973). Cultural diagnosis of Indian polyporaceae III. Genera *Deadalea, Favolus, Ganoderma, Hexagonia, Irpex, Lenzites, Merulius* and *Poria*, Indian forest Records, *Forest Pathology*, 2: 277-304.

Sharma, J.R. (1995). Hymenochaetaceae of India. Calcutta, India, Botanical Survey of India. pp. 219

Slippers, B., Coutinho, T.A., Wingfield, B.D. and Wingfield, M.J. (2003). The genus *Amylostereum* and its association with woodwasps: a contemporary review. *South African Journal of Science* 99: 70 – 74.

Slysh, A.R. (1960). The genus *Peniophora* in New York State, and adjacent regions. State University College of Forestry at Syracuse University.Coll. *Forest.Tech.Publ.* 83: 1-95.

Staleu, F.A. (1983). International Code of Botanical Nomenclature. *Regnum Vegetabile* 97:1-457.

Stalpers, S.J. (1978). Identification of wood-inhabiting Aphyllophorales in pure culture. Studies in Mycology, *Baarn*, 16:1-248.

Sundararaman, S. and Marudarajan, D. (1925). Some Polyporaceae of Madras Presidency. Madras Agricultural Department. Year Book, pp 69-75.

Sydow, H.P. and Butler, E.J. (1906). Fungi Indiae orientalis Part I. *Ann. Mycol.* 4: 424-445.

Sydow, H.P. and Butler, E.J. (1907). Fungi Indiae orientalis Parts II. *Ann. Mycol.* 9: 372-421.

Sydow, H.P. and Butler, E.J. (1912). Fungi Indiae orientalis Parts IV. *Ann. Mycol.* 10: 243-280.

Sydow, H.P. and Butler, E.J. (1916).Fungi Indiae orientalis Parts V. *Ann. Mycol.* 14: 177- 220.

Talbot, P.H.B. (1951). Studies of some South African resupinate Hymenomycetes. *Bothalia* 6 (1): 1-116.

Talbot, P.H.B. (1954). Micro- morphology of lower Hymenomycetes, *Bothalia* 6: 249-299.

Talbot, P.H.B. (1958). Studies of some South African resupinate hymenomycetes. Part II. *Bothalia* 7 (1): 131-187.

Talbot, P.H.B. (1973). Aphyllophorales I: General characteristics; thelephoroid and cululoid families' p. 327-349 in "The fungi. Vol. IV: B.G.S. Ainsworth, F.K Sparrow and A.S. Sussman, eds. Acad. Press New York.

Theissen, F. (1913a). The fungi of India I. *Journal of Bombay Natural History Society* 21: 1273-1303.

Theissen, F. (1913b). The fungi of India II. *Journal of Bombay Natural History Society* 22: 144-159.

Thind, K.S. (1973). The Aphyllophorales in India. *Indian Phytopathology* 26: 2-23.

Thind, K.S. (1975). Recent trends in the taxonomy of Aphyllophorales. In Advances in mycology and plant pathology ed. By S.P. Raychudhuri.

Thind, K.S. and Adlakha, K.L. (1956). Thelephoraceae of Mussoorie hills I. *Indian J. Mycol. Res.* 2: 57-64.

Thind, K.S. and Chatrath, M.S. (1957). The Polyporaceae of the Mussoorie Hills II. Research Bulletin of the Punjab University 125: 431-442.

Thind, K.S. and Dhanda, R.S. (1978). The Polyporaceae of India XI. *Indian Phytopathology* 31: 463-472.

Thind, K.S. and Dhanda, R.S. (1978a).The Polyporaceae of India X. *Kavaka* 8: 59-67.

Thind, K.S. and Khara, H.S. (1968). The Hydnaceae of North Western Himalayas. *Indian phytopath Soc. Bull.* 4: 25-33.

Thind, K.S. and Rattan, S.S. (1968). The Thelephoraceae of the North Western Himalayas. *Indian Phytopath. Soc. Bull.* 4: 15-24.

Thind, K.S. and Rattan, S.S. (1970). The Polyporaceae of India VIII. *Research Bulletin of the Punjab University* 22: 27-34

Thind, K.S. and Rattan, S.S. (1970). The Thelephoraceae of India III. The genus *Tubulicrinis* and *Hyphoderma*. Proc. Indian Acad. Aci. 713:118-131.

Thind, K.S. and Rattan, S.S. (1971). The Thelephoraceae of India IV. The genus *Tomentella. Indian Phytopath.* 24: 32-42.

Thind, K.S. and Rattan, S.S. (1971a). The Thelephoraceae of India VII, *Indian Phytopathology* 24: 290-294.

Thind, K.S. and Rattan, S.S. (1971b). The Polyporaceae of India, V, VII, *Phytopath.* 24, 39: 123-128.

Thind, K.S. and Rattan, S.S. (1972). The Thelephoraceae of India V. Trans. *Br. Mycol. Soc.* 59:123-128.

Thind, K.S. and Rattan, S.S. (1973a). The Thelephoraceae of India VI. *Indian Phytopath.* 26: 485-494.

Thind, K.S. and Rattan, S.S. (1973b). The Thelephoraceae of India VI. *Indian Phytopath.* 26: 528-536.

Thite and Kulkarni (1976). J. Bombay Nat. Hist. Soc. 73: 456-463.

Thite and Patil (1983). J. Shivaji Univ. 21:123-127.

Vaidya, J.G. (1987). Ecological characteristic of wood decay and cord forming fungi from the campus of Poona University, Poona University Press, Pune. pp 109.

Vaidya, J.G. and Bhor, G.L. (1990). Medicinally important wood rotting fungi with special emphasis on Phansomba. *Deerghayu* VI: 1-4.

Vaidya, J.G. and Rabba, A.S.(1993a). Fungi in Flok Medicine. *Mycologist* 7: 131-133.

Vaidya, J.G. and Rabba, A.S. (1993b). Valid names for some common Indian wood Rotting Polypores, their Synonyms and Authenticity II. *Journal of the Indian Academy of wood Science* 24: 35-56.

Vaidya, J.G., Nanda, M.K. and Rabba, A.S. (1991). Community and substratum composition for wood rotting Aphyllophorales from Bhimashankar, Western Ghats. Proceedings of the sixth Engineering Congress on Transidisciplinary premise of Ecology and Environment, Institute of Engineers, Pune: India III\2\2: 56-70.

Wagener, W.W. and Davidson, R.W. (1954). Heartrots of living trees . *The Botanical Review* 20: 61-134.

Wakefield, E.M. (1931). Kew Bulletin of Miscellaneous Information. pp 201-206.

Welden, A.L. (1965). West Indian species of *Vararia* with notes on extralimital species. *Mycologia* 57: 502-520.

Welden, A.L. (1975). *Lopharia*, *Mycologia* 67: 530-551.

Wright, J.E. and J.R Deschamps. (1976-1977). Xylophilous Basidiomycetes from the Argentine Mesopatamia III. The genera *Bjerkandera*, *Gloeophyllum*, *Glocoporus*, *Hirschioponus*, *Hydnoporus*, *Phaeoceroilellus*, *Pycnoporus* and *Xerotinus*, Ibid, XIII (2):27-70.

Wu, S.H. (1990). The Corticiaceae (Basidiomycetes) subfamilies Phlebioideae, Phanerochaetoideae and Hyphodermoideae in Taiwan. *Acta Bot. Fenn.* 142: 1-123.

Wu, S.H. (1992). Notes on resupinate Basidiomycetes of Taiwan I . Yushania. 8: 1-6.

Wu, S.H. (1992). Notes on resupinate Basidiomycetes of Taiwan II Two new record *Porostereum* species . *Bull. Natl. Mus. Natu. Sci.* 3: 267-271.

Wu, S.H. (1993). Notes on resupinate Basidiomycetes of Taiwan III . *Trans. Mycol. Soc. R.O.C.* 8 (1-2): 1-8.

Wu, S.H. (1995). A study of the *Phanerochaete* (Aphyllophorales) with brown subicular hyphae. *Mycotaxon.* 54: 163-172.

Wu, S.H. (1995). Ten Aphyllophorales species new to Taiwan. Fung. Sci. 10: 9_22.

Wu, S.H. (1995). Two new genera of corticioid basidiomycetes with gloeocystidia and amyloid basidiospores. *Mycologia.* 87: 886-890.

Wu, S.H. (1996). Cultural studies of four polypores (Basidiomycotina) collected from Taiwan. *Bull. Natl. Mus. Natu. Sci.* 8: 65-72.

Wu, S.H. (1996). Cultural studies of the Basidiomycotina from Taiwan I.

Wu, S.H. (1996). Resupinate polypores (Basidiomycotina) newly recorded from Taiwan, *Bot. Bull. Acad. Sin.* 37: 151-158.

Wu, S.H. (1996). Studies on *Gloeocystidiellum* sensu lato (Basidiomycotina) in Taiwan. *Mycotaxon* 58: 1-68.

Wu, S.H. (1997). Cultural and ultrastructural studies of *Hyphodontia mollis* and *H. subglobosa*. *Bot. Bull. Acad. Sin.* 38: 191-195.

Wu, S.H. (1997). New species and new records of *Hyphoderma* (Basidiomycotina) from Taiwan . *Bot. Bull. Acad. Sin.* 38: 63-72.

Wu, S.H. (1997). Notes on resupinate Basidiomycetes of Taiwan IV. *Fung. Sci.*. 12: 99-103.

Wu, S.H. (1998). Four Hymenochaetaceae (Basidiomycotina) new to Taiwan. Bull. *Natl. Mus. Natu. Sci.* 11: 129-133.

Wu, S.H. (1998). Nine new species of *Hyphoderma* from Taiwan. *Mycologia*. 89: 132-140.

Wu, S.H. (1998). Nine new species of *Phanerochaete* from Taiwan. *Mycol. Res.* 102: 1126-1132.

Wu, S.H. (1998). Notes on resupinate Basidiomycetes of Taiwan V. *Fung. Sci.*. 13: 121-123.

Wu, S.H. (2000). Six new species of *Phanerochaete* from Taiwan. , *Bot. Bull. Acad. Sin.* 41: 165-174.

Wu, S.H. (2000). Studies on *Schizopora flavipora* s.1., with special emphasis on specimens from Taiwan, *Mycotaxon*. 76: 51-66.

Wu, S.H. (2000). Survey of the Corticiaceae in Taiwan, to 2000, *Fung. Sci.* 15: 69-80.

Wu, S.H. (2002). A study of *Peniophora* species with clamped hyphae, *Bot. Bull. Acad. Sin.* 43: 241-250.

Wu, S.H. (2002). New records of the Corticiaceae from mainland China, *Mycotaxon.* 82: 289-295.

Wu, S.H. (2003). A study of *Peniophora* species with simple-septate hyphae occurring in Taiwan, *Mycotaxon*. 85: 187-199.

Wu, S.H. (2004). Two new species of *Phanerochaete* from Taiwan. *Mycotaxon* 90: 423-429.

Wu, S.H. (2006). *Hyphodontia tubuliformis*, a new species from Taiwan. *Mycotaxon.* 95: 185-188.

Wu, S.H. and Buchanan, P.K. (1998). Species of *Boidinia* (Basidiomycotina) with simple-septate hyphae. *Mycotaxon* 67: 123-128.

Wu, S.H. and Chan, S.J. (1999). Cultural studies of the Basidiomycotina from Taiwan II., *Fung. Sci.* 14: 37-41.

Wu, S.H. and Chen, H.C. (2002). Preliminary survey of wood-decaying Basidiomycota in Lanyu, Taiwan., *Mycotaxon* 15: 77-80.

Wu, S.H. and Chen, Z.C. (1990). New record of Corticiaceae (Basidiomycetes) collected from the National Taiwan University campus, Taiwania. 35: 64-68.

Wu, S.H. and Chou, W.N. (1995). Four Basidiomycotina new to Taiwan. *Bull. Natl. Mus. Natu. Sci.* 6: 139-146.

Wu, S.H. and Losi, C. (1995). *Phanerochaete parvispora* sp. nov. (Aphyllophorales) from Venetian lagoon, Italy. *Mycologia* 55: 543-545.

Wu, S.H. and Ryvarden, L. (2003). Polypore fungi newly recorded from Taiwan, *Mycotaxon* 85: 101-108.

Ying, J.Z. (1980). A preliminary study on the resupinate Aphyllophorales from China. *Acta Bot.Yunnanica* 2:241-274.

Website references

1. http://www.punjabenvironment.com/bd_listhtm
2. http://www.mycobank.org/MycoTaxo.aspx?link=Tand
3. http://www.indexfungorum.org/Names/Names.asp

GLOSSARY

The glossary given here is intended to help you with the terminology primarily by acting as a source of reference. The terminology is an essential part as a mycology useful as a convenient vehicle for communication, which help to put it into proper perspective of understanding the basic concept of present text. The mycological terms that are used are taken from "Glossary of Mycology" by Snell and Dick, 1957, Anisworth & Bisby's "Dictionary of Fungi" (Kirk *et al.,* 2001).

Abhymenial: Opposite the spore, producing surface.

Abrupt: As if cut off transversally; truncate.

Acanthophysis: A sharp pointed process; a spine.

Acuminate: Gradually narrowing to point

Acute: Pointed; less than a right angle.

Adnate: (of lamellae or tubes) - joined to the stipe; if lamellae, proximal end not notched - (of sinuate); sometimes restricted to lamellae widely joined to stipe, tightly fixed to the surface.

Aerial mycelium (am): Vertically growing hyphae, not to the agar surface.

Agglutinate: Fixed together as if with glue.

Aggregated: Crowded.

Allantoid: (esp. of spores), slightly curved with rounded ends, sausage like in form.

Amyloid: Stains blue or black in Melzer's iodine solution.

Aphyllophorales: Having macroscopic basidiocarps in which the hymenophore is flattened, club like, or has the hymenium.

Applanate: Flattened.

Appressed: Closely flattened down.

Arcuate: arc - like.

Aseptate: Without a crosswall

Asperlate: Delicately asperate.

Asterosetae: Stellately branched setae.

Basidiocarp: A fruit - body which bears basidia.

Basidiomycetes: A member of the Basidiomycotina (basidium - producing fungi).

Basidiomycetous: From a subdivision on fungi in which meiospores (basidiospores) are formal exogenously from a meiosporangium (basidium).

Basidiomycotina: Fungi are characterized by the presence of a basidium bearing basidiospores.

Basidiospore : Spore produced on a basidium.

Basidium : (pl. Basidia) - A structure characterized of the Basidiomycotina on which basidiospores are arranged.

Binding hyphae : A thick - walled, much branched often non - septate, hypha found in certain basidiocarps to which it gives strength by binding the tissue together.

Blunt: Of the base, rounded.

Bracket: A fungus which projects ± at right angles from the substrate, usually hard or leathery.

Brown rot: Wood decay process in which cellulose is degraded and lignin is left as brown rot.

Catahyminium: In which hyphidia are the first formed elements and the basidia embedded at various levels elongate to reach the surface and do not from a palisade.

Cellophane paper: Special cellulose paper, used for spore prints.

Chlamydospores: A thick - walled, thallic, terminal, intercalary spores, generally not readily detachable and dispersed.

Clamp connection: A bridge like hyphal connection between adjacent compartments of hyphae of certain basidiomycetes.Usually characteristic of the dikaryon or secondary mycelium.

Clavate: Club - shaped.

Colonization: Process by which a mass of individuals, generally of one species, live together.

Colony: (of mycelial fungi) A group of hyphae (with or without conidia), which arise from one spore or cell. Colony appearance varies e.g. - velvety, floccose (cottony), funiculose (hyphae aggregated into strands), fasciculte (in little groups or bundles), granulose, powdery, synnematous (compact groups of erect and sometime fused conidiophores).

Connate: Joined by growth.

Consistency: Holding together and retaining form.

Constricted: Narrowed somewhere along its length.

Context: the hyphal mass between the superior surface of the subhymenium and the trama of basidiocarps.

Coriaceous: Of leathery texture.

Cosmopolitan: Belonging to all parts of the world.

Cottony: Rather long, single mycelial hyphae spreading in all directions.

Crust: A thick and hard layer on the outer surface of pileus and / or stipe.

Concolorous: Of the uniform color.

Cuticle: The outer layer consisting of compressed hyphae parallel to the surface.

Cyanophilous: Readily absorbing a blue stain such as cotton blue or gentian violet.

Cylindric: Of equal diameter throughout the height.

Cystidium: A sterile body, frequently of distinctive shape, occurs at any surface of a basidioma, particularly the hymenium from which it frequently projects.

Daedaloid: With elongate, tortuous, maize - like pores.

Decay: The destruction of plant or animal material by fungus and other microorganism.

Dendroid: Tree - like in shape, i.e. many branched.

Dentate: Having teeth which are ± torn and acute.

Dendro - hyphidia: Same as dendroid binding hyphae.

Dextrinoid: Stained yellowish - or reddish brown by Melzer's iodine; pseudoamyloid (Singer); cf. amyloid.

Dichotomous: Dividing into 2± equal branches.

Digitate: With figer - like processes.

Dimeidiate: Appearing to lack one half, or having one half every much smaller than the other; (of a pileus), without a stalk and semi - circular; (of lamellae), stretching only halfway to the stipe; (of perithecial wall), having the outer wall covering only the top part.

Dimitic: Having hyphae of two kinds (generative and skeletal hyphae) - which are thick - walled, branched or unbranched, aseptate, straight or slightly flexuous, and have thin - walled apices.

Dissepiment: A portion, e.g. that between the pores of a polypore.

Distant: widely separated.

Echinulate: Describes cell wall surfaces (e.g. of conidia, or conidiogenous cells), with small pointed processes or spines.

Effuse: Stretched out flat, ± irregularly, over the substrate with no pileus [=resupinate].

Effuso - reflexed: (of basidiocarps) - Mostly resupinated but turned away from the substratum in places.

Ellipsoid spores: Elliptical in optical Section.

Ellipsoid: Ellipse - shaped.

Encrusted: Forming a crust on the substrate.

Encrusted Cystidia: Moetuloid Cystidia.

Entire: Not dentate, denticulate or serrate.

Farinaceous: Mealy, Powdery.

Felty: Cottony or wooly Mycellium, which has become matted or packed; emerges hyphae absent.

Ferruginosus: Reddish brown.

Fibril: A very small fiber.

Fibrillose: Bearing radially (longitudinally, of the stipe) aligned superficial fibrils.

Fibrous: Made up of loosely woven distinct hyphae parallel with the long axis of the thallus.

Filamentous Fungus: A term used for a fungus with a hyphal or mycelial organization, as distinct from single - celled forms (e.g. yeasts).

Filiform: Slender and thread - like.

Fimbriate: Distinctly fringed with cystidia.

Flabelliform: (of a basidiocarp) - Fan - shaped; in the form of half - circle (flabellate0.

Floccose: With cottony particles (flocci).

Flocculose: With small cottony particles.

Fringed: Border of short threads or cords.

Fructification (fruit body) : A general term for spore bearing organs in micro and macro - fungi.

Fruit - body: A plectenchymatous structure bearing or containing fungal spores.

Fulvous: Reddish yellow.

Fungus :A eukaryotic, heterotrophic, vegetatively simple organism, plant - like in possessing cell walls (but these not usually containing cellulose) and lacking locomotive powers, but lacking photosynthetic pigments.

Fusiform: Spindle - like.

Generative Hypha: An unspecialized hyphae present in basidiocarps, generally thin - walled and septate, often with clamp connections, which may give rise to specialized strengthening binding and skeletal hyphae.

Glabrous: Not scaly, scurfy, fibrillose, etc., but not necessarily smooth.

Globose: Globe - like, e.g. very young Field Mushroom.

Glossy: A superficial lusture or shine; a false or deceptive appearance.

Granular: Of a surface with a minutely pimply appearance.

Guttulate: Having one or more oil like drops inside.

Heterobasidiomycetes: In which the metabasidium is divided by primary septa.

Hirsute: Having long hairs, shaggy.

Hispid: Having hairs or bristles.

Hollow: With a lumen.

Homogenous: Of the same kind, uniform.

Homobasidiomycetes: In which the metabasidium is not divided by primary septa but may some times become adventiously septate.

Host: A living organism harbouring a parasite.

Hyaline: Colourless and transparent.

Hydnoid: With long slender, +/ - conical teeth.

Hymenium: (pl. Hymenia) - A fertile layer containing asci or basidia.

Hymenophore: The surface on which the hymenium is borne.

Hymenomycetes: Basidiomata typically gymnocarpous or semi - angiocarpous; basidiospores ballistospores.

Hypha: (pl. hyphae) - (vegetative) Filament of a mycelium, without or with crosswalls intercalary - between two cells, between the apex and base (of cells, spores etc.).

Generative hypha: The basic hypha from which arise other elements and other hyphal types:

Binding hypha: Thick - walled much branched hypha which binds generative and skeletal hyphae.

Skeletal hypha: Thick - walled hypha, usually unbranched and aseptate.

Hyphae: One of the filaments of the mycelium.

Hyphal Interference: Process whereby contact between two different hyphae leads to death of one or both of the cells involved.

Hyphal system: The description of the hyphal structure, thus:

Monomitic: Of generative hyphae only.

Dimitic: Of generative and skeletal hyphae.

Trimitic: Of generative, binding and skeletal hyphae.

Imbricate: Where many pilei are close and partially cover each over or overlapping.

Incised: As if cut into; esp. of a pileus margin or lobes of a foliose lichen.

Irregular: Of trama, with no particular disposition of the hyphae.

Karyogamy: fusion of haploid nuclei to form diploid nuclei during sexual or parasexual reproduction (after plasmogamy).

Labyrinthiform: Daedaleoid.

Lacerate: As if roughly cut or torn.

Lamelloid: Made up of thin plates.

Lateral: At the side if the cap.

Lignicolous: Living on or in wood.

Lumen: The central cavity of a cell or other structure.

Membranaceous: Like a thin skin or parchment.

Monomitic: (of a basidiocarp) - Containing only generative hyphae.

Mycelia: Mass of hyphae.

Mycelium: (vegetative) Mass of hyphae; thallus (vegetative) body of the fungus.

Non-amyloid: Non - Un stained with Melzer's iodine solution.

Oblong: (of spores) Twice as long as wide and having somewhat truncate end.

Obtuse: Rounded as blunt.

Ochraceous: Pale brown yellow.

Ornamented: Not smooth but with warts, ridges, and /or spines etc.

Ovate: Egg - shaped.

Palisade : Of the cuticle, where the elements are ± parallel and elongate - clavate.

Parasite : An organism living in or on another organism (its host) from which it obtain its food without benefitting the host in any way.

Pellicular or subfelty : Covered with thin, low coherent mycelium.

Perennial : Living for a number of years.

Persistent : (of spores), Non - deciduous; remaining firmly attached to the spore after liberation.

Pileate : Having a pileus - the hymenium - supporting part of the basidioma of a non - resupinate higher fungus , esp. a Hymenomycetes; cap.

Pileus: (Pl. Pilei) - Upper portion or cap certain of ascocarps and basidiocarps.

Pilose: Covered with hairs.

Plicate: Folded into pleats.

Plumose: Mycelial tuffs with short or long hyphae or group axis often fan like arrangement.

Pointed: At the stem base.

Polypore: General term describing a basidiocarp with a poroid hymenophore, but not nowadays including the boletus.

Poroid: with pores.

Protuberances: Having short projections.

Pruinose: Covered with a very fine powder.

Pubescent: With short, soft fine hairs.

Pulvinate: Cushrion like in form.

Reflexed: Turned upwards.

Regular: Of the trama, with the hyphae parallel.

Resinous: Covered with resinous material or oil drops.

Resupinate: Lying flat on the substratum with the hymenium on the free surface.

Reticulate: With a net - like ornamentation.

Rigid: Not flexible.

Rimose (of pileus surface) : Cracked; originally, cracked in all directions; frequently cracked by radial fissures.

Rimulose: Having small cracks.

Scared groove: Dev - rai, protected vegetation due to religious beliets.

Saprophytic: Saprobe using dead organic material as food and commonly causing its decay.

Secondary mycelium: The dikaryon of heterothallic basidiomycetes.

Septate: With septa.

Septum: A transverse wall in a hypha or other element.(basidium, spore).

Serrate: Edged with saw - like teeth.

Sessile: Having no stem.

Seta (Pl: setae): A short, stiff hair, a bristle.

Setae: A sterile hyphal end, brown thick walled, darkening in KOH, found frequently projecting from the hymenium in xanthochoric basidiomata.

Silky: With long, radiating parallel hyphae or hyphal bundles, more or less prostrate, often glossy.

Simple: Of pores, without subdivision.

Sinuate: Usually taken to be synonymous with emarginate but also of a gill with an abrupt narrowing near stem, or a waxy gill.

Sinuous: Notched at the proximal end at junction with stipe.

Skeletal hyphae : A thick - walled, unbranched , usually non - septate hypha present in certain basidiocarps to which it gives strength by re - inforcing them.

Smooth: Without any form of roughness.

Snuff brown: Powdered tobacco for snuffing up.

Soft - rot: A decomposition of plant parts by fungi or bacteria resulting in the tissues becoming soft.

Spathulate: Spoon - shaped or spatula - shaped.

Spatial: In relation to space or size.

Spore: Any small propagative, reproductive or survival unit which separates from a hypha or sporogenous cell and can grow independently into a new individual e.g. chlyamydospore, ascospore, zoospore, basidiospore.

Sterigma (pl. Sterigmata) : A small projection on which a basidiospore is borne. Sometimes also used for similar projections bearing sporangia or conidia.

Stipitate: Stalked.

Stratified: Where the hymenophores of successive years are layered on each other.

Stroma: A mass or matrix of vegetative hyphae, with or without tissue of the substrate, in or on which fructifications can be produce.

Subiculum: Crust - like growth of mycelium under fruit bodies.

Submerged: Embedded.

Substrate: The material on or in which an organism is living.

Substratum (pl. Substrata) : The material on or in which a fungus grows. Preferred to substrate which has a special meaning for enzymologists. Resource is an alternative word.

Subulate: Thickest at the base and taper even towards the top.

Subulate setae: Slender and tapering to. a point ;awl shaped.

Sulcate: Distinctly grooved.

Superior: Positioned above mid - height.

Tenuis: Slender.

Texture: Arrangement of constituent parts, structure, representation of surface of tissue.

Tomentose: Having a covering of soft, matted hairs; downy; corious like leather in texture.

Toothed: With teeth like projections.

Trama: Tissue underlying the hymenium in a lamella.

Trimitic (of a basidiocarp): Containing generative, skeletal and binding hyphae.

Tuberculate: Having small rounded warts on the sulci.

Tuncinate: A thin white membrane, round the peridiole.

Umbonate: Having an umbo, a central swelling like the boss at the centre of a shield, esp. on top of a pileus above the stipe.

Undulated: Wavy - showing alternate contrary curves.

Ungulate: Shaped like a horse's hoof.

Utriform: Bag like.

Velutinate: Thickly covered with delicate hairs; like velvet.

Velvety: Velutinate.

Ventricose: Swelling out at the middle in a more pronounced way than in 'fusiform'.

Verriuc-ose (- ulose): Having small rounded processes or 'warts'.

Warted: With wart like projections.

White rot: Wood decay process in which lignin is degraded and cellouse is partially degraded.

Woolly: Fairly long interwoven, hyphae or groups of hyphae, somewhat matted, resembling woolen cloth.

Wrinkled: Furrow like crease or ridge in the skin or other flexible surface.

Xanthochoric: (of a Hymenomycete basidioma), having a reddish - or yellowish - brown context which darkens on treatment with KOH.

Zonate: With concentric bands of different colour or texturerg.

HOST- SPECIES INDEX

	"	*Wrightoporia africana*	228-230
21	*Dimorphocalyx lawianus* Hook.	*Pycnoporellus alboluteus*	212-213
	"	*Tyromyces undosus*	224-226
	"	*Trichaptum versatile*	222-224
22	*Drypetes venusta* (Wright) Pax. & Hoffm.	*Gloeocystidiellum flammeum*	66-67
23	*Eugenia jambolana* Lam.	*Lepidomyces subcalceus*	107-108
24	*Ficus bengalensis* Linn	*Gloeocystidiellum lacticolor*	74-75
	"	*Lopharia fulva*	110-112
25	*Ficus benganina* L.	*Leucogyrophana mollis*	108-110
	"	*Perenniporia gomezii*	192-194
	"	*Phanerochaete laevis*	133-134
	"	*Phanerochaete tuberculata*	138-139
	"	*Gloeocystidiellum convolvens*	63-65
26	*Ficus religiosa* L.	*Vararia sphearicospora*	159-162
	"	*Wolfiporia cocos*	226-228
27	*Gliricidia sepium* (Jacquin.) Steud.	*Amylocystis sericeomollis*	169-170
	"	*Gloeocystidiellum citrinum*	59-63
	"	*Cejpomyces terrigenus*	54-55
	"	*Perenniporia medulla-panis*	194-195
28	*Grevilla robusta* A. Cunn.	*Xylobolus apricans*	164-165
29	*Jacaranda mimosifolia* D.Don .	*Perenniporia medulla-panis*	194-195
	"	*Phanerochaete sanguinea*	135-136
30	*Livistona chinensis*	*Hymenogramme javensis*	184-185
31	*Mallotus philippensis* (Lamk.) Muell.	*Aleurodiscus cremeus*	45-46
	"	*Diplomitoporus hondurensis*	177-178
	"	*Hymenochaete rubiginosa*	90-91
	"	*Phanerochaete sanguinea*	135-136
	"	*Phanerochaete sordida*	136-138
32	*Mangifera indica* Linn	*Ramaricium polyporoideum*	150-151
	"	*Oxyporus cervino-gilvus*	188-190
	"	*Hypochnicium eichlerie*	99-100

	"	*Schizopora roseotigens*	219-220
41	*Swietenia mahogany* (L.) Jacq	*Schizopora paradoxa*	218-219
	"	*Botryohypochnus anomalus*	51-52
	"	*Phellinus chryseus*	200-202
42	*Syzygium cumini* (Linn.) Skeel.	*Hymenochaete cinnamomea*	88-89
	"	*Hyphodontia alienata*	95-97
	"	*Peniophora gladiola*	119-120
43	*Tamarindus indica* Linn.	*Peniophora nuda*	122-129
	"	*Lopharia fulva*	110-112
	"	*Junghuhnia crustacea*	186-187
44	*Tectona grandis* Linn.	*Oxyporus latamarginatus*	190-192
45	*Terminalia crenulata* W.& A. Prodr.	*Hypochnicicum cymosum*	97-99
46	*Xanthophyllum rhetsa* (Roxb.) DC.	*Hyphoderma radula*	93-95

INDEX TO GENERA AND SPECIES